Farm M[illegible]

-

Tractors

A Collection of Articles on the Operation, Mechanics and Maintenance of Tractors

By

Various Authors

Contents

The Tractor as a source of power

THE APPLICATION of power to the tasks involved in farming has been a main factor influencing the history of farm machinery. So long as the only sources of power were human labour or that of the ox, the mule and the horse, development of farm machinery was necessarily restricted. Even in favourable conditions, draught animals could not produce sufficient power unless used in unwieldy groups. The steam engine, which 100 years ago seemed to hold out great promise as a source of power for most purposes, proved to have limitations when applied to farming; steam tackle in its day was too ponderous and the provision of fuel for steam engines was always difficult. The perfecting of the internal combustion engine, however, made possible an efficient and economical farm tractor, and this in turn has led to such developments as the power take-off and the hydraulic lift, with consequent changes in the design of many farm implements.

Excluding the early experiments in the application of steam power to the land, tractor history began about 1890, when one of the first tractors powered with an internal combustion engine was made and used in the United States. A British oil-engined tractor followed in 1897. Built by Ruston-Hornsby, this was awarded the Silver Medal at the Agricultural Show of that year. Since then there has been a continuous development.

There were, of course, many problems to be solved in developing a tractor that would work efficiently and economically under farm conditions. As a self-propelling power and traction unit it needed an engine sufficiently robust to stand up to the most difficult conditions and its general construction had to be suited to travel over very uneven ground. This last factor was one which gave considerable trouble when tractors first began to be used in the field, for it was found that the rigid construction originally used gave rise to what is known as 'frame whip'. The remedy adopted was to do away with the frame and to bolt the engine to the transmission, making a single unit. This was known as unit construction and together with the practice of suspending the unit at three points, overcame the difficulties of travelling over uneven ground. This design is now almost universal.

To-day the range of tractors is very wide. Petrol, kerosene and diesel oil engines are available, and a wide choice exists between high-powered tracklayers at one end of the scale and four-wheeled tractors, small tracklayers and two-wheeled walking tractors at the other end of the scale. The most significant development during the last few years has been the steady advance in the use of implements mounted on the tractor as opposed to implements merely trailed behind it.

The modern tractor is a versatile power unit. In addition to direct pulling from the drawbar, it is capable of delivering power at the belt pulley, at the power take-off and through the hydraulic mechanism used to lift tractor-mounted implements. In addition, it may be fitted with a winch. If it is mounted on wheels, its capacity for heavy work can be increased by exchanging the wheels for half-track or rotaped equipment. Starter motors and lighting equipment are becoming standard, and a wide range of gears is regarded as essential. Gear ratios range from very low, for row-crop tractors, to high ratios for transport work. Special types have been developed for horticultural work, with the engine placed behind to gain improved visibility for the operator.

The Engine

A tractor engine derives its power from an internal combustion engine. This engine may be:—

(1) A *spark-ignition* engine, using as fuel:

- (a) petrol;
- (b) kerosene.

(2) A *compression-ignition* (diesel) engine, using heavier fuels.

The operation and maintenance of these various types of engine are discussed in detail in chapters 18, 19 and 20.

Transmission of Power

The power produced by the engine is used in various ways—either to move the tractor, to drive the belt pulley, the power take-off shaft or the hydraulic mechanism.

The Meaning of Power

Before discussing the problems involved in the transmission of power, it is well to define what power means, and how it is measured. *Work* is expressed as the product of a force times the distance through which it moves. *Power* is the rate of doing work. It is the amount of work done in a given time. The unit measure is so many *foot-pounds* of work done per minute.. This unit is too small when large measurements are to be made and so the unit of *horsepower* is used. The horsepower is a unit devised by James Watt and is equivalent to 33,000 foot-pounds of work per minute.

Power from the Belt Pulley

The first point at which power from the engine of a tractor becomes available is at the belt pulley. The output can be estimated

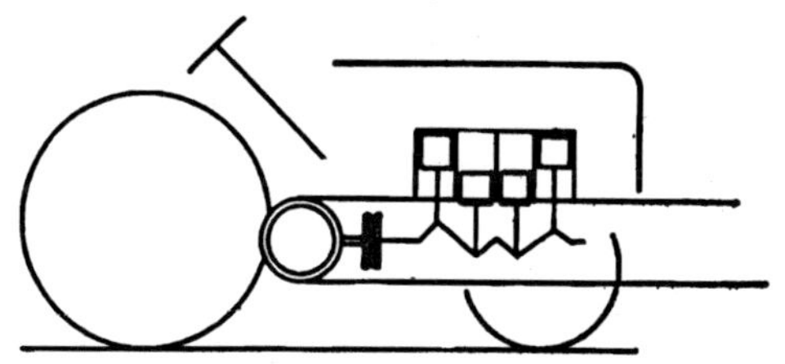

BELT PULLEY

Fig. 15.

by means of an electrical absorption dynamometer, a prony brake, a transmission dynamometer, or an equivalent power-measuring device. When planning work for the belt pulley consult the manufacturer's specification to make sure the tractor has the capacity for the job. Also make sure that the proper operating speed for the driven machine will be obtained.*

* The following formula for belt speed and pulley size can be used.

$$S \times D = s \times d$$

Where S = revolutions of tractor pulley
s = revolutions of machine pulley
D = diameter of tractor pulley
d = diameter of machine pulley.

A revolution counter should always be used for belt pulley work.

Excessive slippage of the belt may be due to too narrow, or too loose, a belt. The belt should ride on the crown of the pulleys and be no wider than either pulley. Operate in a direction that will keep the top side of the belt slack. With a rubber-tyred machine static electricity generated by the belt should be grounded by means of a chain or wire from the frame of the tractor to the ground. Belt drives are discussed in detail in chapter 14.

Power from the Take-off

The power take-off shaft is not assessed in terms of horsepower, but in terms of the number of revolutions per minute.* Some tractors are equipped with a separate power take-off clutch. This is a definite advantage. If the machine or implement driven from the power take-off becomes overloaded, the forward travel of the tractor and the coupled machine can be stopped by disengaging the transmission clutch. The power take-off goes on working and clears the machine. Also, of course, the forward speed of the tractor can be varied by means of the normal gears without stopping the operation of the power take-off shaft. If a separate clutch is not fitted the power take-off shaft ceases to turn when the transmission clutch is disengaged.

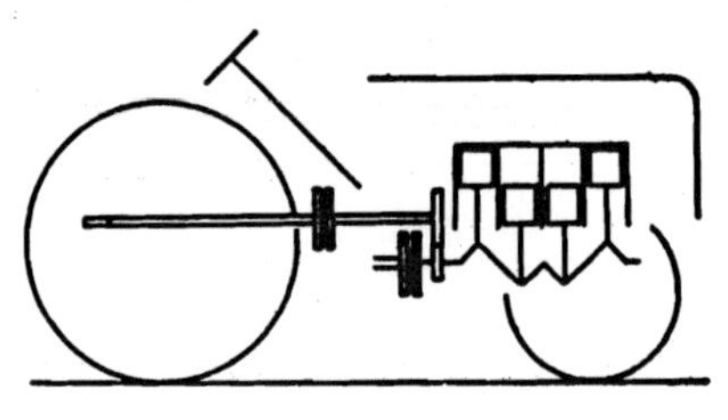

POWER TAKE OFF
Fig. 16.

Machine-cut gears and anti-friction bearings provide an efficient means of transmitting power. There is very little loss therefore between the maximum engine power as developed at the flywheel and what is available at the belt pulley and the power take-off. Any loss is due to friction and the churning of oil in the transmission.

* Standards have been laid down for power take-off speeds and for the machining of the ends of the shaft where coupling to an implement takes place.

Transmission of Power

Power at the Drawbar

Most of the work of a tractor, however, is done at the drawbar, and it is here that tractors are least efficient.

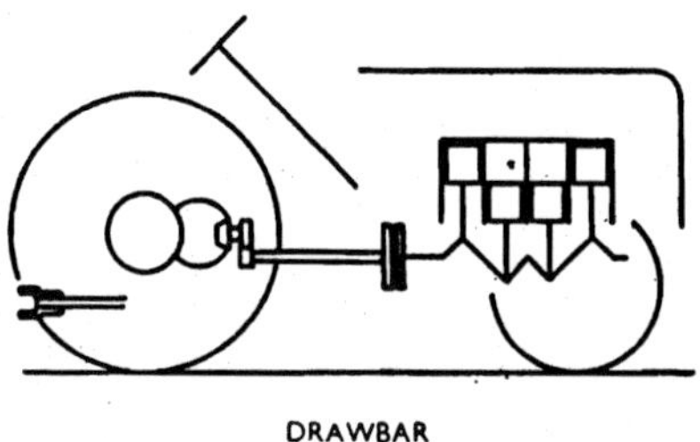

Fig. 17.

Tractor performance at the drawbar begins with the contact the wheels or final drive assembly make with the ground. The cable tackle of last century, with a stationary engine and wire ropes to draw a plough or other implement, was the most efficient means of changing engine power into a pull. The problem of getting a grip on the soil has still not been solved satisfactorily. Suppose a tractor to be equipped with rubber-tyred wheels. The tread of the tyre imprints its pattern on the surface of the soil so as to form a rack. As the wheels turn, it is the resistance to shear (or breaking) of the soil imprisoned between the tread bars, plus friction between the tyre and the soil, that prevents the wheel from slipping. To every force there is an equal and opposite force. As the wheels turn, the force exerted by the tread bars against the sides of the rack, plus friction between the tyre and the soil, causes an equal force to act in the opposite direction at the axle. This force is the drawbar pull. A better grip is obtained on the soil if more weight is added to the rear wheels, and in tractor design allowance is made for weight to be transferred to the rear wheels from the front wheels when the tractor exerts a drawbar pull. This weight is transferred from the front wheels, which have a tendency to be lifted owing to the force representing the drawbar pull having a turning effect (a *moment*) about the point of contact of the rear wheels and the ground.

When a hitch is inclined upwards from an implement to the tractor the vertical component of the resistance to drawbar pull offered by the implement acts downwards, thus adding to the weight already on the rear wheels. Mounted implements also provide some downward force on the rear wheels, plus of course their own weight. Provision is sometimes made for wheel weights to be carried on the rear wheels and it is common practice for water ballast to be used with rubber tyres.

To gain more efficient use of engine power various final drive assemblies have been developed. There are all manner of strakes (wheel grips) and chains to be fitted over rubber tyres to increase their grip on the soil. There are steel wheels with lugs, rotaped equipment, half-tracks and full-tracks. There are advantages and disadvantages with each type.

Wheels

The grip of wheels and the rolling resistance they offer in their passage over the soil both vary with the size, the shape, and number of lugs or bars on the tread, with the overall diameter of the wheels and with the weight upon them. Still more does the grip vary with the structure, moisture content and type of soil. Pneumatic tyres are useful for light work. Steel wheels with lugs have a high rolling resistance but cannot be used on tar-sealed roads without road bands. They have, however, greater tractive efficiency and a low first cost, and need no maintenance.

Full-tracks

A full-track crawler or track-layer provides the most efficient means of getting a grip on the soil. The first cost and maintenance are high. There is also extensive daily greasing. Steering is not always satisfactory under full load with some steering mechanisms, which tend to throw the full torque on to one track. There are others which are more efficient in overcoming this problem. Track-layers, also, have a limited capacity for carrying row-crop tools.

Half-tracks

In a half-track assembly the rear wheels are replaced by sprockets and a spring-loaded idler is carried behind and below the rear axle. A crawler-type track is driven round the sprocket and the idler. The track is in compression where it comes into contact with the soil. A large surface of the track is in contact with the ground, and the effect is the same as that of a large rear wheel.

The first cost is less than that of a full-track, and a return can be made to wheels in a more favourable season. Maintenance costs are low and daily greasing is confined to the two idlers. The available pull is usually increased by one-half with this assembly. The change-over, it should be mentioned, takes two men about four and a half hours when aided by a block and tackle.

The Hydraulic Gear

Hydraulic gear is provided to lift modern tractor-mounted implements. The essentials of a hydraulic system are a pump (to provide oil under pressure), a ram (against which the oil acts), and a means of control. The ram operates arms which lift the implements. The pressure throughout a system containing fluid is the same. Thus, if a one-pound weight is supported on a piston contained inside a quarter-inch diameter cylinder and connected to a two-inch diameter cylinder, a weight of sixty-four pounds will be supported on a piston inside the two-inch diameter cylinder.* The area of the large piston is sixty-four times the area of the smaller one and can therefore support sixty-four times as much weight.

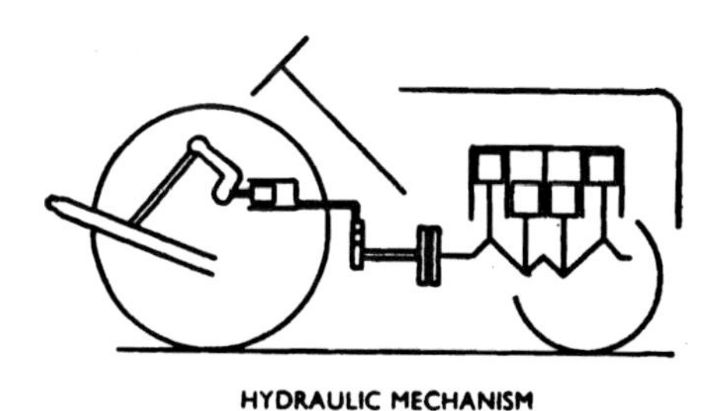

HYDRAULIC MECHANISM

Fig. 18.

In tractor hydraulic systems a large ram and a small pump are used. The primary requirement is a steady supply of oil under pressure. The pumps are sometimes gear pumps. Piston pumps have to be multi-cylinder pumps so that a continuous pressure may be supplied to the system. The pistons are generally arranged in pairs, one on each side of a camshaft. The cams drive home the pistons which return by means of springs. Gear pumps are slightly more

* The area of the small piston is 3.1416 times the radius squared. Similarly the area of the large piston is 3.1416 times the radius squared.

robust than is usual, as the pressures may be very high. To produce such high pressures the pump parts must be very accurately made, and specially hardened. The oil must be kept clean, and in some makes the designer has gone to the trouble of arranging for a magnetic filter element to be incorporated on the intake side of the pump. Where remote cylinders are concerned, care must be taken not to allow dirt to enter the connections. They must also be self-sealing, so as not to lose oil.

The ram is most conveniently arranged so that the pressure of oil pushes it out, and it is returned to its previous position by means of the weight of the implement. More complicated systems are available for delayed-action lifting and there are compensating devices for regulating depth control, details of which cannot be given here. The method of control depends on the intentions of the designer, but it generally consists of a single lever. There may be a clutch to engage the pump, or it may be operating all the time and delivering oil through a by-pass until it is needed.

Overturning of Tractors

The pull of an implement on a tractor transfers weight from the front wheels to the rear wheels of the tractor. If the pull is sufficient the front end will rear. As it rears, it will pivot about the point of contact of the rear wheels and the ground, not about the axle. The force, or pull, has a moment or turning effect about the pivot. This turning effect is, by definition, the product of the force times the perpendicular distance between the pivot and the line of action of the force. If the pull is kept the same, while the hitch or point of application of the force is raised, then the moment or turning effect of the force on the tractor will be increased. Obviously the height of the hitch point is all-important to the amount of weight transferred from the front wheels. Taken to the extreme case, it determines the amount of risk with regard to overturning backwards. The higher the hitch point the greater the risk.

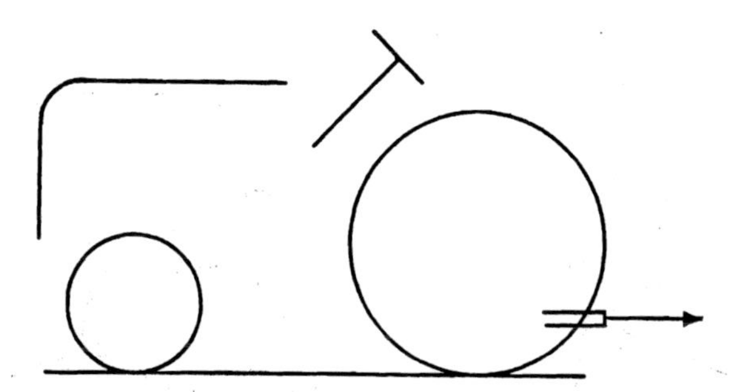

Fig. 19. A load at the drawbar has a turning effect (or moment) about the point of contact of the rear wheels and the ground. The tendency of a load at the drawbar to cause the tractor to rear will be lessened if the hitchpoint (drawbar) is low down.

When the hitch point—that is, the drawbar—is behind and below the rear axle, as it is in standard tractors, then if rearing should occur the hitch point is lowered automatically, thus reducing the turning moment. There is therefore very little danger of overturning backwards due to the pull of an implement on a tractor *when the implement is hitched to a standard drawbar.* The danger, however, lies in hitching to a point other than at the drawbar, as for example, to the rear axle housing. Should rearing then occur, the hitch point would not be automatically lowered.

The resistance to the pull of a tractor by an implement transfers weight from the front wheels to the rear wheels

Too much weight transfer results in front end rearing. A high hitch point results in greater weight transfer

Fig. 20. External forces.

Sideways Overturning

Turning over sideways is more dangerous because it happens more quickly. When a tractor moves in a straight line it tends to keep moving in a straight line. When it begins to turn, centrifugal force comes into action. This force acts outwards from the turning circle and tends to pull the tractor over sideways. The axis over which tipping would occur is represented, in a row-crop tractor, by

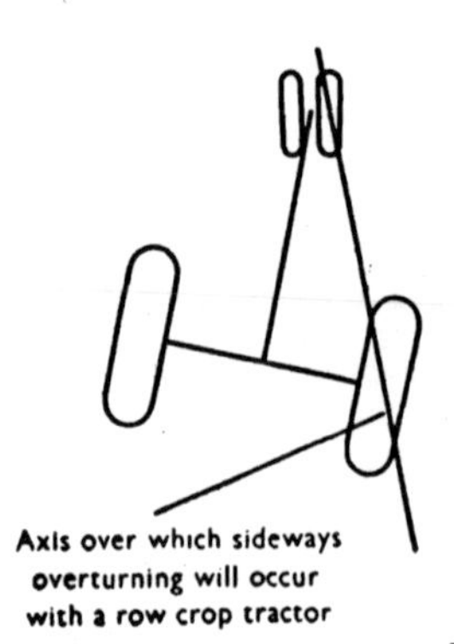

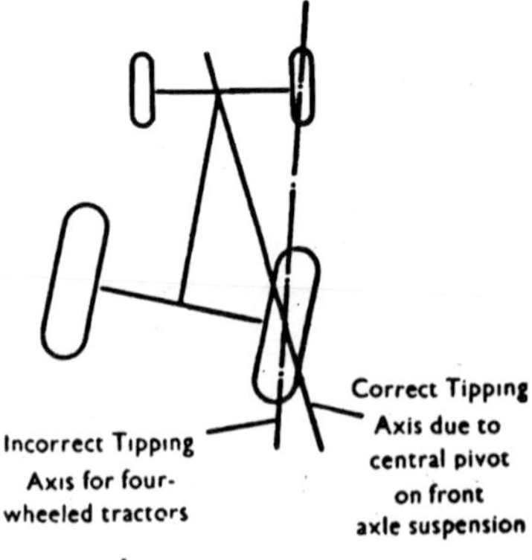

Fig. 21. Sideways overturning.

a line joining the point of contact of a rear wheel with the ground and the point of contact of a front wheel with the ground, both wheels being on the same side. In a four-wheeled tractor, three-point suspension results in the tipping axis being the same as that of a three-wheeled tractor. If the stop at the location of the front axle comes into contact with the trunnion before overturning occurs, then tipping will have to take place over an axis joining the rear wheel with the front wheel. This would be more difficult as the tipping axis would be further from the centre of gravity. However in these matters it is best not to take chances; four-wheeled tractors should be regarded as capable of tipping over the line joining the centre pivot of the front axle and a rear wheel.

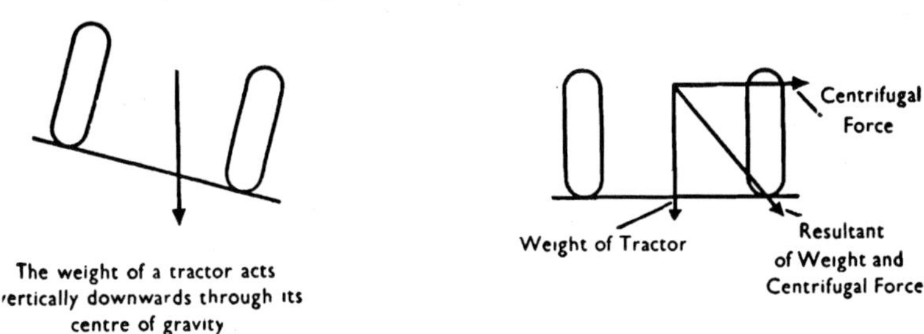

Fig. 22. Sideways overturning.

Figure 22 represents the rear view of a tractor on a turn with the two forces acting on it at the centre of gravity. If the resultant of the centrifugal force and the weight passes outside of the tipping axis then the tractor must overturn. Widening the distance between the rear wheels is the best safeguard against overturning sideways. The width between the front wheels is not important as a measure of safety.

A dangerous practice, when the tractor is not loaded, is to drive along a sideling at speed and then to turn suddenly up the hill. Centrifugal force varies directly with the square of the speed and inversely as the radius of the turn. The speed at which the turn is made is therefore of very great importance.

Internal Forces

So far mention has only been made of the external forces acting on a tractor. The internal forces that are involved with overturning are those associated with *torque reaction*. Torque reaction is a twisting force acting inside the tractor. It arises from the application of engine torque against the inertia and other resisting forces of the drive mechanism. Torque is an engineering term and is a measure of the effect produced by a twisting force. Engine torque is a measure of the capacity of the pistons to twist the crankshaft. Inertia is a property which tends to keep still bodies at rest and to keep those that are moving in uniform motion. For example, when a car is at rest, it takes a greater effort to start it moving than is necessary to keep it moving once it has started. This is partly due to inertia. The use of a low gear and a high engine speed when starting and then a change to a higher gear and a lower engine speed, is a common everyday example of the extra effort that must be provided to overcome the inertia, friction, and rolling resistance of a car.

Torque reaction is most easily understood by considering the driving pinion and the ring gear inside the differential housing. When starting a tractor or when accelerating, there is a tendency for the driving pinion to climb the ring gear. This is due to the inertia of the ring gear and the other drive mechanism to circular motion, and to the inertia of the tractor as a mass to forward motion, plus rolling resistance and friction. This reaction to the application of engine torque tends to cause the rear axle housing to twist in the opposite direction to the drive wheels, and if it is sufficient—that is, if torque is applied suddenly—it may result in the rearing of the front end of the tractor. It should be noted here that wheel weights or rubber tyres filled with liquid, would add considerably to the inertia effect, due to their extra weight, and therefore to the increased difficulty to start them moving. Mention of this is not meant to be an argument against adding weight to

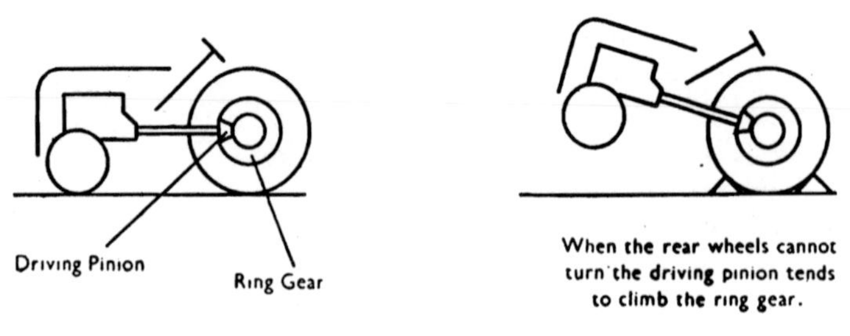

Fig. 23. Internal forces.

the rear wheels. The additional weight is very necessary in some circumstances. It is the intention to point out that additional care must be taken only when either extra weights or water ballast are being used. Also it must be realised that it is only with high engine torque and a sudden engagement of the clutch that the reaction encountered due to inertia and other resistances reaches high enough proportions to become dangerous.

To overturn a tractor in this way the centre of gravity of the front end must be raised to a point vertically over the axle. The energy necessary to do this can only be supplied by an engine with a wide throttle opening (that is, at high engine torque), and in a low gear. It has been ascertained theoretically that there are some tractors which can be wound round their rear axles when standing on a flat surface if high engine torque is applied suddenly.

Counteracting the rearing of the front end by means of the driving pinion is the weight of the front end of the tractor acting through its centre of gravity. It must be pointed out that the turning

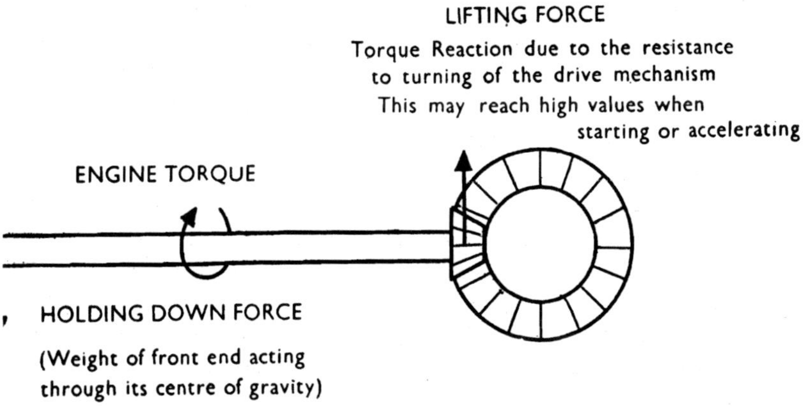

Fig. 24. Torque Reaction

effect of the force holding down the front end of the tractor has a great advantage over the turning effect of the lifting force exerted by the tendency of the driving pinion to climb upwards. It is readily seen that the distance from the centre of gravity of the front end to the rear axle (the point where pivoting is going to take place) is much further than the distance from the ring gear teeth to the rear axle. The advantage clearly lies with the holding down force. The front end also has inertia which has to be overcome.

Should the front end rear, then the higher it rises, the more dangerous does the position become. This is due to the line of action of the holding down force (the weight of the front end acting through its centre of gravity) approaching the rear axle, or pivot, as the front end is lifted. As it gets nearer, so its turning effect is reduced, while the turning effect of the lifting force remains the same. As a result of this, the tractor tends to rear relatively slowly at first and then speeds up as the front end rises higher and its inertia has been overcome.

Apart from the sudden application of engine torque and the reaction resulting from the inertia of the moving parts, there is another set of conditions which can be even more dangerous. When the driving pinion turns, either the ring gear must turn, together with one or both of the rear wheels, or else the driving pinion, *if it keeps turning* will have to ride up the ring gear. What happens generally, is that one or both of the rear wheels will start to slip—most likely one. One wheel slipping is enough to provide an outlet for the energy supplied by the driving pinion. Again it must be emphasised that a good deal of energy is needed to cause overturning in this way. If the clutch is engaged at low engine torque, when both wheels cannot turn, the engine will probably stall. The danger lies at high engine torque (wide open throttle) and of course in low gear.

In general two conditions on the farm may cause both rear wheels to stick. The driver may have encountered a sticky patch and, in the effort to get out, the rear wheels may have dug in sufficiently to prevent them from turning. A similar condition could arise if the rear wheels got stuck in crossing a ditch. The second condition could arise from the common but dangerous practice of jamming a length of wood, or similar object, underneath the front of the rear wheels to prevent them slipping. Under such, or similar conditions, engage the clutch carefully: better still, back out if possible, or else fetch another tractor. On slopes the conditions outlined would be more dangerous, depending of course on the angle of slope. The front end would then already be raised.

Drawbar Loading and Torque Reaction

A load at the drawbar helps to keep the forward end down *against torque reaction* if the drawbar is behind and below the rear

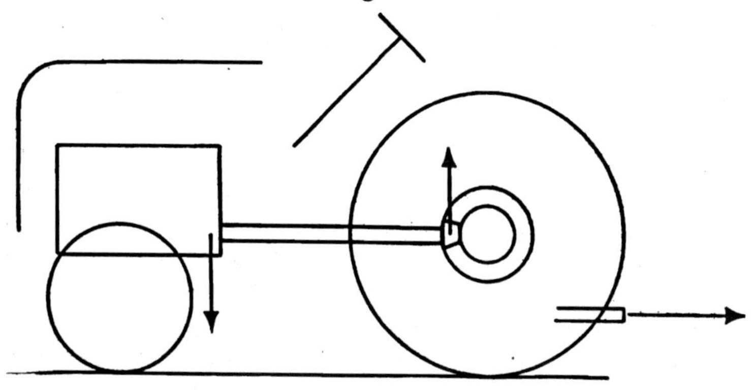

Fig. 25. A load at the drawbar, if the drawbar is below the rear axle, opposes the effect of torque reaction about the rear axle. Torque reaction must be guarded against when a tractor has no load at the drawbar, or is coupled to a load at a point level with or above the rear axle.

axle. A moment's reflection will show that the turning effect of the pull at the drawbar is in the same direction as the holding down force, which is the weight of the forward end. Hitching above the rear axle, or level with it, could be dangerous. The result would be to increase the turning effects of the forces tending to turn the tractor over backwards. Previously the drawbar pull was considered for its external effects on the tractor as a whole. Regarded in this manner a load at the drawbar has a turning effect about the point of contact of the rear wheels and the ground. Considering the internal effects, and taking moments about the rear axle, it becomes obvious that the drawbar pull has a turning effect opposed to that of torque reaction, if the line of action of the pull is below the rear axle. (See Fig. 25.)

Crawler Tractors

In general, crawler tractors are safer on hillsides than wheeled tractors. This is due to the low centre of gravity, to the extra weight of the larger engine and the track mechanism, and to the low setting of the drawbar. The drawbar is placed low so as not to transfer too much weight to the rear of the track. The weight should be evenly distributed over the length of the track in contact with the ground.

The remarks on torque reaction in wheeled tractors apply also to the crawler tractors. But with crawler tractors, which are so much heavier in front than wheeled tractors, the effect of this reaction should not be so dangerous. However, when going up a steep slope with no load at the drawbar a sudden opening of the throttle of a very powerful engine might cause the tractor to overturn backwards. The reaction in this case is due to the inertia of the moving parts and of the tractor mass moving at uniform motion—an inertia which has to be overcome before the whole will move at a higher speed. A sudden increase in slope might cause the operator to open the throttle suddenly. The reaction resulting from the sudden increase in engine torque, might be sufficient to cause overturning. For similar reasons never apply high engine torque suddenly to a tractor that is at rest facing up a steep slope.

The tendency of a crawler tractor to overturn sideways over an axis represented by the outside edge of a track depends on the height of the centre of gravity, and upon the width between the tracks. The greater the width between the tracks and the lower the centre of gravity, the more difficult would it be to turn over sideways. If the tracks themselves are wide, a large surface area is provided to carry the weight of the machine. This large surface with its relatively light weight per unit area may cause the crawler to slip sideways down a slope. A narrower track, however, would tend to dig in because a smaller surface area is carrying the same weight. If a crawler tractor, when sliding down a slope, catches an obstruction on the edge of the lower track, the momentum may well be enough to turn it over.

An Unsuspected Danger

Another sometimes unsuspected danger arises when the driver of a crawler tractor, which is moving down a slope, tries to turn it on the slope. This applies to tractors which are steered by means of a clutch and brake. A turn to the left, for example, is made by disengaging the clutch controlling the drive to the left-hand track, so that this track is no longer driven by the engine. The right-hand track, still being driven, turns the tractor to the left.

When going down a slope, however, with the engine idling and acting as a brake for the tractor, and the tracks driving the engine against the compression in the cylinders, the position is different. In these circumstances, disengaging a track clutch to make a turn

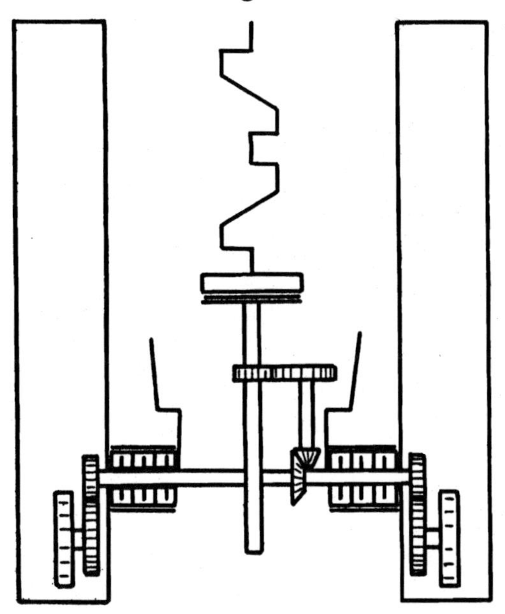

Fig. 26. Steering of a tracklayer by clutch and brake. When going down a steep hill an idling engine will be driven by the tracks. The engine will act as a brake. The steering clutches, under such conditions, will have a reverse effect on directional movement.

has the opposite effect from that which is produced when driving on the flat. Suppose a driver going down a steep hill wishes to turn to the left. He disengages the clutch controlling the left-hand track. Being now free to turn under the downhill momentum of the tractor, this track gathers speed, while the other track, not being disengaged, is still braked by the engine. The result is that the tractor swings to the right in a direction opposite to that intended by the driver. If the intended left-hand turn was being made to avoid a gully or some other obstruction the result of this unexpected turn in the wrong direction could easily be disastrous. Obviously, to make the desired turn under the circumstances outlined, the brakes would have to be used as well, or else the opposite clutch would have to be disengaged.

The trained engineer will realise that the problem of tractor stability is a good deal more complex than would appear from the foregoing analysis. The problem is a dynamic one and not a static one. Momentum, the reaction of the engine to the delivery of its own torque, and the reaction to the impact of the wheels on a rough surface would have to be taken into account for a complete analysis. Some of these factors will depend on particular conditions at a particular moment.

All that has been attempted in this brief outline has been the sounding of a warning by picking out the three major forces concerned with tractor stability—centrifugal force, drawbar pull and torque reaction. It is hoped that the dangers of over-simplification have been avoided.

Hints when buying a Tractor

A TRACTOR must be considered in relation to the farm as a whole—the soil and contours, as well as the type and range of work it is expected to perform. There are a number of general decisions to be made before deciding between different makes. These are the type of fuel used, the horsepower, the kind of wheels or tracks, and the range of gears needed.

It is not a difficult matter to become over-mechanised on a farm. Before purchasing, ask yourself, will the tractor earn money, or will it save money? Will it increase the acreage that can be handled by existing labour? Will it increase income (as, for example, by making it possible to do contract work), or will it just tie up capital that could be used more profitably elsewhere?

The possible purchase must be considered from a number of angles. First of all there are the costs, both direct and indirect. Direct costs are made up of initial cost and running costs. Running costs are labour, maintenance, repairs and renewals. The indirect costs are made up of interests on capital invested (which is an annual charge on the machine) and depreciation, which is an estimate due to the lapse of time and not wear and tear. It is a figure obtained from the new cost, plus capital additions, minus the discard value, divided by the number of years' use. It may in times of rising prices be less than nil, and in times of falling prices be very great. The rates now current are as follows—machinery and plant, 7½ per cent of the diminishing value; tractors, headers, cars, and trucks, 20 per cent of the diminishing value. Indirect costs also include any other equipment required, such as sheds to house tractors, drums for storage, etc.

Other angles from which the purchase must be considered are annual use, adaptability, technical efficiency of construction, and improved performance over existing types. In these last two matters use should be made of the tests undergone by tractors at Nebraska, at Silsoe in England, and also in Australia.

To decide between buying or hiring a machine on contract, make an inventory of the total estimated cost of the machine during operation for one year. Depreciation, maintenance, repairs, housing, interest on capital invested, insurance, fuel and oil supplies and wages. The cost of working per hour comes down as the machine is used more often. More tools may increase the use of the tractor. It is helpful to make a graph showing the cost of operating as against the number of hours worked. The cost per hour diminishes by increasingly small stages as more hours are worked. Somewhere on the graph is the price you would be prepared to pay for contract work. If the number of hours work per year you have for the machine is more than the number of hours associated on the graph with the price you are prepared to pay for contract work, then it is more profitable to buy a machine. If the number is less, then it is more profitable to hire a machine.

If it is decided that the tractor is really needed, then there remains the choice between different makes of tractor.

Examine the tractor from the point of view of the operator—its operation, such as manoeuvrability, turning circle, ease of steering, and vision backwards and forwards. Is the seat comfortable and is it easy to mount and dismount? Can the operator stand to drive, and is there enough leg room? What about noise, vibration, and fumes? Is there sufficient platform space to carry tools, shackles, chains, etc.? What safety measures are there, and what about the brakes? Will they hold the tractor, plus a load, on a reasonable slope? How long does it take to pull up from maximum forward speed? What equipment goes with the standard model? The pulley and power take-off may be extra. What provision for mounting implements is there? If they are directly attached, what kind of fitting is there and how does the lift work? If you intend to buy two kinds of wheels how easy is it to change wheels? Is there provision for back axle weights and can they be easily fitted? How high is the drawbar above the ground and is it swinging or fixed? Note whether the section dimensions of the bar are suitable for your implements. Is there an adequate waterproof sheet to cover the engine?

You, as operator of the machine, will be called upon to service it. Obtain the instruction book, see what you will be expected to do at all the various service periods, and try to estimate how easy or difficult your job will be. Small things, such as whether or not you are able to use a large enough drum under the oil drains to take all the oil during a change, increase in importance with the number of times they are done.

Note the capacity of the fuel tank and estimate if you will be able to do a day's work without refuelling, say at two-thirds throttle. Could spillage from the tanks cause fire by dropping on to a hot exhaust or cause damage to a magneto or battery? How easily does the engine start and what alternatives are there to manual methods? Lastly, look at the tool kit that goes with the standard model and see if the tools are adequate for all the jobs you will be called on to do. Does the kit include a wheel puller if it is necessary for changing the wheels, and any special tools for spark plugs and track adjustment?

Having finally decided to buy a certain make, go to a dealer in your local town who has adequate equipment. With him you will be able to arrange a demonstration on your farm, and you will find it easier to obtain spares, repairs and advice. These things are not so easy if you have gone to a distant town for your tractor, passing over your local dealer.

Tractor Testing

Tractor performance is tested at Nebraska in the U.S.A., at Silsoe in England, and also in Australia. The tests provide a ready means by which a prospective buyer may select the machine most suited to his power requirements before he asks for a demonstration on his particular farm.

Tests can be divided under two headings—engine performance and field performance. Under engine performance the power output and the fuel consumption can be measured with accuracy over a wide range of engine speeds. Tractor engine power is most conveniently measured at the belt pulley. The pulley is coupled by a belt to an electrical adsorption dynamometer, prony brake, transmission dynamometer or an equivalent power-measuring device. Corrections for belt slip are made, and as the measurements are made under controlled conditions a high standard of accuracy is to be expected. No attempts are made to tune the engine especially for the occasion, testing being carried out on a standard model as it comes from the factory. At Nebraska different carburettor settings are included in the tests.

Fuel consumption tests are related to the output of work, and not to time, as that would give very little indication of how the fuel was used.

It can readily be seen that the kind of soil, its structure and moisture content, play a vital part in the amount of drawbar pull available. No two soils are alike and there is no means of exact classification so as to bring the results to a common denominator of soil conditions. The capacity of a tractor in one field is only an approximation to its capacity in another field. The discrepancy is greatest with tractors fittted with pneumatic tyres. A prepared track can be used and on this a more accurate comparison can be made between the performance of two tractors.

Tractor capacity in the field is usually measured in terms of drawbar horsepower. Although farmers tend to define the power of a tractor in terms of the number of furrows of the plough it will draw, statements in these terms can be most misleading. It has already been shown how difficult it is to obtain an accurate measurement of field performance due to differences in soil type, moisture content, and structure, and their effects on the problem of getting a grip. To introduce in the definition of the power of a tractor the many variables associated with a plough is to make the results less useful for general application. Plough bodies vary, and a standard plough body would have to be defined. Draught or pull can be altered considerably by plough adjustments. These are too difficult to assess with any accuracy. Draught fluctuates also with the differences in the soil and in the depth of ploughing.

The field capacity of a tractor is therefore measured in terms of drawbar horsepower. This is obtained as the product of the drawbar pull and the rate of travel in feet per minute, divided by 33,000 to bring the answer to horsepower. In the field the speed at which work is done has a bearing on the quality of the work. It is therefore important to know the pull and the forward speed at which the drawbar horsepower was calculated.

Hints when buying a Tractor

Farmers who use dynamometers obtain the most value from tractor tests. By the use of the dynamometer they will know how many pounds' pull is required to draw an implement on their farm under the most difficult conditions. They will be able to purchase a tractor accordingly. A dynamometer is essential for economic loading. This is particularly true for plough setting, where draught is frequently increased unnecessarily even to the extent of losing another possible furrow.

The percentage of drive-wheel slip at different loads gives a measure of the efficiency of the wheels in getting a grip on the soil. Apart from being a major source of power loss, wheel slip increases the wear on tyres.

The maximum sustained pull gives an indication of loading limits. This figure, together with the use of a dynamometer, enables a farmer always to give an economical load to his machine.

Gradient tests are of great importance to New Zealand conditions, where a large proportion of the cultivated land is on a slope. There are some slopes on which standard wheeled tractors should never be used. It is very important to find out upon what slopes tractors become unsafe. A very large safety factor would have to be allowed, to provide for uneven slopes and for stones.

Tractor Engines (I): Spark-ignition

INTERNAL COMBUSTION ENGINES are so named because combustion takes place inside the cylinder, whereas in steam engines combustion takes place in the fire box and steam only is led into the cylinder. Internal combustion engines for tractors are commonly designed to burn either petrol, tractor vaporising oil (e.g. kerosene), or crude (diesel) oil. Fuel is burnt in the cylinder, and the gases as they expand under heat, are made to do work on the piston. The engine parts are so organised that as soon as one charge of gas has burnt and performed work on a piston the cylinder is cleared of

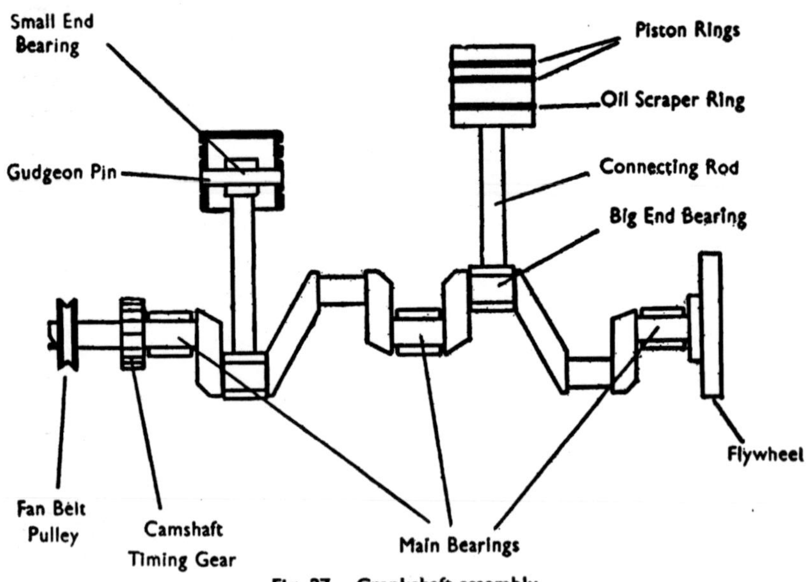

Fig. 27. Crankshaft assembly.

exhaust gases and a fresh charge is introduced. The work is transmitted into rotary motion by means of a connecting rod and crankshaft. To make combustion more rapid and complete a constant proportion of air is mixed with the fuel, and the mixture is compressed before burning.

Fuels

From the engineering point of view petrol as a fuel has advantages over tractor vaporising oil (kerosene). It is better for cold starting, it has a higher thermal efficiency, less lubricating oil is used, and it has a higher power output per unit of quantity of fuel. The compression ratio of an engine determines the type of fuel that it can use. Low-grade fuels, such as tractor vaporising oil (kerosene) are used in low compression engines.

Tractor vaporising oil must be vaporised before it will burn in a cylinder. To this end a vaporiser is used, the heat being provided by the exhaust gases. The engine is started from the cold by means of petrol, the change-over taking place after the engine is hot. The vaporising causes the gases to expand slightly and thus cuts down the volume drawn into the cylinder. Because of poor vaporisation unburnt vaporising oil remains in the cylinder and tends to seep past the piston rings down into the sump, where it dilutes the lubricating oil. Improvements have been made to overcome this.

Low-grade fuels are used in compression-ignition (diesel) engines, which are special high-compression engines. Pure air is compressed, and into this the fuel is injected separately and gradually. This overcomes detonation arising from the compression of a mixture, such as would occur if low-grade fuels were used in engines of the spark-ignition type. In diesel engines detonation occurs, however among other reasons, through ignition lag.

The two main types of internal combustion engines are *spark-ignition* engines and *compression-ignition* or *diesel* engines. In each of these two types of engine combustion of the fuel may be designed to occur in a four-stroke cycle, or in a two-stroke cycle. For convenience, the principles of the four-stroke cycle and the two-stroke cycle are described here only for *spark-ignition* engines.

The Four-stroke Cycle

The four strokes of the piston in the four-stroke cycle are the induction stroke, the compression stroke, the ignition stroke and the exhaust stroke. In the induction stroke the inlet valve opens, the piston moves down the cylinder, and the fuel and air enter to fill the practical vacuum formed. When the piston reaches the bottom of its travel and the cylinder is full of the explosive mixture the inlet valve closes. In the next (or compression) stroke both inlet and exhaust valves are closed. The piston rises and the mixture in the cylinder is compressed into a much smaller space. As the piston reaches the top of its stroke a spark jumps the gap between the electrodes of the spark plug and ignites the compressed mixture. The force of the expanding gases drives the piston down the cylinder. This is the ignition or firing stroke, at the end of which the exhaust valve opens. Then the piston rises again in the exhaust stroke, driving the burnt gases out past the exhaust valve. At the end of this stroke the exhaust valve closes and the inlet valve opens for the induction stroke, and the cycle is repeated.

The Two-stroke Cycle

There are two strokes of the piston in the two-stroke cycle. Every down stroke is a power stroke. Towards the bottom of the power stroke two processes are telescoped into one. The exhaust gases are removed from the cylinder and a new charge is introduced. (In compression-ignition engines, where the fuel is introduced separately, a blower is sometimes used for clearing out the exhaust gases.) Although it was originally believed to be more powerful than the four-stroke cycle, the two-stroke cycle is in fact generally less powerful. This is because two processes are taking place at the same time, so that neither of them is done well. It is difficult to arrange for complete scavenging of exhaust gases without loss of some of the new charge. The two-stroke cycle has some advantages, however, such as mechanical simplicity of design, and the absence of valve gear. More smooth running is obtained than in a four-stroke engine with the same number of cylinders.

So that a spark-ignition engine may function properly, a fuel system, an ignition system, a lubricating system and a cooling system are essential.

THE FUEL SYSTEM

On tractors fuel is carried in a tank placed in such a way as to allow for a gravity feed to the carburettor. Fuel passes out of the tank through a stop-cock to a filter and sediment bowl. To pass out of the bowl the fuel has to negotiate a fine wire gauze. Foreign matter, which can be separated by sedimentation, collects in the bottom of the bowl. Fuel passes on to the float chamber of the carburettor, where it is kept at a constant level by the carburettor float, which operates a valve at the fuel inlet.

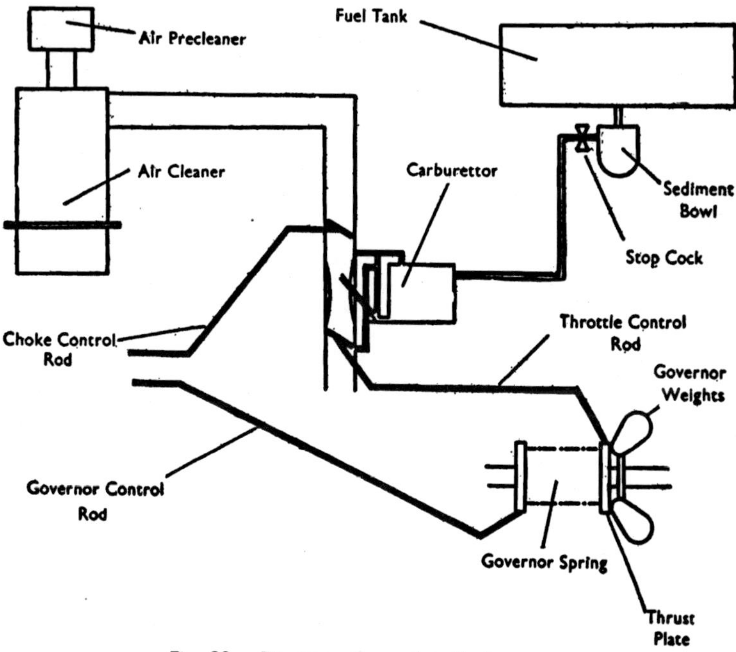

Fig. 28. Diagram of complete fuel system.

From this point it is necessary to trace the air which has to be mixed with the fuel. Air enters the engine through the air filter, which must remove the dust effectively but must not restrict the air flow too much. The filter consists of an element with a large surface area (such as wire wool or very small portions of tubing), together with an oil bath. There may be a pre-cleaner consisting of a series of vanes, which sift out large particles by centrifugal force. This lightens the burden of the main element. Air rushes in to fill the partial vacuum in the cylinders, and is led in to pass over an oil bath, through a filtering element, and into the choke tube. There its passage can be blocked partially by means of the choke valve. This is operated during starting only, and provides for a richer mixture in the cylinder. Air now picks up the fuel which is broken up into small particles in such a way that a constant proportion of fuel to air is maintained. This process is known as carburation.

Engine power depends on the weight of air that can be consumed in a given space of time. Only a short space of time is available for burning to take place in the cylinder. Fuel burns when it comes into contact with air. Small droplets burn more readily than large ones, due to the increase in surface area in contact with the air. For complete combustion, a constant proportion of fuel to air must be maintained.

From this it can be seen that a fine mist of very small droplets of fuel is needed to combine with a constant proportion of air. It is incorrect to speak of complete vaporisation of petrol as this cannot occur to any extent in the short space of time available, nor is it desirable. In order to make a liquid into a vapour quickly, heat must be applied. Hot air occupies more space than cold air and would result in a loss of power due to a smaller volume entering the cylinder.

Inside the cylinder a small flame starts at the spark plug electrodes and then spreads outwards, first of all slowly, then more rapidly. The heat of combustion causes the burnt portion to expand rapidly and force the piston downwards. Pre-ignition of the charge may be caused by plug electrodes, exhaust valves, or glowing pieces of carbon acting as hot bulbs. This gives rise to erratic and uncontrollable power impulses. Severe local over-heating may also cause the metal to warp.

Knocking, detonation, or pinking are caused by the high temperature and pressure conditions which prevail in the last portion of the charge to burn, and at the stage just before it is reached by the flame front. Energy at this point is released at a much higher speed and the violent pressure produces an audible knock. Each type of fuel has a critical temperature and pressure above which knocking will occur. Damage caused in this way includes cylinder head cracks, scarred piston rings (due to failure of lubrication), and burning of the cylinder head or piston top. The resistance of a fuel to knock is an important characteristic and is very difficult to determine satisfactorily. To the purchaser the octane number is a guide: the higher the octane number, the more resistant is the fuel to knock.

The Carburettor

The fuel is maintained at a constant level in the carburettor float chamber by means of a valve operated by a float. As the float rises, so the supply of petrol is cut off. A jet, or a series of jets, supplied from the float chamber with petrol, project into the air flow at a point where a constriction called a venturi is placed. This constriction increases the speed of the air flow at that point and thus lowers its pressure. Petrol issues from the jet, due to atmospheric pressure in the float chamber, and is mixed with the air. Various devices are used to prevent too much petrol from issuing from the jets at high speeds. These are compensating jets

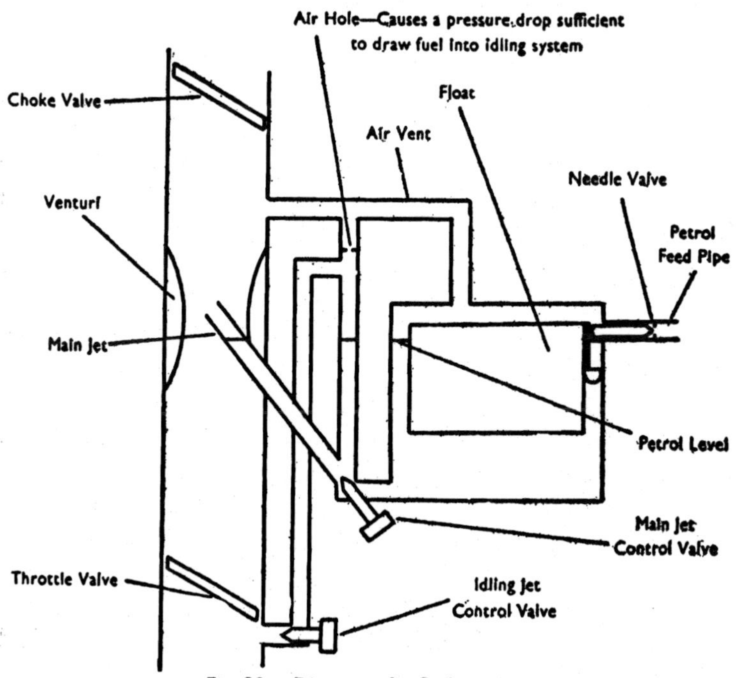

Fig. 29. Diagram of a Carburettor.

(one of them supplying a weak mixture at high speeds), air bleeds and idling jets. Air bleeds and idling jets are the most common. Air bleeds allow air to enter the discharge nozzle of a jet to prevent too much petrol being drawn out at high speeds. Idling jets, which are to be found in most carburettors, come out in the form of a slot beyond the throttle, so that with the throttle partly closed the engine continues to run.

A down-draught carburettor can operate with a slightly larger venturi than a corresponding horizontal carburettor. This is due to the acceleration of the air and fuel droplets by gravity. However, with down-draught carburettors, if there is a faulty float mechanism or an excessive use of the choke, fuel can find its way into the manifold and be a nuisance. A manifold drain valve may be provided for getting rid of any fuel that gathers there.

The carburettor is an adjustable mechanism, and control by the operator is achieved in a number of ways. The throttle valve controls the amount of mixture reaching the inlet manifold. The choke controls the amount of air reaching the main jets, and is only used for starting, in order to obtain a rich mixture. Closing it places the entire space below it under suction. The fuel reaching the main jets is controlled by the main needle valve. For idling, in some carburettors there is control of the air supply to the idling jet, while in others the fuel supply is controlled. To allow for idling a throttle stop prevents the throttle from closing completely.

The inlet manifold is an iron casting bolted to the engine by studs. Considerable care goes into its design to enable an even quantity and quality of mixture to be received by each cylinder. Perfect distribution is difficult, as air is gaseous and fuel is liquid. The fuel tends to go to the outside of a bend and clings to the wall and travels along it. Some evaporation occurs in a manifold, and this absorbs latent heat: the manifold temperature would fall if no heat were added. The main point where fuel tends to cling to the wall and to evaporate is at the part where the mixture from the carburettor impinges on the walls of the manifold. To cope with this it is common practice to add heat to the area. Heat from the exhaust is reliable and easily obtained, so the exhaust manifold is brought down to touch the inlet manifold at this point.

Kerosene Engine Carburation

Engines operating on vaporising oil (kerosene) have a different system of carburation. In order to produce a combustible mixture vaporising oil has to be heated. This is done in a vaporiser, which takes the place of part of the petrol carburettor. The exhaust gases are used for supplying heat and are conveyed to the vaporiser. Air is made to pass over a jet where it picks up the vaporising oil. The mixture is made to pass through a series of passages which are heated by the exhaust gases. A throttle valve controls the supply to the cylinders.

This type of engine uses petrol for starting, and after ten minutes or so, when the engine has heated sufficiently, a change-over is made to vaporising oil. This type of engine is not so efficient as a petrol engine, due largely to the poorer quality of the fuel used. Maintenance costs are higher because unburnt vaporising oil tends to pass the piston rings and dilute the crankcase oil, so that there is excessive wear on the working parts.

With kerosene engines it is essential to maintain a high operating temperature, and the radiator shutters or blinds must be used.

Valves

The valves provide an inlet for the combustible mixture and an outlet for the exhaust gases. They must do two things. They must

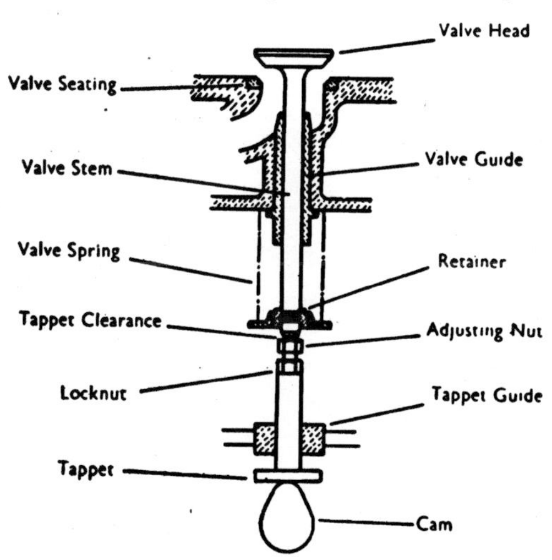

Fig. 30. Typical side valve assembly.

provide an adequate area and they must open and close when needed. For mechanical reasons they cannot close very suddenly, but it is also undesirable that they should do so. When the inlet valve is open a column of high speed gas is rushing past. To halt this column too suddenly may cause damage to the passages. The effect is similar to water hammer in pipes. A rebound might be caused and the back pressure would upset carburation. Also, air moving at high speed tends to keep moving and fills the cylinder more completely. It therefore pays to delay the closing of the inlet valve.

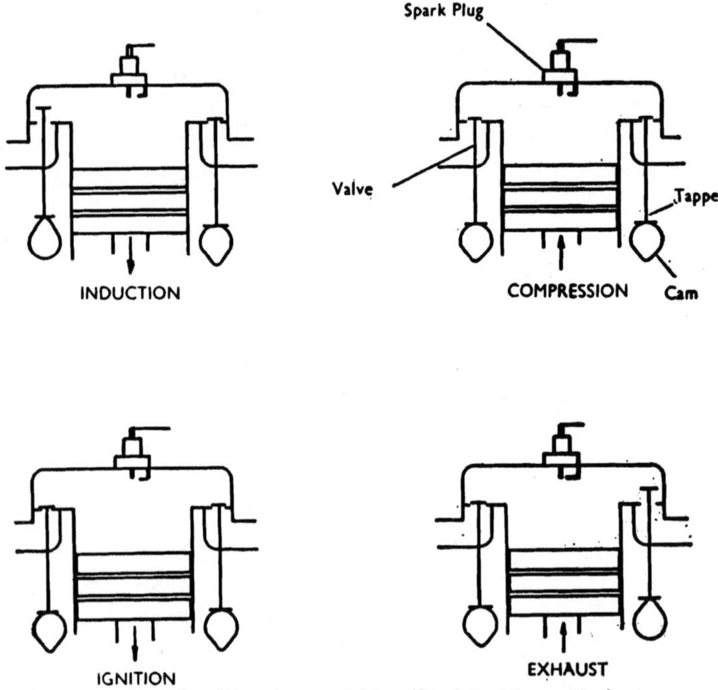

Fig. 31. Side valve assembly. (Spark-ignition engine).

The exhaust valve opens before the piston has reached the bottom of its travel on the firing stroke. This is due to the fact that it takes time to open, and it is desirable to have the pressure dropping a little before the piston starts to rise for the exhaust stroke: with a later opening the energy gained would be lost on the up-stroke. The closing of the exhaust valve is delayed after the piston reaches the top of its travel on the exhaust stroke. The column of exhaust gases in the manifold tends to keep moving and results in a better scavenging effect.

There is a period when the exhaust valve and the inlet valve are open together. This is known as the overlap. It is at the end of the exhaust stroke and the beginning of the induction stroke.

The most common type of valve is a poppet or mushroom valve. The bevelled or conical edge of the valve mates with a corresponding seat. Strong springs keep the valves firmly on their seats, from which they are raised by tappets or cam followers. Push rods are used for overhead valves. The shaft upon which the cams are situated is driven by a train of gears or by a chain, and it is of the utmost importance that the drive should be assembled correctly. For guidance there are generally marks which should correspond with one another.

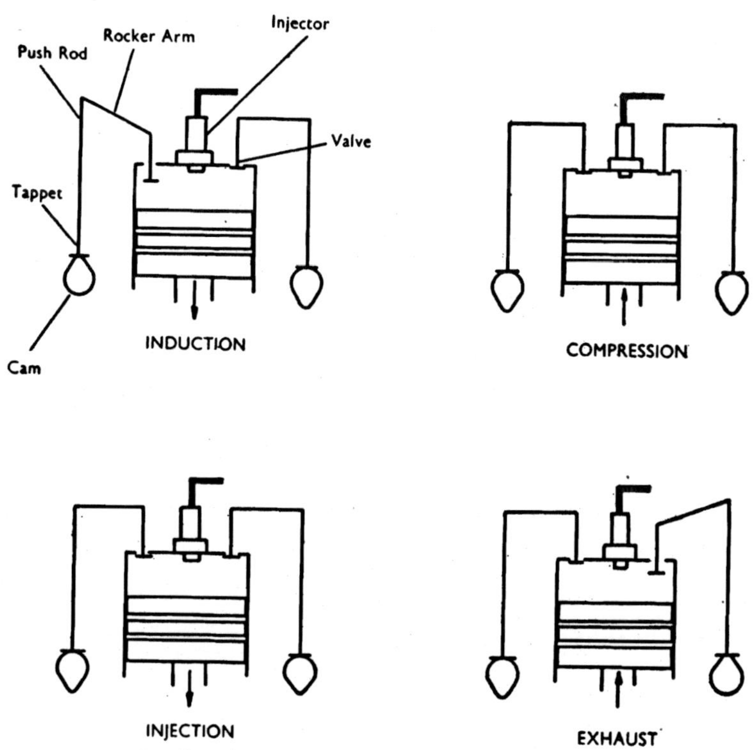

Fig. 32. Overhead valve assembly. (Diesel engine).

Tappet clearance is a small distance in the linkage between the tappet and the valve and this must be maintained to make sure that the valves seat correctly and to allow for expansion on heating. The spring must exert an effort to keep the valves on their seats. Too great a clearance would result in the valves not opening sufficiently, while too small a clearance might result in incomplete closing. Both extremes result in a loss of power.

The Governor

The governor is another accessory to the fuel system. It is a device for maintaining a constant engine speed whether the load is light or heavy. The essentials of a governor are the weights, which fly outwards when the engine revolutions increase. They press against a thrust plate which is held in position by a spring. The movement of the thrust plate is transmitted to the throttle butterfly valve by means of a linkage mechanism. The arrangements are such that when the weights are at rest the throttle is wide open. When the engine starts with the throttle wide open the governor shaft turns at high speed, the weights fly outwards and push against the thrust plate, and this moves the linkage mechanism to close the throttle. The governor spring is compressed by this action. Control is provided for the operator by altering the tension on the spring. This is usually done by altering the location of the end of the spring by means of a system of levers. Vacuum governors are also in use.

THE IGNITION SYSTEM

The behaviour of electricity is best likened to that of water in a pipe; to make it flow along a wire a difference in electrical pressure is needed.

Electrical pressure is measured in *volts*. The rate of flow or current in *amperes*. The unit of power is the *watt* and is the product of the volts times the amperes. The *kilowatt* is a more practical unit and is equal to 1,000 watts.

Considerable electrical pressure (or voltage) is needed to force a spark across a gap inside a cylinder. Under favourable conditions about 4,000 volts are needed. Under unfavourable conditions about 10,000 volts may be needed. An ordinary battery provides 6 to 12 volts. Clearly some special equipment is needed to develop voltage or electrical pressure. Two kinds of systems are used. The magneto system and the battery and coil system.

The Magneto

The operation of the magneto depends on two fundamental principles of electricity. The first one is that when a coil of wire is rotated in a magnetic field an electric current is induced in the wire. The current is proportional to the number of turns of the wire. The second principle is that when an electric current passes through a wire a magnetic field is set up around the wire.

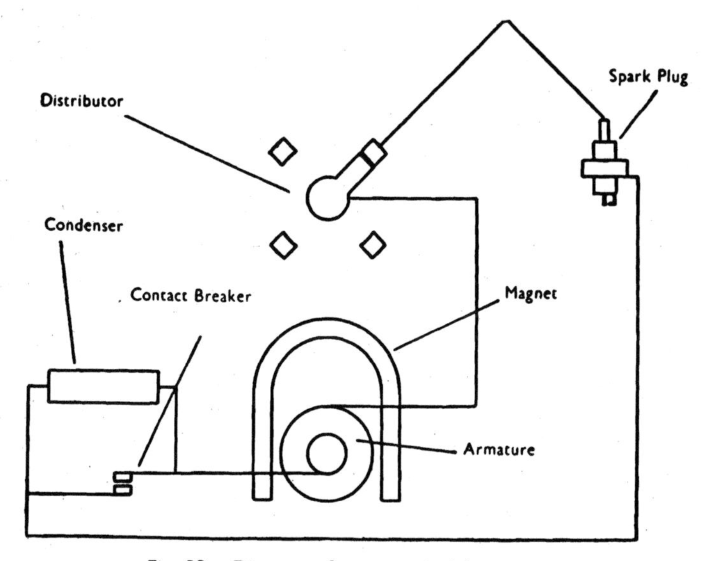

Fig. 33. Diagram of magneto ignition system.

A coil of wire is wound round a soft iron core and is rotated in a magnetic field formed by a magnet. Soft iron is chosen because more lines of force pass through soft iron than through any other substance. The current formed in the wire has not enough pressure to jump the gap in a spark plug but a powerful magnetic field is formed around the coil. A secondary coil of many turns of thin wire is wound round the primary coil. The two coils are insulated from one another. By rotation a current is made to pass in the primary circuit. By breaking the primary circuit and making it again, a secondary current of greatly increased voltage is induced in the secondary winding. This current, which has enough electrical pressure to jump the gap of a spark plug, is taken to a distributor, which connects it at the appropriate moment to the electrode of a spark plug.

The arrangement of a secondary winding and a contact breaker in the primary circuit provides a better control for the spark. A condenser is an essential part of the primary circuit to prevent sparking at the points of the contact breaker.

The soft iron core with its two windings is called an armature. It rotates at half crankshaft speed because one spark is needed for every two revolutions of the crankshaft. There is no reason why in a four-stroke engine the armature should not rotate at crankshaft speed and thus give a spark at the exhaust stroke as well.

The high-voltage current has to be picked up from the secondary winding, which is revolving. This is done by means of a slip ring, connected to the secondary winding, and by a carbon brush. The brush is spring-loaded to maintain contact with the ring. At the other end another brush, also spring-loaded, makes contact with the distributor pencil around which the distributor rotor revolves. The rotor makes contact with the electrodes of each spark plug at the appropriate moment. The primary current must be interrupted by the contact breaker at the moment the spark is desired.

Combustion is not instantaneous and takes time to develop. For this reason the spark is timed to occur just before the piston reaches the top of its stroke. Particularly for starting, and for small throttle openings, it is not desirable to have the spark timed in this way. The contact breaker provides a good means of altering the timing of the spark. The contacts are opened by means of a cam and closed by means of a spring. Provision for altering the position of the cam can be made. It may be manually adjusted or it may be done automatically. Automatic adjustment to the needs of the moment is usually effected by weights on the distributor shaft. These can fly outwards the same way as in a governor, and their movement is transmitted to the contact breaker cam.

To obtain a suitable spark the armature must be rotated at a fairly high speed. During starting, particularly by hand, some device is needed to increase the speed of rotation. The device used is known as an impulse coupling. The armature is held while the drive to the magneto rotates and winds up a spring.

When a certain tension is reached the energy stored in the spring is released and gives the armature a flick over at high speed, thus producing a suitable spark. As soon as the engine makes a few revolutions the coupling no longer functions, being thrown out of action by centrifugal force. The drive is then solid from the crankshaft to the magneto.

Deterioration of the magneto occurs in two main ways—through dampness and through loss of magnetism. Dampness will affect the wiring of the armature shaft, so that every effort should be made to keep the magneto dry. To prevent loss of magnetism, should the magneto be taken down, it is advisable to place a keeper across the poles of the magnet. In an assembled magneto it is possible to obtain the same effect by turning the pole pieces of the armature until they are opposite the poles of the magnet. This in fact is how the armature comes to rest when the magneto is assembled in the tractor, due to the flexibility of the drive provided by the impulse coupling.

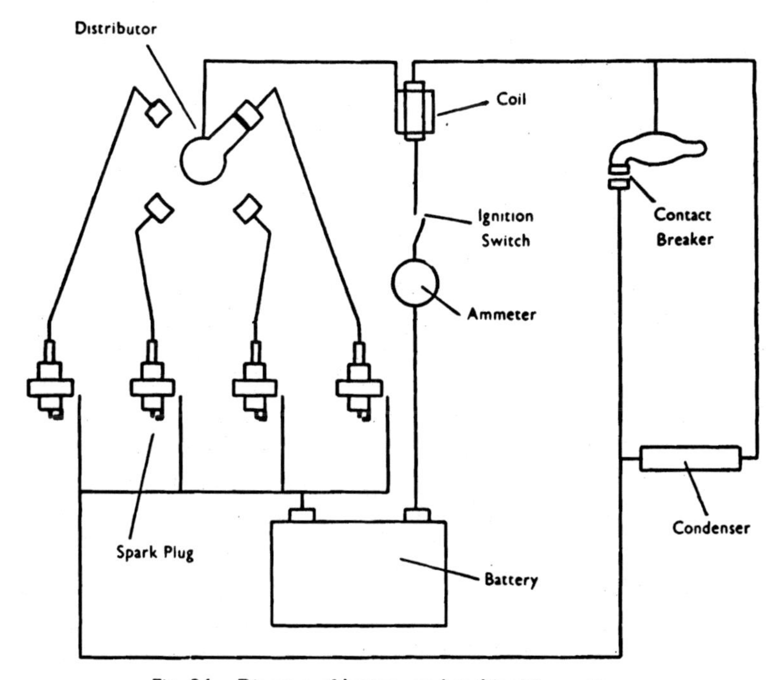

Fig. 34. Diagram of battery and coil ignition system.

The Battery and Coil

In the battery and coil system the primary current comes from a battery. The windings of the coil are the same as in the magneto but they do not rotate. The condenser, which is an essential, is frequently included in the contact breaker assembly. The coil usually has a core made from a bundle of soft iron wires, with the secondary winding wound round first and then the primary outside. The whole may be encased in a metal cover. This is partly for protection and partly to increase the number of lines of force. The secondary winding is of very fine wire and has many turns. All its turns are in the magnetic field created by the primary winding. When the primary current is interrupted, so that its field collapses, the lines of force cut across each turn of the secondary winding, thus adding voltage to the neighbouring turns, and resulting finally in a very high voltage, sufficient to jump a gap. A contact breaker, distributor and condenser are needed, as in the magneto system. It should be mentioned that in both systems a safety spark gap is provided in the high tension circuit in order to protect the insulation. By this means the current is earthed. An impulse coupling is not needed in the battery and coil system.

With magneto ignition the primary current is generated as long as the engine runs. In order to stop the engine the primary current is earthed. In a battery and coil system the primary current is switched off from the battery.

The Spark Plugs

From the distributor the current passes along the high tension leads to the spark plugs. The function of the spark plugs is to create a gap in the circuit inside the cylinder, across which a spark can jump. A plug consists of a central electrode insulated from the body and from the earth point by a porcelain insulator. Some plugs are of the detachable type, where the central electrode and its insulator can be detached from the plug body to facilitate cleaning.

Spark plugs differ in a number of characteristics, three of which are of special importance.

(1) *Thermal Characteristics*

The heat flow in plugs is governed by the shape of that part of the insulator which is exposed directly to the heat of combustion and also by the thickness of the electrodes. If the insulator makes contact almost at once with the sides of the plug body, heat can then flow readily to the cooling water in the cylinder head. If contact is made further up the insulator then the dissipation of heat is slower. The desired heat characteristics depend on engine design and the type of fuel.

(2) *The Reach*

There exists a choice in the length of the threaded portion of a spark plug. This is known as the reach. Some engines have deeper plug holes than others. Difficulty in starting as well as sooting of the plugs is a common fault which can be traced to too little or too much reach.

(3) *The Threading*

Several standard sizes of threading are available, and care in choosing the right one is obviously needed.

On all these points reference should be made to the instruction book.

The Generator

A generator will be incorporated where battery and coil systems are used, or where there is an electric starter motor. The generator works on the same principle as that used for generating the primary current in a magneto—namely, that of a wire coil rotating in a magnetic field. A simple generator is that in which a coil of wire is made to rotate between the poles of a permanent magnet. If the two ends of the coil are connected to a conductor circuit the voltage created in the coil, as it rotates and cuts through the magnetic field, will cause a current to flow in the circuit. This effect can be increased by using a stronger magnet such as an electro-magnet. In this way two coils would be used, one being for the field winding. The current generated in the revolving winding has to be connected to the external circuit. The two ends of the coil are connected to a commutator to produce direct current. The commutator reverses the connections between the armature windings and the outside circuit in time with the reversals of the current in the windings.

Only direct current can be used to charge a battery. An automatic switch is provided to isolate the battery from the generator at all times except when the generator voltage is slightly greater than the voltage produced by the battery. An electro-magnet pulls a small lever in one direction and a spring returns it to the original position. A pair of contacts is provided for closing the circuit, one being situated on the lever. The electro-magnet has two windings: one, of a few turns only, comes from the generator and the other from a terminal of the battery. Current flows from the generator, and energises the magnet, which pulls the lever down and keeps the contacts firmly together until the battery voltage rises above the generator voltage. Then the current starts to flow in the opposite direction along the other winding from the battery, weakens the magnet, and, with an appropriate design of spring, the contacts are parted and the battery is isolated from the generator.

The Battery

The battery stores electricity so that it can be used for a battery and coil ignition system, for a starter motor or for electric lighting. Tractor batteries are generally of the lead-acid type and consist of a series of units known as cells. The current forced into a battery is used up in causing a chemical change to take place. This change is capable of reversal. The materials in the battery revert to their original form if the terminals are connected to an outside conductor circuit.

Each lead-acid cell has two plates, the positive and the negative. They are kept mechanically apart, but are joined chemically together by a weak solution of sulphuric acid which is called an electrolyte. Each plate consists of a lead grid containing a chemical called lead sulphate. This is the discharged state. When current is forced into the cell it flows through the electrolyte from one plate to another. The lead sulphate contained in the positive plate is converted into lead peroxide, the sulphate part going to strengthen the acid. The lead sulphate in the negative plate is broken up, the sulphate portion going again to strengthen the electrolyte. When all the lead sulphate has been broken up the battery is fully charged. After this more electricity only causes the water in the electrolyte to split up into its ions, resulting in loss of water. For this reason distilled water has to be added occasionally to make up for this loss due to over-charging. On complete discharge all the chemical in both plates will have been once more converted into lead sulphate. If a cell is kept in a discharged state for long the lead sulphate hardens and careful treatment is needed before recharging. The battery may even become a total loss.

THE COOLING SYSTEM

A cooling system is essential to internal combustion engines to prevent damage due to excessive heat. They may be water-cooled or air-cooled. Air cooling is usually confined to small engines, an extra surface being provided in the form of ribs cast in one piece with the cylinder head. Air may be blown over this arrangement by means of a fan or led there by ducts to achieve the same object.

With water-cooled engines water jackets are provided in the cylinder heads and in the cylinder block. The jackets are connected at the top and the bottom by a flexible connection to the radiator. The radiator consists of series of upright tubes with fins joining them. Water passes through the tubes and the increase in surface area gained by the fins causes rapid cooling. A fan is provided to draw air through the radiator. The hot water, rising, passes to the top of the radiator by convection currents, cools as it passes through the radiator, and enters the bottom of the cylinder block. To aid the circulation an impeller is used and is driven off the fan belt shaft. In addition there are generally radiator shutters or a blind to control the amount of air being drawn through by the fan. Variable pitch fans have been developed in America. This results

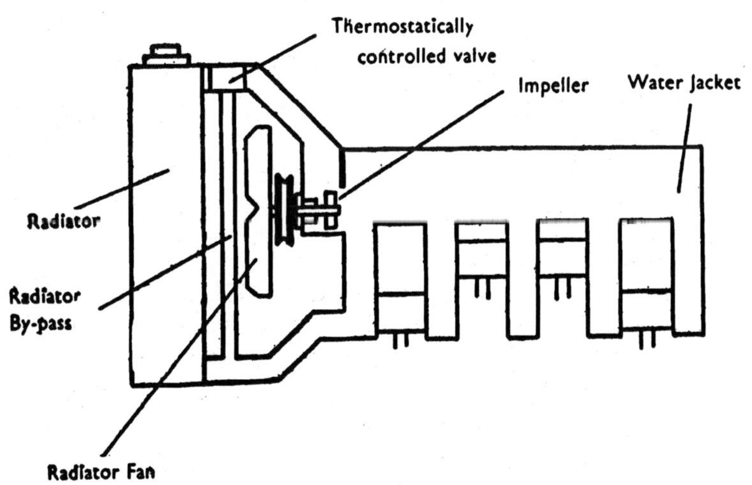

Fig. 35. Typical cooling system.

in a saving of power, as the pitch automatically suits the revolutions of the engine. Further refinements consist in the exclusion of the radiator from the water circuit until the temperature of the water has been built up around the cylinder block. This results in more rapid warming up of the engine and is achieved by means of a thermostatically controlled valve. After a certain temperature has been reached the valve opens and the radiator is included in the circuit. A thermometer is usually fitted and is visible to the driver.

Considerable attention should be paid to the operating temperature of an engine. Engines are more efficient when operating with the cooling water just below boiling point. This is particularly true for engines operating on vaporising oil. Proper vaporisation cannot occur when the temperature is too low and unburnt vaporising oil passes the piston rings and dilutes the crankcase oil. The control for the radiator shutters or blind should be within reach of the driver and should be used in conjunction with the temperature gauge. The cylinder and piston wear increase as the temperature drops. This wear is thought to be caused by moisture condensing from the exhaust gases when the cylinder wall temperature drops below about 212 degrees Fahrenheit. The main causes of friction in an engine lie in the piston rings. Lower oil viscosities due to high operating temperatures reduce this noticeably.

THE LUBRICATION SYSTEM

The performance possibilities of an engine are largely determined by the fuel on which it is designed to operate: its service life depends to a large extent on the lubricant used. An adequate and properly directed supply of lubricant is second only in importance to the supply of fuel and air in the operation of a modern internal combustion engine.

Functions of Engine Oil

Engine oil has multiple functions. It keeps the engine clean, it lubricates, and it dissipates heat. Lubrication consists of the inserting and maintaining of a film of oil or grease between two bearing surfaces. Metal to metal contact causes serious wear and generates heat. An oil film must be formed as quickly as possible. For this reason it is advocated that an engine should be run immediately after starting at fast idling speed to circulate the oil as rapidly as possible. Any impediment to quick circulation will reduce efficiency and service life. With high speeds and heavy loads, bearing surfaces will generate heat, even presuming a complete oil film separates the surfaces. In such cases oil acts as a coolant and carries away sufficient heat to keep the surface temperature within safe limits. The running-in period allows tight bearings to wear away excessive material so as to provide adequate clearances for high speed operation, and the maintenance of an oil film.

Where operating speed and high pressures make it difficult to keep and maintain an oil film, and where design would not exclude dust and water and would result in oil leakage and high consumption, grease lubrication is used.

The place where it is most difficult to maintain an oil film is on the cylinder walls. This is due to the action of the piston, which rises, speeds up, slows down, and stops, and then reverses the procedure and the direction. Most cylinder wear occurs during the first few minutes of running. The choke should be used as little as possible as excess petrol in the combustion chamber tends to wash the oil film from the cylinder walls.

Oil will not maintain its efficiency for long periods. It becomes contaminated with abrasive dust, with products of incomplete combustion, with particles of metal worn from the working parts, and so on. It must therefore be changed at stated intervals. Oil changing is a good preventive maintenance practice which will in the end prolong engine life, lower maintenance costs and allow the engine to give a better performance. Thus it is of vital importance to keep a log book, and make the necessary changes at the times indicated by the makers in the instruction book. Oil and air filters and crankcase ventilator filters do not remove all the dust and dirt. Farm conditions are particularly conducive to contamination of the engine oil.

The lubrication system of an internal combustion engine consists of a reservoir or sump from which a circulation of oil begins. Circulation is achieved by one of three methods—the splash, the pressure feed, and the splash and pressure feed combined. In the splash system the big ends dip into trays which are in the sump and are kept full of oil. The lower part of the bearing carries a small scoop. As it dips into the tray once every revolution it splashes oil round the crankcase, forming a mist inside, and covering all the engine parts with a thin oil film. Oil passes up the scoop and passes through a hole in the bearing and thus lubricates the big end. In this system the flywheel may be indented, and as its lower portion runs in the oil it will pick up oil in the indentations and deliver it to a pipe known as the gallery pipe. This pipe feeds all the main bearings of the crankshaft with oil.

Some provision for filtering is made, and advantage is taken of the flow, caused by the removal of oil by the flywheel, to interpose a filtering unit. A dam in the rear of the crankcase prevents oil from returning direct to the flywheel for re-circulation, and conducts it through a fine-mesh gauze screen. This removes the particles of dirt and carbon which cause wear. If the filtering element becomes blocked oil builds up on the crankshaft side of the dam and overflows to the flywheel compartment.

Pressure Feed System

In the pressure feed system an oil pump is used. Oil is drawn from the sump through a fine mesh and is pumped through pipes to every part of the engine. Pressure gauges should be included in such a system. This system lends itself to more efficient filtering of the oil. There may be as many as four separate filters. The first one may consist of a sheet of gauze or perforated metal placed in the sump, the second may consist of a wire gauze placed over the end of the pipe leading from the sump to the pump. There will also be a filter in the filler cap orifice to stop any solid particles from entering accidentally when filling the sump with oil. Lastly there is the main oil filter which is designed to remove any deleterious matter from the oil. All the possible points where oil filters may be placed need periodic attention. The gauze filters need washing in the paraffin, rubbing lightly with a brush—not with a rag, which might leave behind some fluff. The main filtering element will not be able to function without cleaning and renewal at the periods stated by the manufacturers. When the oil is cold and thick it will have difficulty in passing through an element that also has to filter efficiently when the oil is hot and running freely. To prevent a dangerous build-up of pressure and, more important, to get the oil to the bearing surfaces quickly, a by-pass system is used. The by-pass comes into operation when the pressure in the system rises slightly above normal. The by-pass valve is operated by means of a spring. There are obvious evils attached to such a system, and a by-pass filter must not be regarded as very efficient for removing impurities from the whole system. Oil-changing and frequent cleaning or replacement of the filtering unit, depending on design, are the only ways of ensuring that the oil will be kept clean. Oil-changing makes the job of the oil filter easier, but remember that oil changes should also be accompanied by occasional cleaning of all the oil-filtering elements in the engine.

In the pressure lubrication system oil is made to pass along pipes and through holes drilled in the parts of the engine in order to reach the bearing surfaces. The overflow drips down to the sump and is re-circulated. This is the most efficient system, but it is also more complicated, and thus increases the possibilities of something going wrong. A pressure gauge provides a means whereby the operator can be warned that the system is not working properly.

Oil pumps may be of the piston type or the gear type. The gear type consists of two gear wheels meshing inside a casing which fits snugly round them. Piston pumps, unless arranged in groups, do not provide a steady flow. Thus, gear-type pumps are more popular. Pumps are generally housed in the sump and driven off the crankshaft. As was mentioned before, the intake of the pump is guarded by a fine wire-mesh strainer. The outlet may have a spring-loaded safety valve to prevent damage to oil pipes if the pressure rises too high as it will do when the oil is cold and thick.

The third kind of lubrication system is a combination of the above two. Usually the big end, little end and pistons are lubricated by force feed and the remainder of the moving parts by splash.

The crankcase breather has an important function with a bearing on lubrication. Burnt gases passing the piston contain a large percentage of water vapour which would condense and contaminate the oil. The gases contain traces of sulphur and lead compounds which, if dissolved in the condensed water vapour, would form corrosive compounds. The gases are hot and would overheat the oil and also build up pressure in the crankcase. This pressure could cause oil leaks through shaft seals and through the crankcase gasket. The function of the breather is to prevent this. The breather tube taps into the crankcase at some point relatively free from splashing oil. Baffled passages or a filtering element, such as wire wool, allow gases to escape but retain oil droplets.

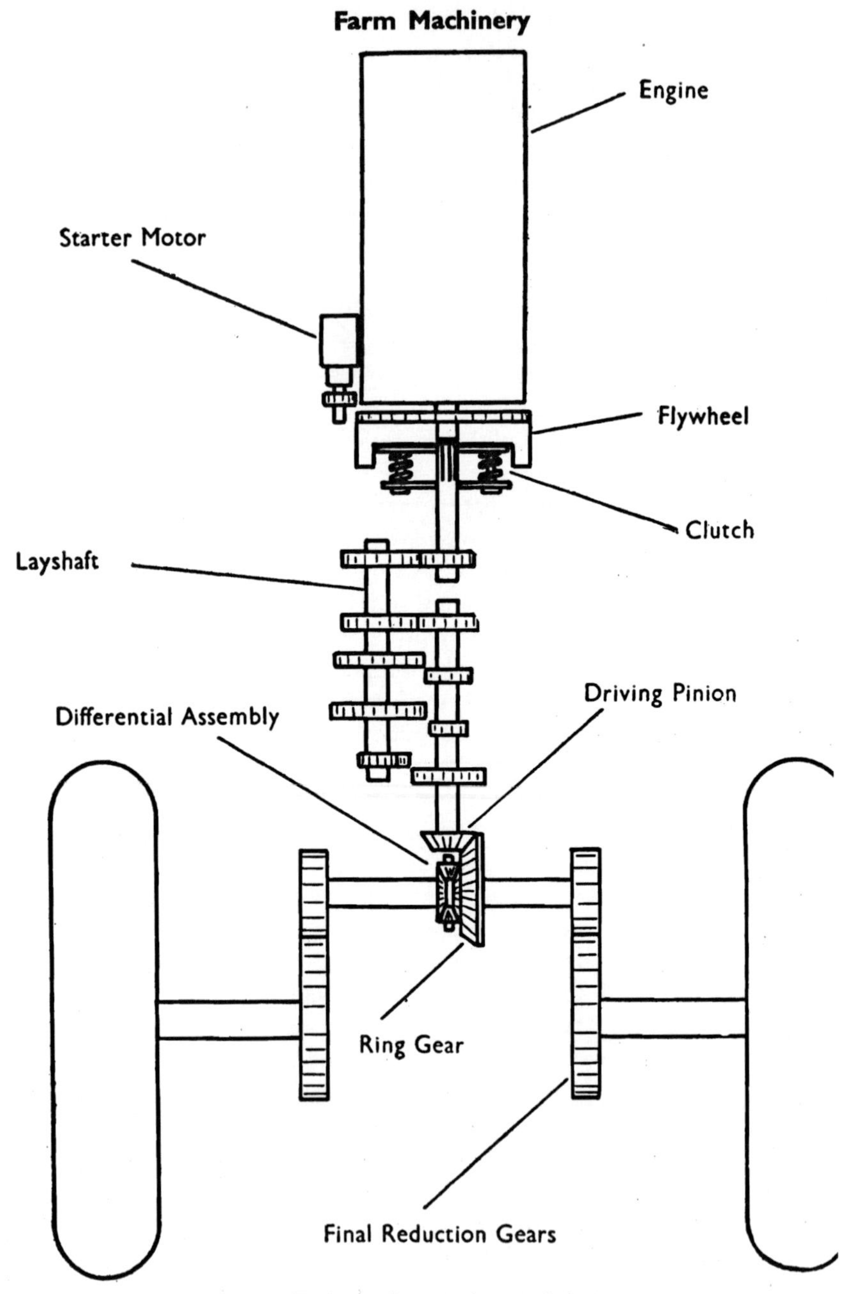

Fig. 36. Diagram of a typical transmission system.

Gear Box and Transmission

The gear box and transmission also receive oil from an oil bath. With gear teeth a certain amount of sliding friction occurs where the teeth make contact with each other. Gearing however can operate so that the wheels dip into an oil bath. There is no need for any special method to aid circulation unless the gears are running at very high speed. This does not apply to tractor engines. For the transmission a different grade of oil is used. No one oil has all the desirable points for a particular situation nor has one oil the properties to suit a number of different situations. In this respect it is essential to follow the maker's recommendations. Oil is graded for this purpose by its S.A.E. number. Buy also from a reputable firm and do not economise by buying a cheaper oil.

Grease

All lubrication is not achieved by means of oil; some of it is done by the use of grease. Greases suitable for lubrication must be soft and yet not be squeezed out easily from between bearing surfaces. Bearing surfaces are either plain, ball or roller. With plain bearings a lining of some special metal of low melting point is usual. This lining can be renewed easily when worn, and also provides a safeguard in case of failure of the lubrication system. Such a failure quickly results in overheating, which causes the lining to melt and so safeguards the crankshaft, which is an expensive item to replace. Ball and roller bearings change sliding friction to rolling friction. Some rubbing occurs between the balls or rollers and the retainer which keeps them properly spaced. Ball races that are to be packed with grease should not have too much grease smeared on them, as too much, due to the violent churning effect, overheats and may fail as a lubricant.

Heavy-duty Oils

Where spark-ignition engines work under heavy-duty conditions it is beneficial to make use of the higher priced heavy-duty oils. Heavy-duty conditions comprise high speeds and temperatures, long hours and heavy loads, all of which impose an added strain on the lubricating oil.

Should a change-over from the conventional oils be made, certain precautions must be observed. It must be realised that loosened

F

deposits may clog oil pump screens or even block small oil passageways. If the engine has operated for a long time using conventional oils it is necessary, after a short period with the heavy-duty oil, to remove the sump and clean out any deposits, paying particular attention to the oil filters.

New or clean engines should be able to operate directly on special heavy-duty oils without this precaution. Keep a close watch on the oil pressure in both cases after the special oil has been introduced, in case of blocked oil passageways. Oil-change periods cannot be extended by the use of heavy-duty oil. The material normally deposited on the cylinder walls, piston top and valves remains in suspension, and there is obviously a limit to the amount that it is desirable for the oil to retain.

Filter cartridge life is generally found to be greater with heavy-duty oils because the particles removed are in a very fine state of dispersion. In this matter it is best to be guided by actual observed results.

Heavy-duty oils are available in the same grades as the conventional oils. Follow therefore the grades recommended by the manufacturers.

Never forget this slogan: *Lubricating oil is the caretaker of the moving parts of an engine.*

For all points of lubrication follow carefully the recommended procedures in the instruction book and buy only the grades and makes of oil that are recommended by the makers. The importance of buying only the best oil cannot be stressed too strongly. The three most important characteristics in relation to oils are viscosity, adherence to surface, and deterioration. The most sensitive indication of the tendency of the oil to adhere to the engine parts and prevent lubrication failure is given by the piston rings. Partial breakdown of the oil film will cause these parts to become scuffed. Deterioration produces acids, tarry sludges, and resins. Acids corrode. Sludges and resins form coatings and deposits. In this connection piston rings and valve guides are the worst offenders. Buy therefore only the grades and makes recommended in the instruction book.

Tractor Engines (2): Compression-ignition (Diesel)

COMPRESSION-IGNITION (or diesel) engines are steadily becoming more popular for use in farm tractors. They are of heavier construction than spark-ignition engines, due to the higher compression pressures used, and are more expensive. They are more reliable, and for that reason need less maintenance.

In petrol and kerosene engines (as discussed in the previous chapter) a mixture of fuel and air is drawn into the cylinder and ignited from an externally produced spark. In a compression-ignition engine pure air is drawn into the cylinder, where it is compressed to a greater degree than in a spark-ignition engine. The high compression raises the temperature sufficiently to ignite the fuel, which is introduced separately through the injector. Fuel has to be injected at a pressure higher than that prevailing in the cylinder. A pump, incorporating a metering device, is used in conjunction with the injector. The fuel is atomised by passing through small holes in the injector nozzle. A high power is developed with fewer revolutions of the engine, resulting in less wear and tear on the engine parts. A greater fuel economy is another feature of this type of engine.

In compression-ignition engines the compression ratios are from 10 to 1 to 20 to 1. If there is not sufficient compression to ignite the fuel, additional heat can be provided by spraying the fuel on to a hot bulb. This bulb is preheated with a blowlamp for starting and, as it is not being subjected to water cooling, it remains hot during running. The Lanz Bulldog is an example of a two-stroke semi-compression-ignition engine of this type.

The compression-ignition engine has features which make it more efficient than the petrol engine. The air passages are not obstructed by the mechanism of carburation. It has no choke, no fuel jets and no throttle valve. The fuel supply is independent of the air supply, so that this engine is more difficult to stall than a petrol or paraffin engine. It is said to have more 'lugging' capacity. Other problems, associated with carburettors, such as the provision of equal charges to all cylinders, are also eliminated.

There are four-stroke and two-stroke compression-ignition engines. With the two-stroke engine it is not necessary to use the technique of crankcase compression. There are no carburation problems, so that the air can enter through a port directly connected with the atmosphere. On some engines a blower is used to act as a scavenger.

Fuel Injection

A fuel injector is fitted to each cylinder and the injector is attached to a pump. The only satisfactory way of regulating the power developed by an engine of this type is by varying the amount of fuel delivered to the cylinder. Various schemes have been tried, including variation of the stroke of the plunger, but the most widely used method is to divert from the engine, at a particular moment, that part of the fuel which is not needed. Fuel must be injected into the cylinder against the high pressures prevailing towards the end of the compression stroke. The amount of fuel must be exactly the same for each cylinder. This is essential for smooth operation. The combustion rate and character are very sensitive to the fuel spray form and characteristics, to injection nozzle dribbling, and to the timing and duration of the injection period.

The fuel pump consists of a hardened barrel and plunger ground to a very high degree of accuracy. The plunger is operated by a cam, and it forces fuel past a spring-loaded delivery valve through the fuel lines to the injector. The fuel lines are made of steel so as to cope with the very high pressures involved. The upper part of the barrel of the pump is connected with the lower part by means of a groove cut down the side of the plunger from the fuel pump. The cam lifts the plunger, and fuel is forced back through the inlet ports until they are covered, after which it must pass through the spring-loaded delivery valve. Fuel continues to pass through until the inlet port is uncovered by the bottom of the plunger. Fuel then escapes from the top of the plunger down the groove and out of the inlet ports. The time during which the inlet port remains covered, with the plunger delivering fuel, can be altered by turning the plunger round so as to present another part of the helix on the side of the inlet ports. Teeth are cut on the lower part of the plunger; these mesh with a rack, which provides a means of rotation. The rack is connected by a suitable mechanism to a governor assembly similar to that of a petrol or kerosene engine.

In some fuel systems a single-plunger metering pump is provided, and distribution is made by cam-operated spring-loaded valves. In others individual metering pumps are provided and a spring-loaded valve prevents injection until a certain pressure has been built up. The pump and the injector may be one, or separate or grouped.

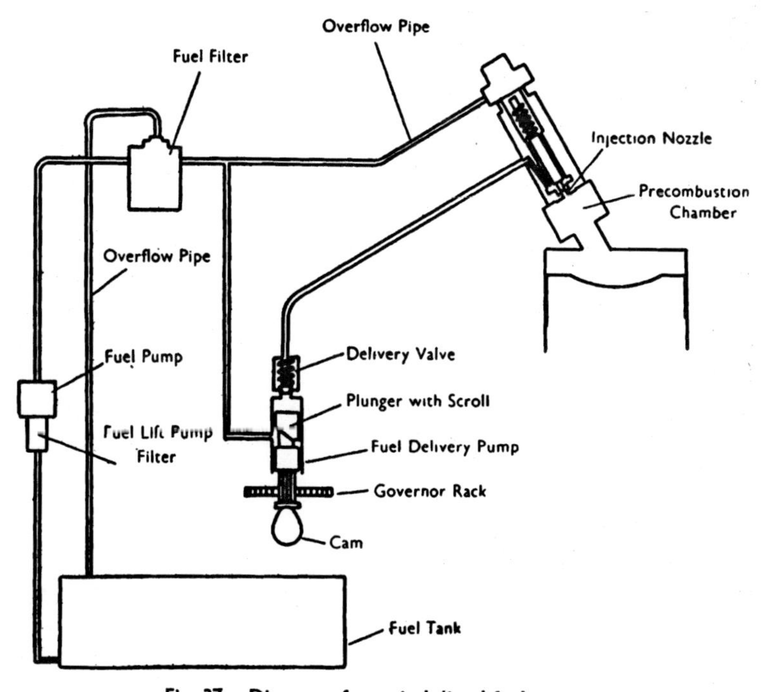

Fig. 37. Diagram of a typical diesel fuel system.

The injector consists of a needle valve held on to its seating by a powerful spring. The pressure of the fuel in the delivery lines is enough to lift it off its seating and drive fuel through the spray hole. The size, shape, and length of these holes determines the kind of spray produced. A fuel drain conducts excessive fuel back into the system. Nozzles fall into three types—simple orifice, multiple orifice, and pintle or variable orifice. The ultimate goal is the same however—the production of a finely-divided spray throughout the air charge. The vital factors are the droplet dispersion and the distance the spray penetrates. The principal aids to dispersion are turbulence and the fluid frictional forces.

Injection takes place over an interval of time, and the fuel burns more slowly than a petrol and air mixture. For this reason the power impulse is more prolonged. Combustion chambers are of many different designs. Some designers favour pre-combustion chambers, while others prefer swirl chambers, and so on. The object is to obtain a thorough mixing of the air and fuel. Turbulence is of great importance, as it influences the degree of mixing of the fuel and air, which in turn determines the amount of fuel that can effectively be burned in a given combustion chamber. The lag in igniting the fuel is a direct cause of diesel knock. When combustion does get started it takes place at a very rapid rate. To prevent this, fuel is needed that ignites spontaneously at relatively low temperatures.

The performance rating of diesel fuels has not received much attention. This is partly because of the smaller demand for them, and partly because of the smaller gain in engine performance, due to the efforts of designers to produce an engine that will run satisfactorily on the cheapest possible type of fuel. Cheaper fuel offsets the higher cost of engine manufacture. Special fuel properties would mean higher fuel costs. Diesel fuels are rated by their cetane number.

Starting a Diesel Engine

Some of the latest diesel tractor engines are started solely by the use of an electric starter motor. Not all compression-ignition engines, however, are started so easily. Various methods are used. One of these is by means of a separate petrol engine engaging with the main engine by means of a clutch.

A second method is to incorporate in the engine an additional system which allows for starting on petrol. In this second case a spark plug and ignition system are added, and when the engine has heated up the change-over is made. The main cylinder is built with a compression ratio suitable for compression ignition. An additional chamber is provided which produces a compression ratio suitable for a petrol engine. The spark plug is fitted in the additional chamber. The induction manifold admits air, either direct to the main chamber or else through a carburettor. When a certain temperature has been reached the small chamber is isolated, and the inlet manifold delivers air to the main cylinder. The magneto drive is disconnected and the compression-ignition cycle starts as the fuel pump commences to operate.

A third method uses a compression release, which lifts the exhaust valve, allowing cranking by hand. To this may be added a means of obtaining higher than normal compression when momentum has been achieved simply by cutting off some of the combustion chamber.

In another design an inertia starter is used. The starter consists of a flywheel running freely. This is cranked by hand and then connected to the engine by means of a clutch.

Other starting methods include the use of hot spots, electrically-heated plugs, and cartridges which are fired off to start the piston moving downwards rapidly, the piston being first of all set in the correct position.

The Cooling System

The cooling system of a compression-ignition engine does not differ in principle from that of a spark-ignition engine (see chapter 18). Always examine an unfamiliar engine with the aid of the instruction book to make sure that any peculiarities of design are fully understood. Trouble can sometimes be anticipated by keeping a close watch on joins, gaskets and bearings, and on the performance of the thermostat, if one is fitted.

The Lubrication System

The lubrication needs of a compression-ignition engine differ from those of a spark-ignition engine. Higher compression pressures and fuel deposits have made it necessary to develop special lubricating oils that will be adequate under conditions in which conventional oils would fail. Very high temperatures can cause incompletely burned fuel oil to combine with soot and lubricating oil to form lacquer deposits which cause rings to stick, with a consequent reduction of compression.

In engines where the underside of the piston is cooled by an oil spray to help reduce lacquer formation, the oil used for cooling is momentarily heated sufficiently to produce severe oxidising conditions, so greatly accelerating the production of sludges, varnishes and acids. To prevent these things happening heavy-duty lubricating oils have been developed (see page 149). With these oils certain chemicals have been added to give increased resistance to oxidation, while other chemicals have been added to give detergent characteristics. Detergency means the ability to clean a surface. The particles removed from the surface are held in suspension in the oil and remain dispersed. If large enough, they are removed by the filter. Other chemical additions provide increased stability, added film strength and anti-foam properties.

For high speed, high output compression-ignition engines special heavy-duty oils are essential. They prolong operating life by reducing piston and valve steam deposits. In the medium and low speed compression-ignition engine justification for the added expense is found in the prevention of ring sticking with its accompanying ills.

Tractor overhaul and maintenance

THIS CHAPTER deals with the general problems of tractor overhaul and maintenance both for the engine and the chassis. Many of these are the same for both types of engine—spark-ignition or diesel.

The instruction book that goes with every machine is the only correct guide to the maintenance of the machine. The present brief summary does no more than call attention to the points that most engines have in common. No attempt is made here to cover any special features, the instruction book is written specifically for that purpose.

Operators of machines are mainly concerned with functional disorders. The designer had certain conceptions which he arrived at with the aid of mathematics and experimental data. The operator must maintain those conceptions by carefully adjusting or replacing those parts subjected to wear and tear. From time to time overhauling must be done to test the condition of the engine and determine the need for specific repair jobs. Finally there is the last step in a repair job, which consists of making sure that the new or repaired part is in correct adjustment with its fellows.

It is very important indeed to keep a log book, noting carefully how many hours' work an engine does every day. Note the amount of oil and fuel used, the breakdowns, repairs and replacements. The instruction book will set out what is to be done at the specified periods, and without the log book those periods could only be estimated by guesswork. The periods will commence with daily checks and progress to major overhauls.

The daily check on a tractor is sufficiently important to outline its essentials. Check all oil levels and the level of water in the radiator. Lubricate where stated in the instruction book, check the fuel and fill up for the day's work. When using a tractor think out the day's work beforehand and make suitable preparations. Take a trailer with spares, fuel, water and oil. Take also shackles, chains, bolts, spare plugs, tools, etc., with you to the field if you are going out for the whole day.

When running always maintain a high operating temperature. The cooling water should be just below boiling. Make use of the radiator shutters or blinds in this connection. Keep the tractor running on full load doing useful work as much as possible. For most efficient running alter the setting of the jets to suit the load, but only if you feel sufficiently competent to adjust the carburettor without endangering the engine. Always open the choke as soon as the engine starts.

Make careful adjustments to your implements to make sure that as little energy as possible is wasted. This is particularly important when ploughing. In other cases use as wide an implement as the tractor will pull without labouring. Use a higher gear and a wider throttle setting where possible, thus allowing the engine to work with full compression, an increased charge being drawn in on every down stroke. Some general suggestions for the overhaul and maintenance of an engine follow.

Compression

To test the compression remove all the spark plugs except the one from the cylinder to be tested. Replace this one with a compression gauge if available. Test the compression by turning over the engine by hand. If you are using a gauge you can read the compression for each cylinder. If you are not using a gauge you can estimate the compression by the resistance offered as you turn the crank. Compression should be the same for all cylinders. Lack of compression indicates a number of possibilities: among them are worn piston rings and cylinder wall, seating of the valves in poor condition, or incorrect tappet clearance. The cylinder head gasket might be damaged or the piston rings gummed in their grooves so that they cannot spring outwards. Eliminate the possibilities until the reason is found. The exhaust tappet clearance will be greater than the inlet clearance due to expansion when heated.

Side valves have their adjustment on the tappet and overhead valves generally on the rocker arm. Use feeler gauges and check the clearance again after tightening up the lock nuts. Apply thin oil to the exposed parts of the valve stem and, on overhead valves, to the rocker arm assembly if no provision has been made for oiling automatically. Replace the valve cover and make sure that the gasket is in good condition.

The Ignition System

The spark plugs should be removed, checked, adjusted and checked again. Never bend the central electrode—only the outside one. Replace plugs with cracked porcelains and badly burned terminals. Use the correct plug. In general, cold plugs are used for petrol engines and hot plugs for a heavier fuel. Too hot a plug would shorten the plug life by cracking and blistering the porcelain and by rapid electrode burning. Too cold a plug generally results in plug fouling.

If the engine is in good mechanical condition, carbon deposits on the plugs would indicate too low an operating temperature, excessive idling or too cold a spark plug. The spark plug gasket is important for the transmission of heat from the plug to the cylinder head. Dirty threads will not conduct heat properly and may throw the plug out of its heat range.

Two-piece plugs, if of glazed porcelain, should not be sand-blasted, as this may break the glaze and alter the insulating qualities. One-piece plugs, however, cannot be cleaned by any method other than sand-blasting. Check and replace defective ignition wires where necessary. The high glaze on the wire acts as an insulator. A cracked or weathered appearance may be a sign of trouble ahead.

The efficiency of the magneto largely depends on the contact breaker, the points of which must be clean and meeting uniformly. If the points are dirty and pitted and meet irregularly, clean them up on an oil stone. If one point needs renewing, renew them both. Check the breaker spring: a weak spring will cause irregular firing and must be replaced. To test the points, remove the spark plugs to reduce compression. Turn the engine until the impulse coupling clicks; then turn the engine backwards slowly with the fan belt until the points are fully open. Arcing at the points may be due to a faulty condenser or to an excessive gap at the plugs.

Lubrication of the magneto is done in the factory, but, if an oiler is provided it is usual to add a few drops of oil for every 50 to 60 hours of use. The contact breaker may be lubricated with a felt pad. If so, remove and work the pad in grease and then replace it. A drop of oil on the rocker arm pivot is the only other point that needs lubrication. If there is no pad, then smear the cam lightly with grease. Do not let oil or grease reach the contacts.

Distributor and Battery

Clean inside the distributor, particularly between the electrodes. If there has been arcing at the electrodes, and there is a burnt path, too large a spark plug gap is indicated. Remove the high-tension pick-up, insert a clean cloth in the pick-up hole and turn the engine slowly to clean the slip ring. Make sure the carbon brushes of the high-tension pick-up make contact with the slip ring and the distributor rotor. They are spring loaded and should work freely in their guides. Make a rough test of the timing of the impulse coupling. When the piston reaches top dead centre at the end of the compression stroke the coupling should click. Remove it, clean and replace.

If the tractor is fitted with a battery and coil ignition, test the level of the electrolyte in the battery and add distilled water as required. Remove any corrosion on the battery terminals and cover with vaseline or light grease. Keep the hold-down clamps tight to eliminate vibration which will shorten battery life. Terminals and ground connections should be kept tight also. Clean the battery top with baking soda now and again and wash off with water.

The Generator

Check the generator commutator and brushes, and add oil to the oil cup regularly. The charging rate of the third brush-type generator is regulated by movement of the third brush. To increase the charging rate loosen the clamp or similar device holding the third brush and move it in the direction of armature travel. Do not push the output beyond the maker's recommendations.

After long service the commutator will become blackened on its surface, and scored slightly. The insulating mica strips may project above the segments. Proper brush contact will not occur. This leads to bad sparking and burning of the commutator and brushes.

A fine hacksaw blade is an ideal tool to remedy this. Withdraw the brushes, unscrew the cover, and withdraw the armature. Clamp gently in a vice with protected jaws. Only a slight undercut into the mica is necessary. Make the bottom of the cut square and not V-shaped. Polish the surface with fine glass paper—never with emery cloth, as the grains would bury themselves in the commutator and rapidly wear the brushes. Bed-in the brushes by passing a strip of fine glass paper with its cutting surface outwards round the commutator. Slip each brush in turn into its holder and move the glass paper backwards and forwards. Remove all the carbon dust which could be the cause of shorts. Excessive spring tension on the brushes will cause rapid wear. Make sure the right brush is used for the generator.

Should a wire lead break away from a brush it can often be reconnected by soldering. It is necessary first of all to copper-plate the upper half of the carbon brush. This can be done quite easily on the farm. Make a solution of copper sulphate in a jam jar and suspend in it a piece of copper. Twist a piece of wire round the brush so that it can also be suspended in the solution. Connect the piece of copper to a positive terminal and the carbon brush to a negative terminal of a small battery. In a few minutes a layer of copper will be deposited on the portion of the carbon brush that projects into the solution. Remove the brush and wash in water. Now soldering of the broken lead can be carried out in the normal way.

Keep the generator belt in proper tension. Too loose a belt slips and too tight a belt causes short generator bearing life and costly repairs.

Cut-out relays and voltage regulators are important devices designed to prevent damage to the battery and loss of charge. They are best left to the expert if anything goes wrong. The cut-out is a safety switch which connects the dynamo to the battery for charging purposes. It prevents the current from feeding back to the generator; it does not disconnect when a full charge is reached. Faults are usually confined to burnt-out windings, burnt or pitted contacts and sticking armature due to a weak spring. Remember to disconnect the leads from the battery before touching the contacts. Turning off the ignition is not enough. A voltage regulator controls the charging rate of the battery, and provides an increased output from the generator to balance the charge taken by the accessories such as the lighting system. Its action is automatic. If a lighting system is fitted, go over the wiring and make repairs to the insulation where necessary. These points will probably be where chafing has occurred.

The Cooling System

Flush out the cooling system every so often. Put a hose in the top of the radiator, open the drain cock at the bottom and allow the water to run out long enough to become clear. To aid in cleaning fill the radiator with a solution of two pounds of washing soda to four gallons of water. Run the engine until hot, then drain and refill with clean water. Use soft water if possible. If leaks are suspected test by putting the system under pressure. Use an adaptor in place of the radiator cap and a foot pump for air. Hose out the radiator from behind now and again to remove chaff and dead bees, etc. Test the fan belt for slack and take up if necessary.

The high engine ouputs and slow forward speeds of tractors make proper cooling essential. With tractor engines proper cooling depends on a forced draught. The fan and the fan belt are very important accessories. The fan belt may also drive the generator and the water pump and must therefore be kept in good order. Too tight a belt wears the bearings unduly. Inspect it frequently for tension and wear.

Some units have special circuits included with the ordinary circuit, such as air compressors, oil coolers, etc. Include all possible systems in the inspection of the main system to make sure they are leak proof, secure and in good condition. Remember to use a high melting-point grease in the water pump bearings.

A chemical combination of air, water and metal produces rust, and in the cooling system the ideal conditions for its formation occur. Corrosion inhibitors are available and should be used. As the radiator suffers from frequent vibration and shocks, inspect it frequently for leakage. Other points of possible leakage are the hose and tubing, the engine water-jacket, the cylinder-head joint and the water pump. With the adjustable gland type of pump normal wear of the packing will cause leakage unless the gland is tightened periodically and the packing replaced when worn. In the new type of packless pump the self-adjusting seals are subject to wear and have to be replaced every so often.

In some engines the radiator circuit is isolated from the remainder of the cooling system until a certain temperature has been reached in the cylinder block. When that temperature has been reached a thermostat operates a valve and adds the radiator circuit to the circuit round the cylinder block. By watching the temperature gauge during warming up it is possible to gain some idea of the functioning of the thermostat and test it.

Some cooling systems operate at atmospheric pressure and some at a higher pressure. The pressure type is completely enclosed and a special cap is used. The principal advantage of the pressure system is that no coolant is lost due to evaporation or to slopping over and higher cylinder wall temperatures are possible without boiling the water.

Alkaline cooling system cleaners such as washing soda cut grease and remove loose rust deposits but are not entirely effective. They do not remove rust and hard scale in the radiator cores and jackets. For this purpose oxalic acid and sodium bisulphate cleaners are used. Acid cleaners should always be followed by a neutralizer. The procedure is as follows. Completely drain the system. Add the cleaner compound and fill the system with water. Run the engine at fast idling speed for at least 30 minutes after the cleaning solution has heated up. It should be maintained at least at 180°F. but below boiling. Completely drain the solution, add the neutralizer and fill with water. Run again at fast idling speed for at least five minutes after the neutralizing solution has heated up. Completely drain the neutralizing solution and fill up with water. Again run the engine at fast idling speed at least five minutes after the engine has heated up, then completely drain to complete the procedure. Keep the radiator covered during the idling periods if a thermostat is fitted. This will make sure that the thermostat is open and the radiator is included in the circuit.

The Fuel and Carburettor System

Inspect and refill the air cleaner oil bath. For the combustion of every gallon of fuel about 9,000 gallons of air are needed. An air cleaner must remove dust effectively but must not restrict the air flow. Frequent servicing is essential to maintain the cleaning properties of the element and to provide a free passage for the air.

The volume of the air flow is altered by the use of an obstructed dirty filtering element, and also by the viscosity of the oil used in the air cleaner.

An obstructed and insufficient supply of air will result in an excessive fuel consumption and loss of power. Constricted air lines and improperly adjusted choke valve may be as much to blame in this respect as a clogged air filter.

Remember that dust can also enter the engine through worn carburettor valve bearings and through defective gaskets. Some dust will enter the engine, and it will end up in one of two ways. It will either mix with the carbon products of combustion to help form deposits on cylinder heads and piston tops, or it will find its way into the lubricating oil. Mixed with the lubricating oil it will cause damage by abrasion or more serious trouble due to the blocking of the oil ways.

Clean the vent hole in the fuel tank filler cap if clogged, as a blockage here will prevent the fuel from flowing. See that all the fuel connections are tight. Clean all the fuel filters and sediment bowls, and locate and clean all strainers. There is one usually where the fuel enters the carburettor float chamber. Drain and clean this chamber and dismantle the carburettor and blow through with an air pump. Direct the air in the opposite direction to the flow of petrol. Leaks of petrol at the jets are caused by loss of float buoyancy, worn float valve seat or valve, and dirt under the seat. Cork floats are apt to lose their buoyancy in time. Tighten the connection between the fuel bowl cover and the fuel bowl. There may be a gasket there because in some makes proper functioning requires an air-tight joint. Set the needle valves according to the instruction book. Wrong setting may cause overheating, loss of power, poor acceleration, damage to the valves, spark plugs and piston rings or excessive fuel consumption. The throttle stop controls the idling speed in some carburettors, while the idling needle valve may control either the air or fuel for idling.

Clean and lubricate the linkage between the governor and the throttle. Binding in the linkage may be a cause of hunting. Warm up the engine and set the governor controls for maximum revolutions. Test the engine revolutions at the belt pulley or power take-off, and check with the instruction book.

Overhaul and Maintenance

The Lubricating System

Test the oil levels frequently. There will be at least two levels to test, one in the engine crankcase and the other in the transmission. Change the oil as frequently as stated in the instruction book. Changing oil is a good preventive maintenance practice. Oil in the transmission and final drive becomes contaminated in a number of ways. Metal particles flake off during contact between gear teeth, particularly with new tractors before they are run in. Condensation causes a gradual accumulation of water, which causes rusting and decreases the lubricating ability of the oil. Dust also enters and reduces the efficiency of the oil. The only sure way to cut down the need for repairs is to drain the oil and flush the housing before refilling with fresh lubricant.

The oil used in the transmission is usually a heavy one which can maintain a film under extreme pressure conditions. Drain after work when the oil is hot, or else add paraffin to drain when cold. To do this tilt the tractor front end downwards by stopping on an incline. On no account allow oil to get on to rubber tyres, and if necessary make a metal trough to act as a runway for the oil. It is best to do this oil-changing procedure away from where the machine is usually parked.

Don't tow a tractor at a faster speed than that at which it normally operates. The transmission gears are not designed for speed work and will overheat the oil if a high speed is continued for a long period. This may result in lubrication failure.

Oil has multiple functions; one of them is to keep the engine clean. Frequent oil changes lower maintenance costs, increase life and give better engine performance. Farm conditions particularly promote oil contamination. Oil filters are too small to be really efficient and frequently work on a by-pass system. Only a small amount of oil passes through them in proportion to the amount that passes through the main circuit. Not only is oil changing necessary at frequent intervals but the oil filters should be taken out and cleaned. Make sure that you know where all the filters are. Some tractors have as many as four separate filters. There may be a sheet of perforated metal in the sump and there will be a filter on the intake side of the pump. The filler cap orifice will have a filter to prevent solid matter entering and then there is the filter proper.

Finally do not forget the crankcase breather. Rinse the filtering element in petrol or paraffin, dip in light oil and replace. Always be guided in your choice of oils by the maker's recommendations.

Special Diesel Engine Maintenance

The care and maintenance of a compression-ignition engine is much the same as for that of a spark-ignition engine with certain differences in the fuel system. The fuel system of a compression-ignition engine is a delicate high-precision mechanism and is best dealt with by an expert. Servicing is confined to the injectors and involves keeping the nozzles and the plunger free from carbon. The plunger spring and adjusting mechanism should not be touched.

It is essential to work on a clean bench. It is a good idea to spread a newspaper on the bench. On no account should cotton waste be used for cleaning, as small particles of the waste are very easily left on the parts and on re-assembly cause trouble. To check the injectors, if trouble is suspected, run the engine at the speed at which the trouble is worst and loosen the fuel line nut above the pump sufficiently to cut out one cylinder. If a difference is made to the engine the injector of the cylinder you have cut out is working properly. If no difference is made, remove the injector, couple it up outside the engine and run the engine at full throttle momentarily, by means of its starting device, to see the quality of the spray. A cone-shaped spray of fuel should be observed. An uneven cone indicates blocked holes. It must be realised that the pump delivering fuel to the injector may be at fault. To tell whether the cause of the trouble is in the pump or the injector involves using special equipment which submits the injector to a hand-pump test. It must be remembered that ignition depends on compression, and poor compression gives poor ignition and poor combustion. Poor compression arises for the same reasons as in a spark-ignition engine, plus the possibility of poor injector seatings. There are therefore reasons for combustion defects other than those traced to pumps or injectors.

The state of the engine exhaust is a general guide indicating when the nozzles should be cleaned. Indications that the nozzles may want cleaning are black or dark smoke from the exhaust, loss of power, an increased leakage from the nozzles through the by-pass leak-off, and rough running. To clean the nozzles fill a new dish with paraffin or fuel oil and have some dry soft cloths handy. Never use fluffy cloths. Soak the nozzles in the dish and clean the interior and the spray holes. Use tools of very soft metal. Rub the nozzle valve with a piece of rag soaked in light oil, and before assembling make sure that everything is clean. Smear the valve with a good grade of petroleum jelly, and then re-assemble. Tighten up the connection and replace the fuel line and the drain pipe.

Because of the small tolerances in the fuel injection system only clean fuel must be used. There may be as many as three or four filters in the fuel system. Clean these regularly, for any dirt getting into the injectors or pumps will cause serious trouble.

As carbon formation is more rapid in a compression-ignition engine the instruction book must be consulted, for a special grade of oil will be used in order to prevent gumming of the pistons. In general, as compared with spark-ignition engines, there will also be more frequent intervals between decarbonising and more frequent oil changes. Due to the higher pressures in the cylinder, wear will be greater but the valve assembly will remain in good order for a longer period due to the low operating temperatures.

The biggest contribution towards a clean and trouble-free fuel system comes from careful sedimentation of the fuel before it is put into the tractor. Two or more storage drums are essential for this.

Chassis Maintenance: Wheeled Tractors

To grease the bearings of the front wheels in some tractors it is necessary to remove the front wheels, clean out the old grease in the bearings and repack with new. In others the wheel hubs are fitted with pressure-gun fittings and are lubricated daily. Do this every half-day if working in very dusty conditions.

Check the front wheels for toe-in, which is an adjustment for camber. Wrong adjustment may cause excessive front tyre wear. Check the clutch and brake pedal free travel and take up excessive slack in the steering gear.

Remove the brake housing, inspect the linings and remove any oil or grease that has found its way there. Adjust the brake bands, where necessary, replacing those that are badly worn. Inflate the tyres to the correct pressure. Inflate the furrow wheel a little harder when ploughing and check the inflations frequently. When doing belt work, protect the tyres from the belt. Keep the tyres free from oil and grease and from sprays. Remove embedded flints before they work into the casing. Jack up a tractor fitted with pneumatic tyres if it is not to be used for a long time.

Chassis Maintenance: Crawler Tractors

Chassis maintenance with crawler tractors differs from that of wheeled tractors. Never lubricate the pins and bushes of a track: they should always run dry. Inspect them for wear every so often, and have them turned when necessary so as to present a new wearing face. Grease the rollers, the idler and the track frame pivot every day and adjust the steering mechanism according to the instruction book. The cost of turning the pins and bushes is small compared with the cost of rectifying neglect. Make a point of doing all that is possible to minimise wear. Load the tractor fully at the rated gear. An off-centre hitch increases wear in the steering mechanism due to the constant corrections needed. Extra furrows decrease the offset of the pull as well as increase the load to the capacity of the tractor. A swinging drawbar is ideal but is only possible with a central pull: extra furrows may make that possible. Make sure that the headland is adequate, so that sudden turns are not necessary. Make sure that the track tension is correct according to the manufacturer's recommendations. After turning the pins and bushes, the further term of service will be less than that which caused the original wear. Pins and bushes should not be allowed to wear too much before complete replacement.

Storing the Tractor

When storing a tractor drain the dirty oil and refill with fresh, having first cleaned the filters. Run the engine for a few moments to cover all the parts with an oil film. Drain water from the radiator and fuel from the tank and the fuel lines. Grease all parts fitted with lubricator fittings. Remove the battery or magneto and store indoors. Jack up the tractor if fitted with pneumatic tyres. Every month, remove the spark plugs and pour in a spoonful of oil. Crank the engine a few times to spread an oil film on the cylinder walls. There are special corrosion-inhibiting oils available which maintain an oil film for a long period and make this unnecessary. These are used for the storage of aircraft engines, and could readily be used for farm machinery.

Overhaul and Maintenance

The life of a tractor varies with the type of work that it has to do. An average life is twelve years or 10,000 working hours. This can be increased with proper care and maintenance.

Before dismantling make sure that you know how each part fits with another and mark adjacent parts. Check all nuts and bolts that are likely to be affected by vibration and attend to all small adjustments at once. Gaskets are fitted where a tight joint is needed and there is no movement. See that the gaskets are in good order when replaced. Oil any bearing surfaces when assembling and always replace all forms of locking devices such as spring washers, cotter pins and wiring for screw heads. Make periodic checks over the electrical equipment to make sure that there is no possibility of a short circuit.

Tractors: Development and Principles of Operation

"There appeared no valid reason why locomotive engines should not be made suitable for moving agricultural machinery, whether threshing, ploughing by means of windlasses, or for other purposes for which the farmer requires motive power ; and it was with the view of encouraging the manufacture of such engines that the Society determined this year to offer a prize, not for a mere locomotive, but for 'the best agricultural locomotive engine applicable to the ordinary requirements of farming.'"

"Trials of Traction Engines at Wolverhampton."
Jour. Roy. Agric. Soc. England, 1871.

If early experiments with steam tractors are excluded, the history of the development of the use of tractor power began in 1890, when one of the first tractors powered with an internal combustion engine was built and used in the United States. An early British oil-engined tractor was built by Ruston-Hornsby in 1897, and was awarded a silver medal by the Royal Agricultural Society of England at the Manchester show in that year. This was soon followed (1901) by the "Ivel", a 3-wheeled tractor fitted with a horizontally opposed 2-cylinder 4-stroke water cooled engine. Since that time there has been a more or less continuous development of tractors in this country, but the design of the modern machine owes much to American influence. The first tractor to offer a serious challenge to horses for draught purposes in this country was the Ford, which was introduced from America in 1917. This differed greatly in design from anything that had preceded it, and was the forerunner of the type commonly employed to-day.

After the 1914-18 war there was a temporary loss of interest in tractors in this country, though a large number of "food production" Ford tractors continued to be used until they became derelict. A reason for the decline in interest was the unfortunate experience of many farmers with unreliable and badly serviced

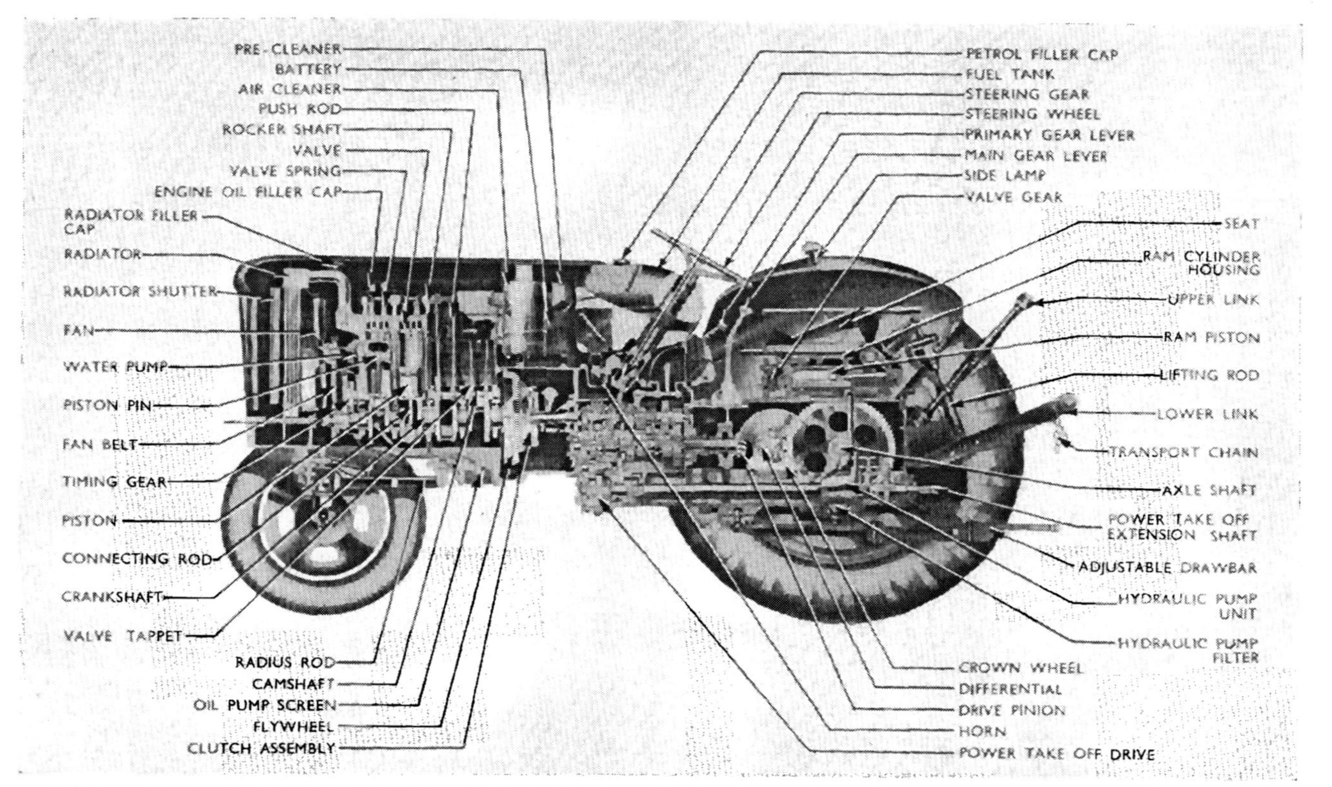

FIG. I.—SECTION OF MODERN MEDIUM-POWERED ALL-PURPOSE TRACTOR WITH HYDRAULIC LIFT. (FORDSON.)

machines during the food production campaign. Progress in America, however, continued to be rapid, and in addition to important advances in general design, such new departures as the production of the original International " Farmall " were made. In more recent years, noteworthy developments have included the introduction of low-pressure pneumatic tyres, together with many other improvements in design and manufacturing precision.

FIG. 2.—LIGHT-MEDIUM ALL-PURPOSE TRACTOR WITH DIRECT-COUPLED PLOUGH CONTROLLED BY HYDRAULIC MECHANISM. (FERGUSON.)

The range of tractors has been widened by the production of high-powered tracklayers at one end of the scale and small four-wheeled, two-wheeled and tracklaying machines at the other. Between these extremes, a wide selection of row-crop and other tractors has been designed for special uses.

The most significant development of recent years has been the steady evolution of tractors which are specially designed for use with " tractor-mounted " or " unit-principle " implements, mounted directly on the tractor itself and raised and lowered by means of a power lift. The " row-crop " tractor which was specially designed for work between the rows of growing crops is gradually becoming merged into an " all-purpose " outfit which retains the essential row-crop features. The modern tractor has, indeed, become a kind of multiple-purpose machine tool on which all manner of attachments may be mounted.

A contrasting development is the evolution of various types of specialized self-propelled machines such as the self-propelled combine harvester and the self-propelled tool-bar frame. While

FIG. 3.—MEDIUM-POWERED ALL-PURPOSE TRACTOR (DAVID BROWN) WITH P.T.O.-DRIVEN DIRECT-COUPLED ROTARY CULTIVATOR. (ROTARY HOES.)

some of these self-propelled machines are extremely efficient at their specialized work and valuable on certain types of farm, it should be realized that the necessity of providing a transmission system with each machine—even if engines were made interchangeable—renders any great extension of evolution in this direction unlikely at present.

Other important advances in tractor design include the introduction of satisfactory " half-track " equipment that can be easily fitted in place of the rear wheels ; better carburettors and governors which lead to improvement in fuel economy ; improved transmissions with a wider range of speeds and better braking ; the more general provision of electric starting and other aids to operating comfort, and a steady improvement in general tidiness of design.

Most modern machines can be relied upon to give long and efficient service. They are, moreover, extremely adaptable power units, in that power may be delivered either to directly mounted tools or at the drawbar, at a belt pulley, or at a "power take-off".

The Use of Tractors in British Farming. In the period between the first and second world wars the use of tractors

FIG. 4.—LARGE WHEELED TRACTOR WITH ENGINE DRIVEN PICK-UP BALER. (MASSEY-HARRIS.)

in Britain increased steadily, but by 1939 the total number in use in England and Wales was still no more than 55,000, as compared with 549,000 working horses. A survey of tractor use in the Eastern Counties during the middle nineteen-thirties showed that tractors represented 42 per cent. and horses 45 per cent. of the total power available in tractors, horses and stationary engines, but that horses did 70 per cent. of the work and tractors only 26 per cent.

By the end of the 1939-45 war the position had been transformed. Tractors had become much more versatile and horses of less importance, both in numbers and in use. By 1951, some 300,000 tractors were in use in England and Wales compared with 249,000 working horses, and to-day there are thousands of farms which rely entirely on tractors for draught work. It seems most unlikely that the change from animal to mechanical power in farming can be arrested or reversed. On the contrary, the steady increase in wages, and the need to obtain increased output from a restricted labour force, make it essential for farm workers to utilize mechanical power on an ever-increasing scale. (See also Chapter Twenty-four.)

It is generally agreed that a man is capable of developing approximately $\frac{1}{8}$th h.p. The cost of this power, with wages at 3s. an hour, is approximately 24s. per h.p. hour. No farmer can afford to rely much on the unaided power of human muscles. The worker who is equipped with a pair of horses can provide power at a cost of about 3s. 6d. per h.p. hour, and with a tractor developing only 15 drawbar h.p. the cost of power at the drawbar is rather less than 6d. per drawbar h.p. hour in average conditions. It may safely be claimed that the future of agricultural progress depends largely on the extent to which mechanical power and machinery can be employed to render labour more productive.

In present conditions, some farmers find it best to do part of their work with tractors and part with horses, partly because there is a limit to what one tractor can do in the peak periods. For instance, a tractor can be used quite satisfactorily for drilling ; but it often happens that the tractor is needed for ploughing or cultivating when the time for drilling comes, and it cannot do both jobs.

There are also many light jobs, where the tractor is necessarily run under-loaded, at which horses can compete successfully with tractors on almost any count. An example of such work is beet hoeing, where the typical small tractor outfit covers only the same width as a one-horse outfit. Thus, while it is possible, with suitable tractor power units, to do practically any job on the farm without horses, there is still on many farms a certain amount of work that is done more economically if horses are employed. In particular, light transport work where a great deal of stopping and starting is involved, and very light cultivations such as harrowing, are often better done by horses. The chief practical questions that arise are just what sorts of tractors are best for the farm and the work, how many tractors are needed, and how few horses it is necessary to keep to do those jobs that the tractors cannot manage or the horses can do better.

In considering the influence of the use of tractors on farming organization the whole farming business must be studied. Tractors have attained their present important position in

FIG. 5.—MEDIUM-POWERED ALL-PURPOSE TRACTOR (NUFFIELD) WITH HYDRAULIC FRONT-MOUNTED MANURE LOADER. (COMPTON.)

British agriculture not merely on account of lower working costs, but also because of the benefits that arise from ability to do cultivations more thoroughly and in better season, with the production of more or better crops. Extra tractor work is frequently justified by a more intensive system of farming and a greater output.

The Cost of Tractor Work. The cost of operating tractors varies from farm to farm and from district to district. Factors influencing costs, apart from the size and type of machine, include the number of days worked annually, the types of work performed, the care the tractor receives, and several other items.

One of the most important factors influencing cost per tractor hour is the amount of use. Low hourly costs are achieved on some farms by using tractors nearly every day of the year and doing all kinds of work with them. On light land it is generally possible to work tractors many more days in the year than on heavy. Low hourly costs achieved by regular use are not, however, necessarily an indication of efficient use, since this may be achieved by using tractors for work that horses could do more economically.

The cost of operating a modern medium-powered paraffin tractor, costing about £450 on pneumatic tyres, for 1,200 hours annually, would be approximately as follows :

	£	s.
Petrol, 95 gallons at 4s. 2d.	19	16
Vaporizing oil, 1,350 gallons at 1s. 4½d. ..	92	16
Lubricating oil, 40 gallons at 8s. 6d.	17	0
Grease	1	0
Cost of fuel and lubricants	130	12
Repairs (at Garage, £25, on farm, £10) ..	35	0
Licence and insurance	3	0
Depreciation $\left(\text{6-years life } \frac{£450}{6}\right)$	75	0
	£243	12

Cost per hour approximately 4s. 0d.

The method of charging depreciation in the above example is open to argument. If the " written down " method were used, depreciation chargeable in the first year would be 28⅛ per cent. on £450, i.e. £126 11s. The sum allowed for repairs is also possibly rather low. Farmers can readily calculate their own running costs by substituting their own figures for fuel, oil, repairs, etc., and a more detailed discussion of methods of costing mechanization processes may be found in another book by the present writer.[1]*

It should be noted that the total cost of operating the tractor considered above must be increased by approximately 3s. per

* 1. Figures refer to list of references at end of Chapter.

hour, since the driver's wages are not included in the calculation. The conclusion is reached that the total operating cost, including driver's wages, for a modern medium-powered tractor, is of the order of 7s. per hour.

The Choice of a Tractor. The best basis for choosing suitable tractors is experience, and practical trial in the conditions in which the machines are to work. Nevertheless, it is helpful to consider some of the principles involved in choice and to mention factors that need to be studied.

TYPE OF TRACTOR. In most instances the tractor will need to be an "all-purpose" machine which can be applied to almost any kind of farm work, including ploughing, cultivations, sowing, row-crop work, harvesting, transport and belt work. This will apply particularly to small farms where one or two tractors are required to do everything. On larger farms there may be sufficient of particular kinds of work to warrant using special types of tractors, e.g., tracklayers for deep cultivations, or light tractors of the self-propelled tool-bar type for drilling and inter-cultivation of root crops. Market gardeners may require tractors specially adpated for work in vegetable crops, and such special needs are briefly considered in Chapter Two, where the principal types of tractors available are described.

ROW-CROP AND STANDARD TRACTORS. Most modern medium-powered tractors are "all-purpose" machines which incorporate such features as make them suitable for work in all common farm row crops. Some manufacturers offer standard type tractors which are a little cheaper than the row-crop type because fewer adjustments and fittings are provided. Such machines may be quite suitable for use on farms where row-crop features are seldom or never needed. If the standard type has a lower centre of gravity it may be preferable for use on hillsides.

TRACTOR-MOUNTED IMPLEMENTS. If a general-purpose tractor is required, an important consideration is whether the tractor should be equipped with a power lift in order to enable it to use tractor-mounted implements and machines. Practically all modern British medium-powered tractors are or can be equipped with a hydraulic lift, and there is a wide range of efficient mounted implements and machines specially designed for use with them. In general, it may be affirmed that modern medium-powered tractors are most efficient and economic when used as " unit-principle " machines with their mounted equipment, and no farmer should ignore this point. Nevertheless, the use of mounted equipment has disadvantages, one of which is mentioned below.

FIG. 6.—MEDIUM WHEELED TRACTOR (INTERNATIONAL) WITH PAIR OF UNIT TYPE HAND-FED TRANSPLANTERS (ROBOT) ATTACHED TO GENERAL-PURPOSE TOOLBAR. (STANHAY.)

TRACTOR SIZES, AND STANDARDIZATION. One of the most difficult problems in choosing tractors for general farm work is to decide whether all the tractors should be of one make and size, or whether a range of makes and sizes should be selected to suit the various jobs. This is a problem on any farm where more than one tractor is needed, and is perhaps most difficult on medium-sized farms where three or four tractors are used. The advent of tractor-mounted implements has accentuated the problem, for though a group of manufacturers of large/medium tractors have power-lift linkages which will accommodate equipment designed for any one of them, two other important manufacturers produce tractors and equipment which are not interchangeable with the first group or with one another. The British Standards Institution has now introduced British

Standards for the three-link system which will in time help to overcome these difficulties ; but the present situation is that if a farmer wishes to operate two different sizes of unit-principle tractors he also has to equip himself with two distinct sets of mounted implements, each of which can only be used with the appropriate tractor. This leads to a high expenditure on equipment, and lack of flexibility in operation. So long as these conditions hold, there is much to be said for sticking to one make of tractor on small and medium-sized farms, and in these circumstances choice will frequently depend as much on the

FIG. 7.—LIGHT ROW-CROP TRACTOR WITH SMALL ENGINE-DRIVEN COMBINE HARVESTER. (ALLIS-CHALMERS.)

range of mounted equipment that is available to go with the tractor as on the size or other features of the tractor itself.

On larger farms, there is considerable advantage in having more than one type and size of tractor, since this arrangement allows the matching of the tractor with the job in hand, and permits choosing the best equipment from more than one manufacturer. Operating more than one type of tractor necessitates keeping more spare parts on the farm, and requires the man responsible for maintenance to have a wider knowledge ; but these are minor disadvantages compared with the greater operating efficiency where both tractor and equipment really suit the job.

Where it is decided to stick to one make and size of tractor, the decision as to which is most suitable must embrace a study of all the operations that will need to be carried out. Among the most important are ploughing and the basic cultivations, which must be done thoroughly and in good season, and will serve to illustrate the problems. The power needed for ploughing and other cultivations varies greatly according to the nature and condition of the land. The " ploughing resistance " of various types of soil may be indicated by a figure in lb. per sq. in. which, when multiplied by the total sectional area (sq. in.) of the furrow slices, gives the drawbar pull needed. Ploughing resistances range from about 5 lb. per sq. in. for very light blowing sand to over 15 lb. per sq. in. for heavy clay, an average figure for medium loam being about 10 lb. per sq. in. of furrow section. Thus, in average conditions, the drawbar pull needed to operate a 3-furrow plough with furrows 10 in. wide and 7 in. deep would be $10 \times 7 \times 3 \times 10 = 2{,}100$ lb. The drawbar horse-power required may be calculated by using the following formula :

$$\text{Drawbar H.P.} = \frac{\text{speed (m.p.h.)} \times \text{drawbar pull (lb.)}}{375}$$

At a ploughing speed of 3 m.p.h. the drawbar horse-power required for the above conditions would therefore be

$$\frac{3 \text{ (m.p.h.)} \times 2{,}100 \text{ (lb.)}}{375} = \text{approximately } 17 \text{ H.P.}$$

Such a performance is just within the capabilities of a medium-powered tractor of about 25 b.h.p., provided that conditions for traction are satisfactory. For heavier land, however, especially if the tractor is to operate on pneumatic tyres, it would be necessary to consider either a 2-furrow plough or a more powerful tractor.

Within limits, securing a satisfactory performance from a tractor is more a question of selecting suitable implements and using suitable gears than one of buying tractors of different sizes. Few of the jobs that modern all-purpose tractors are required to undertake need the full power of even the small/medium sizes, and two small/medium tractors are often more useful than one large one. On the other hand, where a large tractor can be given a full load there may be a considerable saving of labour cost by using it—a fact which may often be taken advantage of on large farms and by contractors.

TYPE OF ENGINE. Many tractors can be obtained with petrol, paraffin or diesel engine. As we have seen earlier, the total cost of operating a medium-powered all-purpose paraffin tractor for about 1,250 hours annually is made up approximately as follows : fuel and oil, just under 30 per cent. ; depreciation and repairs, just under 30 per cent. ; labour (driver's wages), 45 per cent. Owing to the high rate of tax on petrol, using petrol instead of vaporizing oil increases the fuel cost enormously, and compared with this increased cost, any savings in depreciation, lubricating oil and repairs are quite insignificant. While, therefore, the use of petrol may be necessary for tiny engines whose fuel consumptions are low, it can only be justified in very exceptional circumstances for medium-powered tractors. The choice between a vaporizing oil engine and a diesel engine is a difficult one. The first cost and depreciation on the diesel engine are considerably higher, but this is more than balanced by the saving on fuel cost. (The higher efficiency of the diesel (see Appendix Six) results in appreciably lower fuel consumption). The diesel engine usually costs more to repair when an overhaul is needed, and maintenance of the fuel pumps and injection nozzles is still not as easily carried out as that of the carburettors and electrical ignition equipment of the paraffin engine. The diesel engine usually costs more for electric starter batteries.

On balance, there is very little to choose between a good diesel engine and a good paraffin engine for a medium-powered tractor, and the increasing popularity of the diesel is at least in part due to the fact that many of the paraffin engines of the past have not been as up-to-date in design as the diesels with which they are compared. A good paraffin engine, with a bi-fuel carburettor that permits an immediate switch over from petrol to paraffin or vice-versa, and a manifold which will deal with vaporizing oil really efficiently, can be quite free from the starting and operating troubles which were so common in earlier models ; and the choice between paraffin and diesel is now largely a matter of personal preference.

WHEEL AND TRACK EQUIPMENT. For general farm work, where the medium-powered tractor is to be regularly used for transport as well as for all kinds of field work, pneumatic-tyred wheels are now almost an automatic choice. The tyres are so expensive, however, that some thought would be given to the possibility of changing to steel wheels when long spells of heavy ploughing or other heavy cultivations lie ahead. Modern tractor wheels are easy to change, and many farmers over-estimate the time taken for this simple job. The essential

FIG. 8.—MEDIUM-POWERED DIESEL-ENGINED TRACKLAYER WITH MULTI-FURROW PLOUGH. (FORDSON COUNTY.)

requirement is a flat, firm floor and a good hydraulic jack. Some tractors employ the tractor's built-in hydraulic lift mechanism and a simple linkage for jacking up the rear wheels. Where the tractor is regularly required to run on hard roads the advantages of fitting retractable strakes or wheel girdles for heavy work should be considered. Technical details are given on pages 55-56.

For very heavy tillage operations, and for operations on heavy land where damage to soil structure may be caused by tractor wheels, tracklaying mechanisms are most efficient. In really difficult conditions a tracklayer may operate reasonably efficiently where any type of wheel fails completely. A track-layer enables field work to continue late in the autumn and to start very early in spring, while at other times of the year tracklaying devices enable the tractor to translate a high proportion of the engine power into effective work at the draw-bar. Full-track machines are, however, expensive in first cost and maintenance, and non-versatile ; and half-track mechanisms which are easily interchangeable with pneumatic-tyred wheels offer attractive possibilities for medium-powered tractors. In general, it is a sound rule to stick to the cheapest type of tractor that will do the job satisfactorily, and only to buy expensive special-purpose tractors when the need has been clearly established. In this connection, the possibility of getting a

FIG. 9.—MEDIUM-POWERED ALL-PURPOSE TRACTOR (FORDSON) WITH HALF-TRACKS (ROADLESS) PULLING HEAVY DISC PLOUGH.

certain amount of work done by contract should always be kept in mind.

With regard to price, the cheapest tractor is, naturally, not always the best. Some of the higher priced outfits offer value for their extra purchase price in the form of fittings which on other tractors are extras. A cheap tractor is one that is cheap in operation ; and in order to achieve economical operation the tractor must be suited to the needs of the farm. Other factors to be considered are low fuel consumption, low maintenance cost, ease of adjustment and repair, dependability, adaptability, ease of operation and service facilities. There is now a good tractor to suit the special requirements of almost every farmer, and the prospective purchaser in making his choice should carefully weigh the advantages and disadvantages of each type of machine for his own farm, bearing in mind points of construction and operation dealt with in this and the following chapter.

Tractor Power or Capacity. The measure of a tractor's power or capacity for field work is its drawbar horse power. (See Appendix Four.) Tractors are, however, sometimes described, especially by American manufacturers, according to the number of plough furrows they can pull in " average " land at a speed of about 3 m.p.h. In such conditions approximately 5 drawbar h.p. is required to pull a single furrow of normal depth and width, and a tractor of 15 drawbar h.p. may be described as a " three-plough " tractor. Since ploughing conditions vary so greatly it is probably more satisfactory for those who understand the meaning of drawbar horse power to use the horse power figure to indicate capacity.

When it is known how many furrows a given tractor will normally pull on any particular soil, the width of other standard implements that the tractor may be expected to handle can be approximately estimated. A tractor that is capable of pulling three 10-inch furrows can generally handle three times the width of cultivator or disc harrow (7 ft. 6 in.), six times the width of heavy harrow (15 ft.) and five times the width of drill or binder (a 12-ft. drill or two 6-ft. binders).

If it is desired to calculate the rate of working of implements other than ploughs, the following formula, which allows 20 per cent. for time lost in turning at the headland, etc. may be used :

$$\text{Rate of working (acres per hour)} = \frac{\text{Working width of Implement (ft.)} \times \text{speed (m.p.h.)}}{10}$$

This formula is not applicable to ploughing or to other implements with which long stops are necessary for setting, marking out, filling up, etc.

TRACTOR TESTING. In 1920 a scheme for the compulsory testing of all types of tractors marketed in the State of Nebraska, U.S.A., was inaugurated, and these tests, which are carried out at the University of Nebraska College of Agriculture, have come to be recognized throughout the world as a reliable guide to the capabilities of the machines tested. Among the important figures obtained from the tests are the rated belt horse powers and the rated drawbar horse powers of the tractors. The rated powers are determined according to a Standard Farm Tractor Rating Code, drawn up by the American Society of Agricultural Engineers (A.S.A.E.) and the Society of Automotive Engineers (S.A.E.). Without entering into the technicalities of how the maximum powers are measured, it may be briefly stated that the *rated belt horse power* must not exceed 85 per cent. of the maximum belt horse power developed continuously on test, and that the *drawbar rating* must not exceed 75 per cent. of the maximum drawbar horse power.[2]

There are many good reasons for rating tractors below the maximum powers developed on test. For example, the tractors are tested for drawbar power on tracks which are level, straight, and have good surfaces for adhesion ; and the maximum load is gradually applied by electrical and other special devices. Working conditions in the field are always much less favourable, and tractors cannot generally develop their maximum test horse powers continuously when in practical use.

In Britain, tractor testing is not compulsory, but all the leading tractor manufactures make use of the testing facilities provided by the National Institute of Agricultural Engineering. The N.I.A.E. carries out two main types of tractor test in addition to any special test which may be undertaken to assist manufacturers in their development work. The first—the N.I.A.E. test—is a comprehensive one. In addition to the usual belt test and drawbar tests in a range of soil conditions there are also tests on a hillside, and a ploughing test. The drawbar tests are carried out on firm grassland on heavy clay, on stubble on light land, and on loose, freshly cultivated soil ; and the results are shown partly in the form of graphs. These N.I.A.E. tests give a great deal more information than the Nebraska tests, or the British Standard Test referred to below, concerning the pulls and powers that can be exerted by the tractor on farm land. A detailed test report which includes a specification is prepared, and provided the manufacturer agrees, the test report is subsequently published by the N.I.A.E. in the form of a small booklet.[3]

The second type of test—the British Standard Tractor Test—is also carried out by the N.I.A.E. but has a more restricted scope. The test conditions are laid down in British Standard No. 1744 : 1951. The drawbar tests of pneumatic-tyred tractors are carried out only on dry, level tarmacadam, while track-layers and steel-wheeled tractors are tested on dry, level, mown or grazed grassland on heavy clay. The results of the drawbar tests carried out under the British Standard Test regulations are roughly comparable with those produced by the Nebraska testing methods, but the actual performances recorded are obtained from a series of performance curves, and not from "spot" readings.

INTERPRETATION OF TRACTOR TEST RESULTS. It is not possible here to give detailed guidance on how to interpret tractor test reports, but some of the factors that need attention should be mentioned. The reader must first learn to appreciate the significance of drawbar pull, tractor speed and drawbar horsepower—terms which are explained in the Appendix. Maximum drawbar pulls on a test track are usually limited not by the power of the engine, but by wheel slip, and they can be greatly increased at low speeds merely by adding weight to the tractor. With pneumatic tyres on firm ground the addition of 1,000 lb. to the rear wheels will usually increase the maximum drawbar pull by about 500 lb. It is, therefore, important to know how much weight has been added, and whether the addition of such weight would be practicable in farm conditions. It must be realized that it would be possible for a tractor fitted with wheels quite unsuitable for normal farm work to put up an excellent drawbar performance in the Nebraska test or the British Standard Test. It is, therefore, important to study the results of ploughing and other agricultural tests when these are available.

Increasing the maximum drawbar pull by adding extra weight does not necessarily increase the maximum drawbar horse-power developed by a tractor. In the case of pneumatic-tyred tractors, as is shown later, the maximum drawbar horse-power is attained at fairly high speeds and low drawbar pulls, and the limiting factor at these higher speeds is engine power and not wheel slip. The addition of extra weight at these higher speeds will usually, in fact, slightly reduce the maximum drawbar horse-power. The reader of a test report will, therefore, need to study drawbar pulls and drawbar horse-power over the range of working speeds. Fig. 10 shows characteristic test results for a medium-powered tractor in first and second gears when fitted with pneumatic tyres and steel wheels. It illustrates many points referred to elsewhere, e.g. the higher drawbar pull with

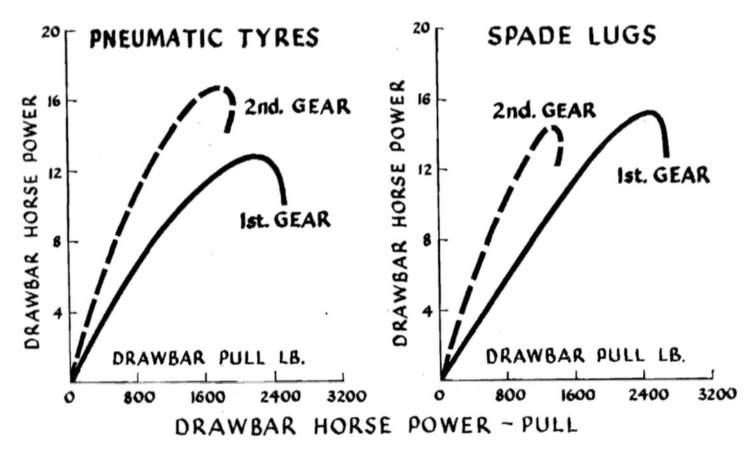

FIG. 10.—GRAPH SHOWING INFLUENCE OF WHEEL EQUIPMENT AND WORKING GEAR ON DRAWBAR PULL AND DRAWBAR H.P. OF A MEDIUM-POWERED TRACTOR.

steel wheels, and the higher drawbar H.P. obtainable in second gear on pneumatic tyres. Other graphs included in test reports may show the effect of added weight on these factors, and the effects of various combinations of conditions on specific fuel consumptions and on wheel slip.

Economic Operation : Loading. So long as a tractor is in use, even though the work is very light, labour has to be employed to drive it and expense is incurred on depreciation, repairs, lubricating oil and fuel. Apart from fuel cost, none of these items is greatly influenced by the load. Thus, labour is a constant

2

charge per hour and it has also been shown that provided there is no overloading of the tractor, depreciation and repairs are very little higher in an engine which works at full load than in one which works at half load, assuming that each works the same number of hours per year. Indeed, the depreciation of a paraffin engine that is not given a full load may be very much greater than that of one working at full load, owing to the serious dilution of the lubricating oil that occurs if the engine is allowed to run too cold. The labour and overhead charges therefore represent an almost constant charge per hour, regardless of how much power is developed ; and on an average, these constant charges represent above two-thirds of the total cost of running the tractor at full load. If the tractor is worked at half load or less, the labour and overhead charges account for much more than two-thirds of the total cost, and the only saving of any importance is the reduction in the amount of fuel used. Thus, suppose that the total cost of running a medium-powered tractor at full capacity is 7s. 6d. per hour ; about 70 per cent. of this amount (5s. 3d.) represents a constant charge, incurred whether the load be great or small. For example, it would cost about 6s. per hour (5s. 3d. constant charge, plus 9d. fuel) to run the tractor without load, and about 6s. 9d. per hour to run it at half load. It is, therefore, important for economical working to provide the tractor with a full load whenever practicable.

The tractor engine is normally fitted with a governor to ensure that the engine maintains a constant speed. A gear-box provides various definite forward speed ratios, and assuming that the governor keeps the engine speed constant and that there are no complications due to slippage, the tractor has definite forward speeds.

For each gear the tractor has a fairly definite optimum drawbar pull, and this optimum is generally about 80 per cent. of the maximum drawbar pull at that speed. The object should therefore be to provide implements which require drawbar pulls approximating to this optimum. For instance, if a medium-powered tractor pulling a 2-furrow plough at 2½ m.p.h. exerts a drawbar pull of 1,500 lb., it is developing only 10 h.p. on the drawbar. In this case a 3-furrow plough could probably be pulled, and the efficiency of working would be much higher for the reasons explained above. If the implement is small, the load may often be adjusted by putting the tractor in a higher gear, or by adding another implement "in tandem".

An adequate load is also of importance from the standpoint of optimum fuel economy with all types of tractors. Typical specific fuel consumptions of tractors recently tested at N.I.A.E. are as follows :—

Type of Tractor Engine	Specific Fuel Consumption (lb. per b.h.p. hour)		
	At full load	At ½-load	At ¼-load
Medium sized petrol ..	0·65	0·90	1·40
Medium sized paraffin ..	0·72	1·05	1·60
Medium sized Diesel ..	0·47	0·58	0·86

Thus, the fuel consumed per b.h.p. hour at half load is approximately 30 per cent. higher than at full load ; while at quarter load it is a little more than double that at full load. The superior fuel economy of diesel engines is more marked at light and moderate loads than at full load.

There are times when it is impossible to load a tractor adequately, but at other times it is possible to increase efficiency greatly by either adding to the load or working in a higher gear. A practical method commonly employed to provide a full load and speed up the work is to hitch implements in tandem. There are often savings in both time and money if operations such as discing and harrowing, drilling and harrowing, rolling and harrowing, etc., are carried out in one journey across the field. There are, of course, operations where it is impracticable to increase either the width of work or the speed, or to hitch implements in tandem. In such circumstances the best solution will be to run the tractor in a higher gear, with the engine throttled down. With a trailer mower, for example, the most economical method of operation may be to run in high gear with the engine throttled down to about half speed.

Wheel Slip. Slipping of a tractor's drive wheels always wastes power and fuel, and with pneumatic tyres this wastage may be serious, even if the wheels do not spin. A simple method of determining wheel slip when a tractor is working is to make a mark on the tractor wheel and then measure the distance the tractor moves forward in, say, 10 revolutions of the wheel, first under load, and then on the same surface with no load. The percentage slip will be :

$$\frac{(\text{Distance travelled without load}) - (\text{Distance travelled when working})}{\text{Distance travelled without load}} \times 100$$

The amount of slip revealed by such a test is often surprising. Fifteen per cent. slip on pneumatic tyres is hardly noticeable, yet it represents an important waste of time and tractor fuel. Slip can never be entirely eliminated, but it can sometimes be minimized by lightening the load and working in a higher gear, while at other times it may be remedied by adding weight, fitting strakes, or fitting alternative types of wheel or track equipment.

Operating Speeds. A three-speed gearbox, giving forward speeds of about 2, 3 and $4\frac{1}{2}$ m.p.h. at the standard governed engine speed, was adequate so long as the tractor was normally fitted with steel wheels ; but to-day, with almost all new medium-powered tractors going out on pneumatic tyres, the four-, five- or six-speed gearbox has many advantages, especially where the tractor can be easily converted from pneumatic tyres to steel wheels or to half-track equipment.

On steel wheels, the maximum drawbar horse power that can be developed generally falls off rapidly at speeds above 3-4 m.p.h., owing to increased rolling resistance.

Practical tests with steel-wheeled tractors have shown that where 3 furrows can be pulled without undue wheel slip in bottom gear, ploughing in this gear gives a saving of both time and fuel over ploughing with 2 furrows at a higher speed. With pneumatic-tyred wheels, however, the situation is quite different. A study of published tractor tests shows that because of the limitations imposed by wheel slip at high drawbar pulls, modern pneumatic-tyred tractors can only exert their maximum drawbar powers at moderately high speeds. With such tractors, practical experience shows that pulling 2 furrows at a speed of 3-4 m.p.h. is on most soils a better proposition than attempting to pull 3 in bottom gear—an attempt which may prove unsatisfactory even after wheel weights or wheel strakes are employed.

On the other hand, it should be realized that arguments against the general use of high speeds for tillage are provided by tests which show that the draught of implements increases with increase in speed. For example, increase of the speed of ploughing from 2½ m.p.h. to 4 m.p.h. may increase the draught by 10 per cent. Unless the increased draught is justified by better work (a rare occurrence), it represents sheer waste of power. Moreover, high speeds cause increased wear and tear of the implements owing to the greater shocks and increased rubbing friction. It should be added that there are certain jobs such as spraying, hoeing, combining, etc., which cannot be performed at above a certain critical speed without causing very poor work.

For ploughing, special high-speed bodies are required for work at above about 5 m.p.h. Normal plough bodies simply will not stay in the ground at high speeds.

A practical objection to the use of very low speeds is that they necessitate high drawbar pulls to provide a full load. It may be inconvenient to provide large enough implements, and in any case, under ordinary field conditions the tractor may be unable to exert very high drawbar pulls owing to adhesion difficulties. The best practical solution is to select implements which, in average conditions, provide a full load for the tractor at about 3-4 m.p.h. so that more difficult conditions may be met by the use of a lower gear.

Light, pneumatic-tyred tractors frequently need to be "ballasted" in order to secure a good drawbar performance at low speeds. Methods of ballasting include the use of liquid filling for the tyres, and addition of cast-iron wheel weights. The effect of added weight on maximum drawbar pulls and drawbar h.p. is illustrated by figures from an N.I.A.E. test report. The rear axle weight (2,856 lb.) of a Fordson Major tractor which was already water ballasted 75 per cent. was further increased by

1,600 lb. This increase of rear axle weight produced increased drawbar pulls in first gear ranging from 800 lb. to 1,600 lb. according to conditions. On light land in a condition favouring the performance of a heavy tractor, addition of 1,600 lb. to the rear axle weight increased the drawbar h.p. in low gear by 91 per cent. It should be added that a heavy rear axle weight on a rubber-tyred tractor is not always desirable. It is especially undesirable when the tractor is used on damp seedbeds or on growing crops.

The value of a high road speed on an all-purpose medium-powered tractor is now generally agreed. A road speed of 15 m.p.h. or so is invaluable for rapidly moving to scattered fields, and also for some forms of road transport.

Engine Speed. The modern tendency is to increase the speed range provided by the gearbox by means of a good, flexible governor control. The governor is made capable of adjustment to give about 50 per cent. more or less than the standard governed engine speed, so that a speed range of say 2 m.p.h. to 10 m.p.h. is extended by means of the governor control to a range of 1 m.p.h. to 15 m.p.h.

Intelligent use of the gears and governor control can result in appreciable fuel economy, as well as in a saving of time.

A modern tractor equipped with an "all-speed" governor and 5-6 gears can work at a normal speed of 3-4 m.p.h. either by use of a low gear and high engine speed, or in a higher gear at lower engine speed. Occasionally there is a choice of more than two gears and engine speeds. N.I.A.E. tests confirm what observant farmers have known for some time—that for fuel economy it always pays to choose a high gear and low engine speed whenever this is practicable. It is, of course, only at light or moderate drawbar pulls that there is any choice, but so much of a tractor's working time is spent on light work that this is a point of some importance.

Maintenance and the Tractor Driver. Farmers do not always appreciate the prime importance of systematic attention to care and maintenance of the tractor. All manufacturers provide a handbook giving detailed instructions on such matters as adjustments, greasing, changing lubricating oil, correct grade of oil for various parts of the tractor, and so on. This handbook is probably the most important part of the tractor's tool-kit, and the instructions contained in it should be carried out as thoroughly as possible.

The driver is assisted and encouraged to attend regularly to maintenance if he is provided with a log-book in which changes of lubricating oil and dates of the various other types of servicing are regularly recorded. Such a log-book, if properly kept, permits a check on fuel and oil consumption, and makes it easy to see at any time whether the tractor is due to be serviced or not. Farmers who have a number of tractors have frequently found that a properly kept log-book enables them to detect faults in operation and to improve tractor efficiency.

A good driver will take care to use clean water in the radiator and drain it out at night in winter. He will avoid getting dirt or water in the fuel, and will keep the working parts of the tractor reasonably clean. He should also be able to keep the engine tuned up by the adjustment of tappets, sparking plug points, etc., and should be capable of detecting trouble in its early stages and carrying out minor repairs and replacements on the farm. He should be able to carry out an annual overhaul and should generally take interest and care in the running and use of the tractor. In addition, he should be a competent ploughman and should be familiar with the working of such machines as the mower and binder.

It is clear, therefore, that high qualifications are called for; and it may be thought that few such men are available. This is, indeed, true; but it is possible to produce one by choosing a young and keen man and giving him a chance to learn by letting him assist the service agent to carry out an annual overhaul and all repairs. If he possesses the necessary aptitude and intelligence he should be able to do these things himself in a short time. It is sound economy to have at least one such man on the farm and to pay him a good wage. In addition to work on the tractor, there is work on fixed engines and all kinds of farm machinery that calls for considerable skill and is expensive if performed by outside labour.

Tractor and implement maintenance are greatly facilitated if a well equipped fuel and spares trailer is provided. Such a trailer should carry sufficient fuel to last several days, lubricating oil and grease ; a semi-rotary pump for filling the tractor or a good funnel and can ; a water-can and water-drum ; a tow-chain and spare shackles ; a good set of tools and spare plough shares, sparking plugs, nuts and bolts, etc. The cost of providing such a trailer is quickly repaid in time saved on servicing and minor troubles. Moreover, the driver is much more likely to attend to routine servicing if he is adequately equipped for the job.

Safety Precautions. Farmers and tractor drivers should never forget that limbs and lives may be endangered unless proper precautions are taken in handling power-driven machinery. With tractors the chief causes of accidents are carelessness in hitching to implements, and failure to fit proper shields over power drive shafts.

Hitching a tractor to an implement is often a difficult job for one man, especially where the implement drawbar is heavy, as with a disc harrow or a roll. Unless a special hand clutch is fitted, the tractor driver should always remain squarely on the tractor so long as it is in gear. The driver who attempts to operate his clutch and also to guide the implement drawbar over the tractor drawplate takes a big risk. His foot may slip ; and if this happens he will be fortunate if he escapes being crushed. Many lives have been lost in this way. Farmers should either see that the driver carries a handy block of wood or a jack to support the implement drawbar, or should send a second person to help with the hitching.

Power drives are seldom treated with the respect they deserve. The square shafts and universal joints, if not properly covered, readily catch up any loose clothing ; and if this happens, the operator is lucky indeed if it is only his clothing that suffers. Tractor drivers and all other workers who have to go near machinery in motion should avoid wearing loose clothing.

In the past, accidents have been caused by tractors which reared and overturned backwards if the clutch was too rapidly engaged. This now only occurs if the tractor is prevented from going forward by having its rear wheels in a grip, and is easily avoided by disengaging the clutch. Sideways overturning, however, still causes a number of fatal accidents, the most frequent causes being careless operation on steep hillsides and running too near the edge of a ditch. Where a tractor is required to work on steep hillsides the wheels should be set out as wide as possible, a sharp lookout should be kept for local hollows in the ground surface, and turns should be made slowly.

Save in very exceptional circumstances no attempt should be made to adjust or lubricate a machine that is in motion. The temptation to save time may be great; but on the farm, as in the factory, the motto should be " Safety First ". It simply does not pay to take unnecessary risks with any power-driven machinery.

REFERENCES

(1) Culpin, C. *Farm Mechanization: Costs and Methods.* Crosby Lockwood & Son, London, 1951.

(2) " Nebraska Tractor Tests." The University of Nebraska, Lincoln, Nebraska, U.S.A., Bulletin 397. January 1950 and supplementary data sheets on current models.

(3) N.I.A.E. Tractor Tests. Reports of Tests on individual tractors and other farm machines are published periodically by National Institute of Agricultural Engineering, Wrest Park, Silsoe, Beds.

(4) Influence of Engine Loading on Tractor Field Fuel Consumption. N.I.A.E., Silsoe, Beds. (1951).

(5) Hine, H. J. " Tractors on the Farm." *Farmer & Stockbreeder,* London, 1950.

(6) Jones, F. R. *Farm Gas Engines and Tractors.* (An American text-book.) McGraw-Hill, 1950.

The following deal with all types of farm equipment:

(7) N.I.A.E., Wrest Park, Silsoe, Beds., publishes reports of research work, and agricultural engineering abstracts.

(8) Wright, S. J. " Farm Implements and Machinery." *Jour. R.A.S.E.* (An annual review of publications on farm power, implements and machinery.)

(9) *Farm Implement and Machinery Review* (London). A monthly journal containing news of new machines and methods.

(10) *Farm Mechanization* (London). A monthly agricultural engineering journal.

(11) *Power Farmer* (London). A monthly agricultural engineering journal.

(12) *Agricultural Machinery Journal.* (London.) A monthly agricultural engineering journal.

Tractor Types: Constructional Features

" Our business was to award the prize in Class XVII, which was to be given 'for the best agricultural locomotive engine applicable to the ordinary requirements of farming '.

We had, therefore, to judge of the merits of the engines when used to replace portable engines, as a mere implement for driving farmyard machinery, of their merits when used as locomotives upon the high road, and of their merits when used as locomotives upon farm roads, or upon the surface of fields where there were no roads."

" Trials of Traction Engines at Wolverhampton."
Jour. Roy. Agric. Soc. England, 1871.

The Principal Types of Tractors. While a high proportion of the vast numbers of tractors now used on British farms have four wheels and a petrol-paraffin engine of some 20-30 b.h.p., account must also be taken of the other important types and sizes. There is an almost infinite variety of types and sizes between the tiny motor hoes and the giant tracklayers, and to add to the difficulties of classification, some tractors can appear as wheeled, half-track or full-track machines, and may be fitted with petrol, paraffin or diesel engines. The classification shown in Table I is, therefore, an arbitrary one, its aim being merely to convey a general picture of the wide range of tractors available. The "row-crop" or "all-purpose" tractors, being now the most important type, are dealt with first.

Row-crop or All-purpose Tractors. This group includes both 3-wheeled and 4-wheeled machines specially designed for row-crop work. The medium-powered size (20-30 b.h.p.) is deservedly popular because it is powerful enough and adaptable enough to tackle satisfactorily almost any kind of farm work, and because its manufacture in large numbers in Britain has resulted in an excellent product at a moderate selling price. The modern row-crop tractor possesses the following distinctive features which enable it to work between crops drilled in rows : (1) high ground clearance ; (2) wheels with narrow rims ;

TABLE I.—*Range of Tractor Types.*

Type of Tractor	Power Group and Approximate Power Range (B.H.P.)	
		H.P.
Market Garden ..	Motor Hoes	½ – 3
	2-wheeled Ploughing Tractors ..	3 – 8
	Baby Tracklayers	6
	Self-propelled Toolbars	5 – 17
" Row-crop " or " All-purpose " ..	Small	12 – 20
	Medium	20 – 35
	Large	Over 35
Standard wheeled ..	Small	12 – 20
	Medium	20 – 35
	Large	Over 35
Tracklayers ..	Small	Under 25
	Medium	25 – 40
	Large	Over 40

(3) wheels adjustable for various widths of rows ; (4) a small turning radius ; and (5) special fittings for the attachment of various tools. Other features found on modern row-crop tractors are power lifts for the tools and rear-wheel steering brakes.

The tricycle type row-crop tractor has advantages for working some crops, not only owing to its smaller turning circle, but also because forward or middle tool-bars can be more easily fitted. Moreover, the single front wheel needs no adjustment for working rows of various widths. The proportion of tricycle type tractors is high in the Fens, where they are found particularly suitable for work in potatoes and sugar beet. On the other hand, most British farmers outside such areas prefer four wheels. A few tractors are fairly easily convertible from four wheels to three and vice versa.

Ease of adjustment of the wheel track widths on a row-crop tractor is an important feature ; for it is easy with some tractors to waste half a day on this job. In general, where only open-field work has to be considered, the use of a sliding wheel hub which is fixed to the axle shaft by means of an easily loosened clamp is easiest to adjust. The type which necessitates re-arranging a dished wheel centre on an offset rim has, however, the merit of relative cheapness, and is rather more foolproof mechanically. There is everything to be said for planning the row-widths of crops in such a way that the changing of tractor wheel widths is reduced to a minimum.

FIG. 11.—THREE-WHEELED ROW-CROP TRACTOR WITH TOOL-BAR WHICH CAN BE FITTED EITHER AS SHOWN, AT THE FRONT, OR AT THE REAR. (JOHN DEERE AND STANHAY.)

Small row-crop tractors (i.e. those of 12 to 20 b.h.p.) are indistinguishable in their general features from machines of the medium-powered group. As a rule, however, it is only on market gardens that a tractor of the small size is considered large enough for general work. On most farms where such machines are used they are employed for particular specialized jobs such as inter-cultivations, and heavy work is left to larger tractors. Machines of just over 20 b.h.p., on the other hand, are used for almost all kinds of work on all sorts of farms.

As previously stated, the row-crop tractor is gradually becoming merged with the unit-principle tractor into an all-purpose machine which is designed to carry implements of all

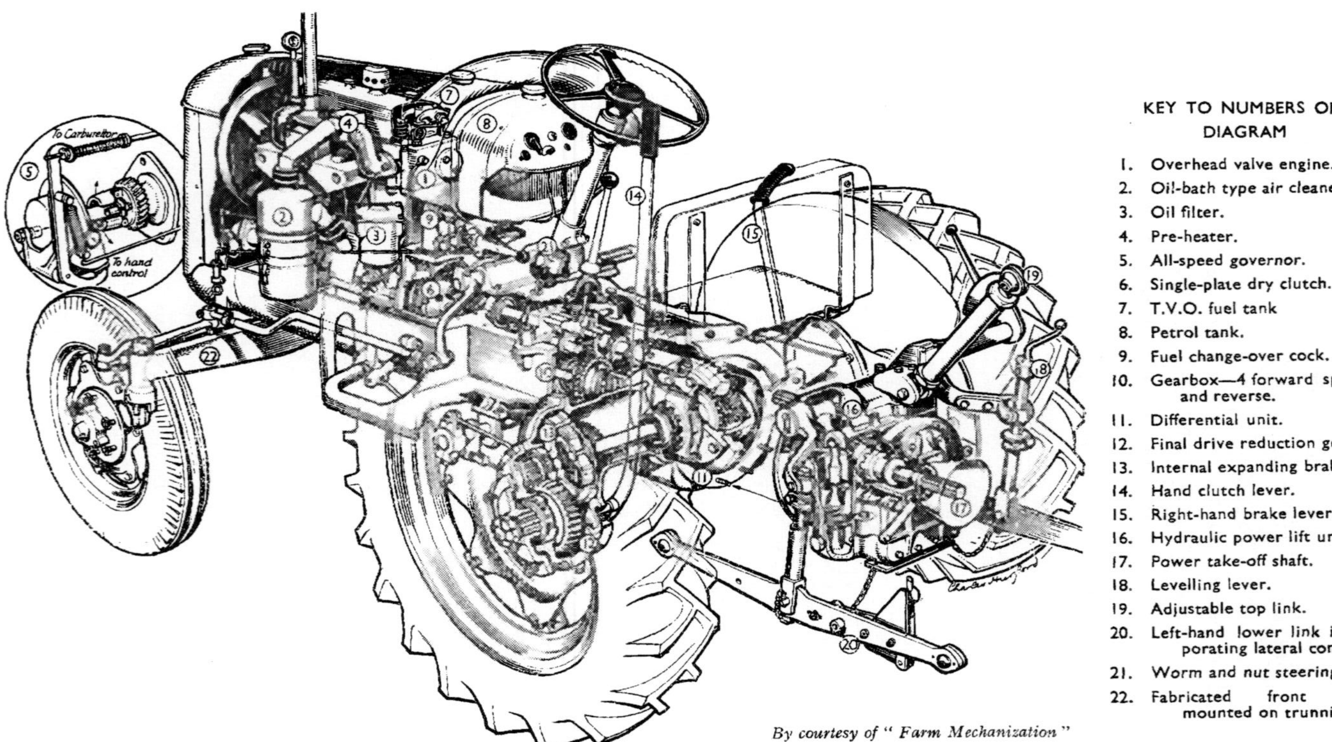

By courtesy of "*Farm Mechanization*"

FIG. 12.—SECTIONAL DRAWING OF FOUR-WHEEL ALL-PURPOSE TRACTOR WITH POWER LIFT SHOWN DETACHED. (DAVID BROWN.)

KEY TO NUMBERS ON DIAGRAM

1. Overhead valve engine.
2. Oil-bath type air cleaner.
3. Oil filter.
4. Pre-heater.
5. All-speed governor.
6. Single-plate dry clutch.
7. T.V.O. fuel tank
8. Petrol tank.
9. Fuel change-over cock.
10. Gearbox—4 forward speeds and reverse.
11. Differential unit.
12. Final drive reduction gears.
13. Internal expanding brakes
14. Hand clutch lever.
15. Right-hand brake lever.
16. Hydraulic power lift unit.
17. Power take-off shaft.
18. Levelling lever.
19. Adjustable top link.
20. Left-hand lower link incorporating lateral control.
21. Worm and nut steering.
22. Fabricated front axle mounted on trunnion.

kinds, e.g. ploughs, hoes, potato diggers, etc., directly coupled to it. With most outfits the implements so attached have ground-engaging depth-control wheels of their own ; but with others, notably the Ferguson, the whole range of adjustment is effected by means of a hydraulic control on the tractor, and adjustments in the linkage between tractor and implement.

The unit principle has many advantages. In the first place there is a saving of material on implement wheels and control gear generally, and this makes a complete outfit of unit-principle implements cheaper to manufacture than the trailed kind. A further important point in favour of mounted tools is the ease of handling and manœuvrability. For example, when a wheeled tractor with a trailed plough enters a boggy patch of land it may take half an hour to extricate the outfit. In similar circumstances a tractor with a mounted plough can get through without real difficulty if intelligent use is made of the hydraulic lift control. There are, on the other hand, occasional instances of exceptionally hard soil conditions where a heavy trailed plough will work satisfactorily and a light type of directly coupled plough will fail to penetrate. In soft soil, tractor wheel adhesion may be increased by some mounted ploughs because both the plough weight and downward soil forces pull downwards on the tractor wheels, but in hard ground wheel adhesion may be seriously reduced. In general, the advantage is with the tractor-mounted implement, and there now seems no doubt that there will be a gradual switch over to this type of outfit.

Standard Type Wheeled Tractors. " Standard " tractors are 4-wheeled machines which have no special provision for doing row-crop work. Little or no adjustment of wheel track width is provided, and the aim is usually to have a low centre of gravity and a cheap, short and sturdy tractor, rather than to secure a high ground clearance, with plenty of room for fitting hoes, etc., between front and rear wheels. Many of the tractors sold in the early nineteen-forties were of this general type, and there are still many farmers, e.g. on hill land, who are not concerned with row-crop work and prefer the general shape of the standard type. Nevertheless, this type is almost obsolete.

As with row-crop tractors, the most popular size is 20-30

b.h.p. High-powered tractors (with engines of over 35 b.h.p.) are much more widely used in some other countries, e.g. U.S.A. and Canada, than in Britain. In Britain they are used mainly on large farms where wide implements can be employed, and tracklayers are not needed. They are also frequently equipped with large pneumatic tyres and a winch, and used for driving threshing machines on contract work in hilly country.

The Transmission. The ordinary 4-wheeled tractor has the engine, clutch, gear-box, propeller shaft and drive axles all mounted in a rigid iron or steel frame. The two rear wheels are the traction members and the front wheels provide directional control. The rigid frame gives freedom from mis-alignment of the transmission, and may at the same time provide a casing that is dust-proof and water-proof. In well-designed machines, the engine, clutch, etc., are all easily accessible. The power is usually transmitted through an easily-adjustable plate clutch to a simple clash-type gearbox with 3-6 forward speeds and 1 or 2 reverse.

The usual type of final drive to-day is by spiral bevel gears to a high-speed differential shaft on which the steering brakes are mounted, and thence by spur gears to the axle shafts (Fig. 1). The whole of this gearing is usually enclosed in the main gearbox casing, but in a few tractors the final reduction gears are situated at the ends of the axle housings (Fig. 12).

Exposed chains were used in some of the early tractors but proved unsatisfactory, mainly because it was impossible to keep them clean or to lubricate them effectively. The roller chains (Fig. 311) on the Case tractor, however, are entirely enclosed, and run in an oil bath. There is much to be said in favour of this type of drive, but it would be surprising if there were any great extension of the use of chains.

Tractors with horizontal engines, such as the John Deere and the Marshall, have a straight-through drive by spur gearing. The belt pulley is carried on the end of the crank-shaft, and the only bevel drive used is that to the power take-off. This type of transmission is very simple and efficient. A conical clutch is used on some tractors. Transmission systems for tracklayers are described on pages 79-80.

Whatever type of final drive is employed there is, of course, a

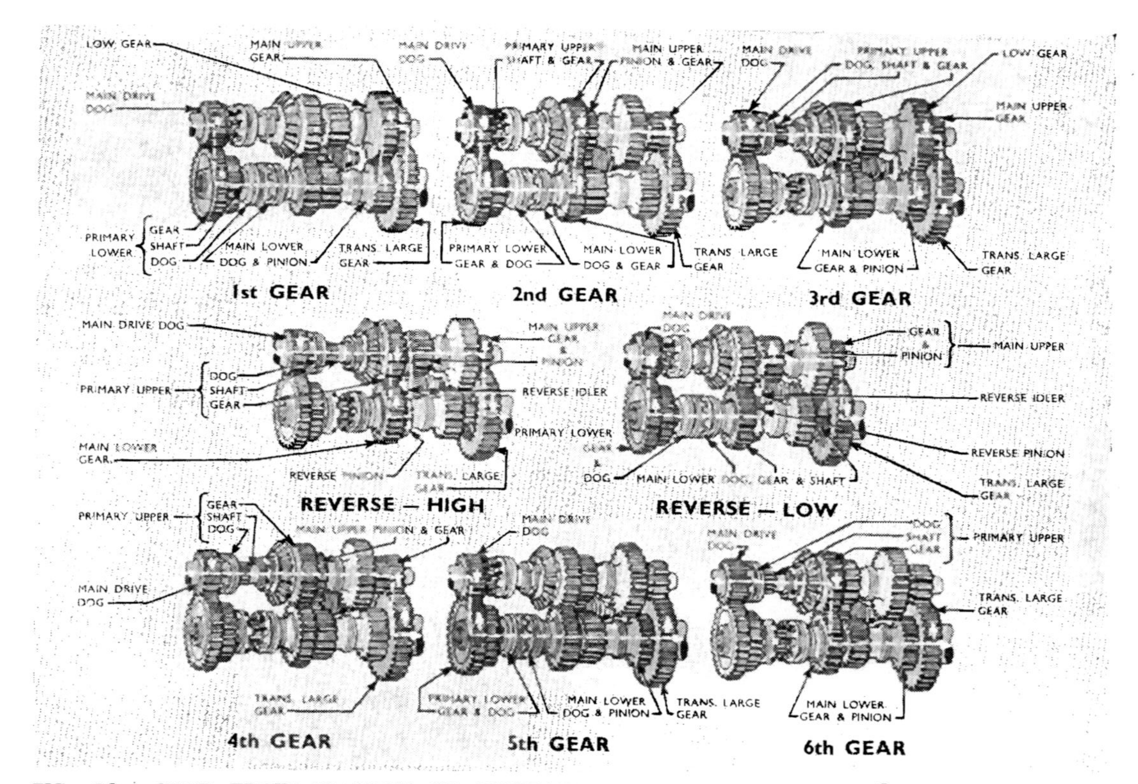

FIG. 13.—GEAR TRAIN DIAGRAM OF CONSTANT MESH GEAR-BOX WITH 6 FORWARD SPEEDS AND 2 REVERSE, CONTROLLED BY TWO LEVERS. (FORDSON.)

differential gear (see Appendix Two) to facilitate turning without any skidding or undue strain. The differential is sometimes a nuisance in field work, because of the fact that the major portion of the tractor power may be delivered to the wheel meeting the least resistance. Thus, when one tractor wheel enters a greasy patch of land which gives poor adhesion it may skid and the tractor becomes bogged because most of the power is delivered to the skidding wheel. Use of a "differential lock" to overcome this difficulty has often been advocated and occasionally tried, but no entirely fool-proof system of providing a solid axle for straight work and a differential for turns has yet been put into production for farm tractors.

The Belt Pulley and Power Take-off. Modern tractors are equipped with a belt pulley for driving stationary machinery such as threshing machines, grinding mills, etc. The pulley shaft is normally driven through bevel gearing and a separate clutch, the pulley usually being situated at the off side, just in front of the rear wheel. With tracklayers the belt pulley is generally situated at the rear, the drive being taken by bevel gearing from an extension of the propeller shaft. The standard speed for belt pulleys adopted by the British Standards Institution is 3,100 ft. per minute, plus or minus 100 ft. per minute.

The power take-off consists of a shaft passing backwards from the gear-box, which can be fitted with universal joints and telescopic sections so as to drive the mechanism of machines hauled by the tractor. It is an efficient means of applying power to such machines as a binder, for up to 95 per cent. of the power delivered to the shaft may be transmitted, where the shaft is fairly straight, so that little energy is lost in the universal joints. The take-off usually rotates at about half engine speed, and the British Standards Institution has adopted a standard speed of 536 r.p.m., making it possible to design implements so that they can be driven by any make of tractor conforming to this standard.

Power drives should always be adjusted to run as straight as possible. The maximum working angle of drive for a double universal joint should be limited to 22 degrees, but it is sometimes necessary to run power driven mounted machines at greater angles for short periods when the machine is being lifted out of work.

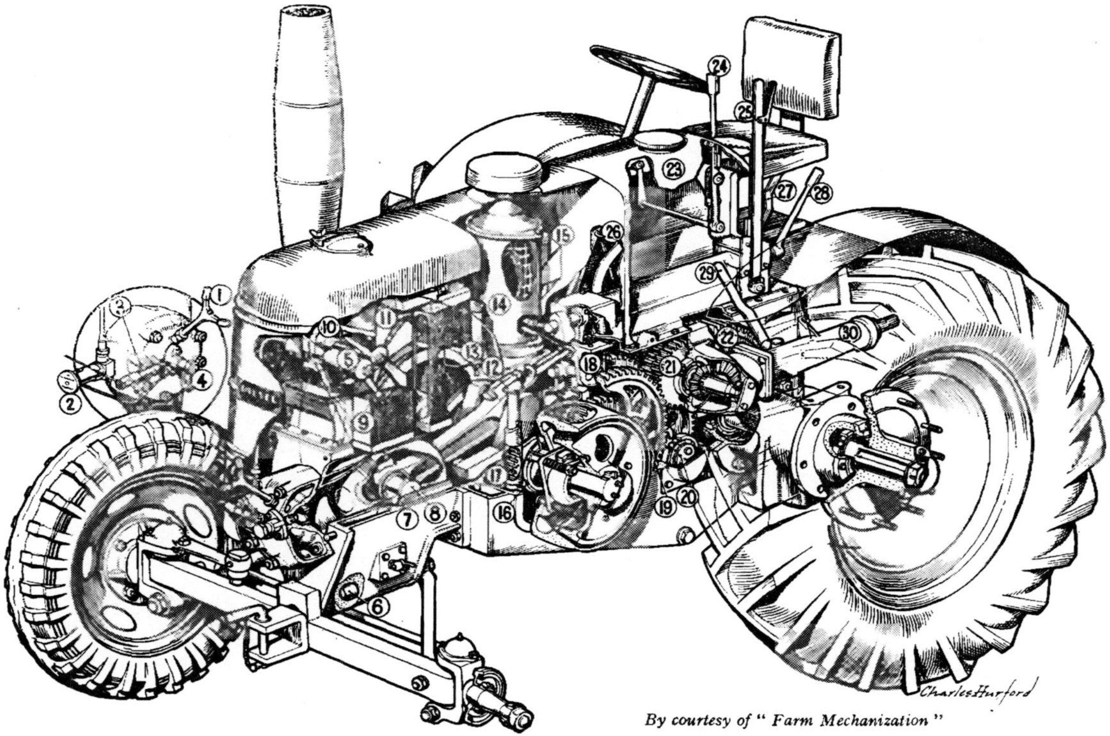

By courtesy of " Farm Mechanization "

FIG. 14.—SECTIONAL DRAWING OF TRACTOR WITH SINGLE-CYLINDER HORIZONTAL DIESEL ENGINE. (MARSHALL.)

KEY TO NUMBERS ON DIAGRAM

1. Ignition paper holder.
2. Handstart valve.
3. Compression release valve.
4. Fuel injector.
5. Cartridge starter.
6. Front axle pin.
7. Transfer port.
8. Piston pad.
9. Radiator.
10. Fan drive.
11. Fan.
12. Flywheel.
13. Oil filter.
14. Air cleaner.
15. Fuel filter.
16. Clutch operating fork.
17. Crankshaft pinion.
18. Gear selectors.
19. First motion shaft.
20. Second motion shaft.
21. Power take-off drive.
22. Differential and final gear.
23. Fuel tank.
24. Fuel control.
25. Brake.
26. Clutch foot control.
27. Clutch hand control.
28. Change gear lever.
29. P.T.O. control.
30. P.T.O. unit.

The Drawbar and Hitch. For many purposes, a tractor delivers the power of its motor through a fixed drawbar. When the engine is exerting its power in driving the rear wheels, the front of the tractor tends to rise owing to the "torque reaction", i.e., the tendency of the drive wheels to remain stationary while the engine winds the tractor round the back axle. This transfers weight to the rear wheels, the front tending to rear off the ground. The drawbar must be placed below the axle, so that the reaction of the drawbar pull has an *opposite moment* (*vide* Appendix One), tending to bring the front wheels back on to the ground. Distribution of tractor weight and location of the drawbar must be such that there is always sufficient weight on the front wheels to make steering possible. The British Standards Institution has adopted various standards for drawbar hitch locations, one of which is that the vertical distance between the ground line and the top of the drawbar shall be 12 to 18 in. Most manufacturers provide a range of both vertical and horizontal adjustments to facilitate hitching to all kinds of implements.

The Wheels. The 4-wheeled machine normally has 3-point suspension, the front of the frame being centrally attached to the axle by means of a horizontal trunnion. This enables the wheels to follow the contours of the ground quite freely. The design of the traction wheels should aim at securing the best possible adhesion with a minimum absorption of power. The tractor often works on loose surfaces, and the power expended by it in propelling itself over the land is considerable. In general, the less the disturbance of the soil, the lower is the rolling resistance. Other things being equal, the greater the diameter of the traction wheels the less the wheels sink into the soil and the better the grip. But very large or wide traction wheels have many disadvantages, and the normal type is about 4-5 ft. in diameter and 8-12 in. wide.

Adhesion is increased by increasing the weight, but Table II (col. 3) shows how the weight per horse power has been reduced since 1920 while, as column 4 shows, the ratio of maximum pull to test weight steadily increased until the influence of fitting pneumatic tyres caused it to level out at about 66 per cent.

There are many reasons why weight should be kept as low as possible, and much attention has been given to devices such as strakes and spade lugs for increasing adhesion.

TABLE II.—*Progress in the reduction of Tractor Weight and Fuel Consumption*

Period in which tested.*	Fuel consumption on drawbar rated load. (lb. per h.p. hour)	Test weight per maximum drawbar horse power. (lb.)	Ratio of maximum pull to test weight
1920	1·63	349	0·44
1923-6	1·15	275	0·62
1927-30	1·01	213	0·69
1935	1·05	215	0·70
1946	0·73	213	0·66

* Figures based on all wheeled tractors tested at Nebraska in the periods stated.

Spade lugs are normally arranged in two staggered rows on a steel rim about 8 in. wide, the lugs being set out so that their tips project beyond the edges of the rim. The efficiency of such wheels depends on the points of the lugs penetrating into a firm layer of soil. They may fail to secure good adhesion if the soil is very loose, so that the lugs cannot reach a firm layer without sinking in. A more usual cause of trouble is sticky soil which builds up between the lugs and eventually completely covers their tips. The first condition is best met by using pneumatic tyres, but the latter are no remedy for sticky soil. To meet this difficulty, many " open ", " tip-toe " or " skeleton " types of wheel, which have no broad rim, have been developed. The absence of a broad rim permits the soil to pass between adjacent lugs and so enables the wheel to keep reasonably clean. A further advantage of this type of wheel on heavy land is that it leaves a level seed-bed and avoids producing a wide strip of puddled, impervious soil in the wheel-marks. Such

FIG. 15.—SKELETON TYPE TRACTOR DRIVE WHEEL (ALLMAN.)

wheels are therefore popular for all kinds of cultivations on many clay farms, and are particularly valuable for seed-bed preparation, drilling and inter-cultivations. On land which is very soft to a great depth the absence of a broad rim may result in skeleton type wheels allowing the tractor to sink up to its axles. They are therefore unsuitable for such conditions.

Pneumatic Tyres for Tractors. The use of pneumatic tyres on tractors began about 1930, and they have become so efficient and popular that to-day, they are the standard fitting on almost all small and medium-sized tractors.

The tyres used for tractor drive wheels are generally low-pressure types, pressures of 10-15 lb. per square inch being employed. Several types of tractor tyres have been tried, some having a very deep tread designed to secure good adhesion in boggy conditions. The present tendency is towards the " wide-base " type with a broad steel rim, walls rather short in proportion to the width, and a deep tread.

The walls are made fairly light, and as the load increases the tyres flatten and cover more ground. Compared with steel wheels, they have a lower rolling resistance under most conditions. Frequently, however, they do not provide as good adhesion as steel wheels equipped with suitable lugs, and the drawbar pull for a given tractor may sometimes be reduced by the substitution of rubber tyres for steel. The adhesion between the rubber tyre and the soil is chiefly frictional, and therein differs from that of a steel wheel in which the lugs behave rather like the teeth of a gear. In some conditions, such as loose sand, rubber tyres may permit higher drawbar pulls than steel wheels; but generally, the drawbar pull which can be transmitted is limited to a fairly well defined maximum.

Most drive-wheel tyres are marked with an arrow to indicate the direction in which they should travel. When correctly fitted, the flexing of the tread bars which occurs when the tractor is at work, assists in freeing soil from the spaces between the bars, and so produces a self-cleaning effect. If the tyres are fitted the wrong way round, and the conditions are at all sticky, the spaces between the tread bars rapidly become filled with soil, and wheel spin occurs.

Where conditions for adhesion are difficult, tyre pressures should not be too high. The lower the pressure, the greater the area of tyre in contact with the soil and the better the grip. Pressures must not, however, be reduced below a certain limit —generally 10-12 lb. per sq. in., since further reduction results in serious tyre "wrinkling" and rapid disintegration of the tyre walls.

The adhesion may be increased by the addition of weights to the wheels. Another method of weighting the tractor is to use water as ballast in the drive wheels. The tube, fitted with special valve equipment, may be half, three-quarters or almost completely filled with water, the air being kept at the normal pressure. In frosty weather it is necessary to ensure that the water does not freeze, and a concentrated solution of calcium chloride may be used. Instructions for the use of such liquid fillings are provided by all tractor tyre manufacturers.

American studies of 100 per cent. liquid filling show that diffusion through the inner tube walls is so small that the rate of loss of pressure is greatly reduced compared with a 75 per cent. liquid filling or 100 per cent. air. On the other hand, wide changes of temperature have a big effect on the pressure of tyres that are completely liquid-filled, and adjustment of pressures may be necessary where temperature changes are extreme. A disadvantage of complete liquid filling is that resistance to bruising when a tyre strikes an obstruction is decreased. The effects of 75 per cent. filling and 100 per cent. filling on riding qualities are complicated. In general, 75 per cent. filling gives a cushioning effect, and riding quality on rough ground is good. With 100 per cent. liquid filling, riding quality is definitely lowered, but the effects are not serious except on rough roads at high speed.

One hundred per cent. filling is advantageous where heavy tractor-mounted implements are employed. When such an implement is lifted, an extra load is placed on the tyres, any air that is contained in them is compressed, and the deflection of the tyre where it is in contact with the ground may be so excessive as to cause the tyre walls to break down. Where 100 per cent. filling is used, the additional deflection caused by raising the implement is much smaller, since water is practically incompressible ; so the danger of overloading the tyres is lessened.

Unfortunately, 100 per cent. filling is not easily effected. It requires special motor-driven apparatus which exhausts the air from the wheel before the liquid is pumped in. The filling is therefore usually a job that must be done at a garage, and even when the utmost care is taken the final result falls short of 100 per cent., since it is impossible to exhaust all the air, or to ensure that the liquid introduced does not contain some air.

FIG. 16.—RETRACTABLE STRAKES FOR WHEELS FITTED WITH PNEUMATIC TYRES. (STANHAY.)

There is, therefore, much to be said in favour of a simple piece of equipment which permits filling tyres to about 95 per cent. The only apparatus required, apart from a liquid pump or gravity tank, is a special air-water valve equipped with a flexible extension-piece which allows air to continue to escape from the top of the tube until the level of the water is well above the valve, and the tube is almost completely filled with liquid. When filling is complete the adaptor is removed and rapidly replaced by the air valve. Whenever a substantial amount of extra weight is added to wheels which are wholly or partly air-filled, it is necessary to increase the pressure in the tube in order to limit tyre deflection. Adding a 95 per cent. filling of liquid ballast to an 11 in. × 36 in. wheel increases the weight on each wheel by about 485 lb. and requires pressure to be increased by about 2 lb. per sq. in.

Generally speaking, rubber tyres give a higher drawbar efficiency than steel wheels and lugs at high speeds, while the converse holds at low speeds. For a given tractor with steel wheels, the maximum drawbar horse power generally falls rapidly at speeds above 3-4 m.p.h., whereas with pneumatics the maximum drawbar horse power may progressively increase for speeds up to 5-6 m.p.h. So long as rubber tyres are used *within the limits of their tractive effort*, they are usually more efficient than steel wheels in transmitting power, and this greater efficiency may in specially favourable circumstances lead to economy in fuel consumption of up to 20 per cent. But in considering the relative merits of the two types of wheels, appreciably increased fuel economy cannot be relied upon with certainty.

FIG. 17.—SPIRAL STRAKES FOR PNEUMATIC-TYRED DRIVE WHEELS. (KENNEDY & KEMPE.)

On wet surfaces which are at all greasy, pneumatic tyres may fail completely. They are seen at their worst on ground which is slippery on the surface and hard beneath, so that the " tread " of the tyre cannot penetrate into the soil to obtain a grip.

In soft ground, on the other hand, pneumatic tyres with a deep tread may perform very much better than steel wheels. In those circumstances where rubber tyres fail to obtain a grip, they may be fitted with various types of lug chains or special strakes to give them adhesion; but in such conditions they lose many of their special advantages.

The great advantage of pneumatic tyres is that a tractor equipped with them is a maid of all work, instantly adaptable to transport, haymaking, harvesting or other of the varied jobs the modern all-purpose machine is required to do. For general farm work the increase in cost and depreciation compared with steel wheels is justified by better performance. Moreover, increased depreciation on the tyres themselves may be more than offset by reduced wear and tear on the transmission system.

For the present, many farmers find that the best plan is to have two sets of drive wheels, one rubber and one steel, the steel wheels being fitted only when prolonged spells of heavy work lie ahead. For the future, it may be expected that the alternative equipment will be on the one hand pneumatic tyres, and on the other, easily fitted half-tracks.

Tractor Engines. The most common type of engine used in tractors is the 4-cylinder 4-stroke petrol-paraffin type, but multi-cylinder 4-stroke diesel engines are becoming increasingly popular for medium-powered tractors, as well as for large. Reference should be made to Appendix Six for an account of the working principles and general constructional features of such engines.

Crankcase Oil Dilution. One of the primary causes of wear in paraffin engines is bad vaporization and incomplete combustion of the fuel, which may lead to heavy dilution of the crankcase oil with unburnt paraffin. It is unfortunately impracticable to adjust vaporizer temperature to all variations of load, so a certain amount of oil dilution is inevitable. With careful operation, dilution can be kept down to about 12 per cent., and the viscosity of the engine lubricant recommended by the manufacturer will be such that with this normal dilution, full protection is given to all bearing surfaces.

With careless operation, however, dilution can exceed 25 per cent. and serious consequences may ensue. In order to minimize the harmful effects of oil dilution, the following recommendations should be followed :

1. Where the vaporizer has a temperature control lever, set this to suit the tractor load and the weather conditions.

2. Change crankcase oil regularly, as recommended by the manufacturer.
3. When starting up, get the engine hot before switching over from petrol to paraffin. Get the tractor to work quickly, and do not leave the engine idling needlessly.
4. Use the radiator blind or shutter control to keep the cooling water as near boiling point as practicable.
5. For long spells of light work, close down the adjustable jet on the carburettor (where fitted).

In cold weather it is advisable to run the tractor extra hot towards the end of the day, so as to evaporate fuel from the lubricating oil. It is always advisable to run on petrol for the last two minutes before stopping, so that all vaporizing oil is cleared out of the carburettor and vaporizer ready for re-starting.

Carburettors.* With most paraffin tractors, the petrol and paraffin are fed to the carburettor by means of a 2-way tap. Such a system is not at all well suited to frequent stopping and starting of the tractor ; for all paraffin must be drained from the carburettor float chamber before attempting to re-start, unless the engine is really hot. With a steady increase in the use of electric starters and the utilization of tractors for all kinds of farm work, more convenient carburation systems are of interest.

Some modern tractors employ a " bi-fuel " carburettor which consists essentially of a simple petrol carburettor added to the paraffin carburettor, each having its own independent fuel line and being put into or out of operation by a simple push-pull control from the driver's seat. With the control lever pulled right out, only the petrol starting carburettor is used, and when the control is pushed in the paraffin carburettor comes into operation. Such an arrangement, in conjunction with electric starting, makes stopping and starting almost as straightforward as it is with a petrol engine.

The engine is started up on the petrol carburettor, and when it has warmed up sufficiently the control rod is pushed in to change to paraffin. If, for any reason, the engine fails to run properly on paraffin, it can immediately be switched back on to petrol. This arrangement operates well in practice, and is worth the slight additional expense. It is to be hoped that

* See also Appendix Six.

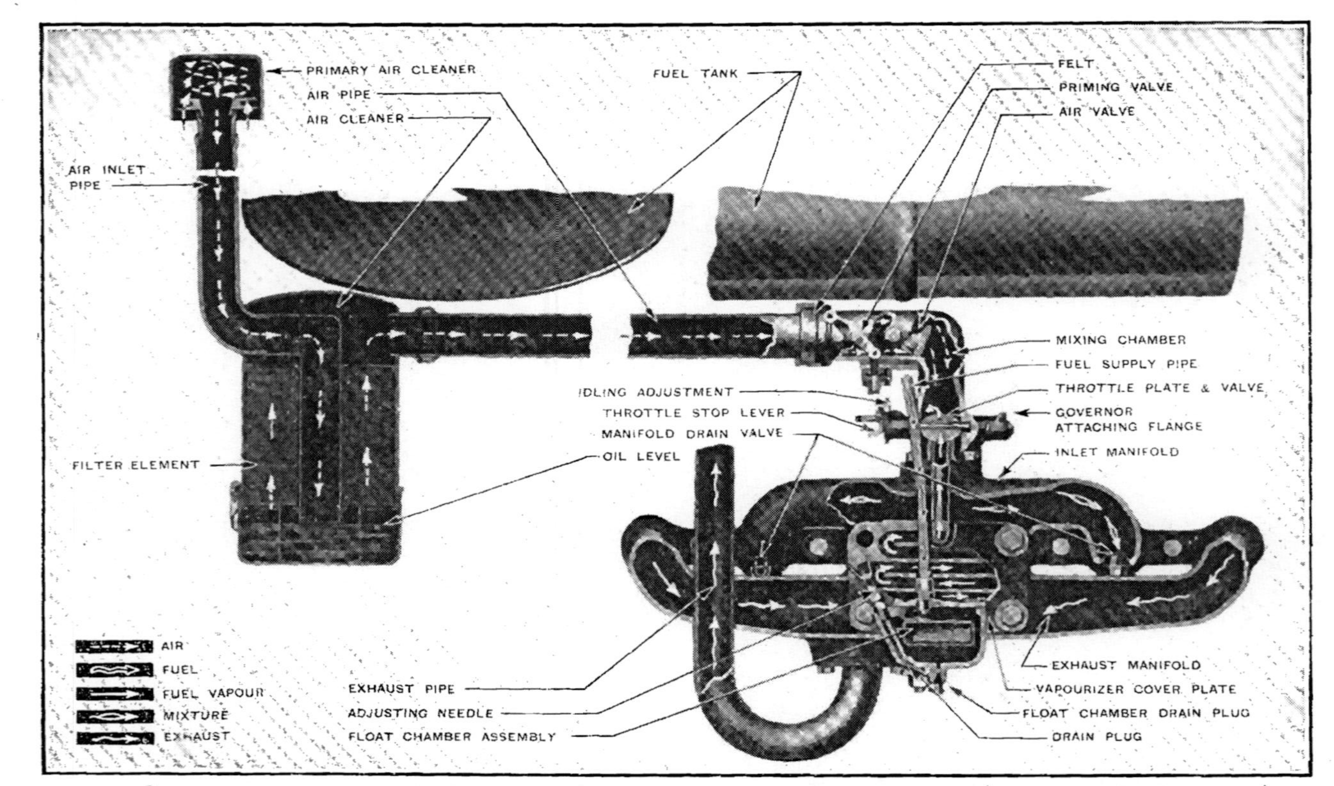

FIG. 18.—AIR CLEANER AND FUEL SYSTEM OF A PARAFFIN TRACTOR ENGINE. (FORDSON BEFORE 1952.)

something of the kind will become generally available for paraffin tractors.

CARBURETTOR ADJUSTMENTS. For efficient operation the carburettor jet must be correctly set, and this setting must be done while the tractor is at work, since the optimum setting varies according to the load. With the carburettor illustrated (Fig. 18) the method of adjusting is to get the engine thoroughly hot and then gradually cut down the amount of fuel drawn through the main jet by screwing in the needle valve until the engine begins to falter. The needle should then be unscrewed slightly, about one-eighth to a quarter turn, until the engine runs normally again. Fuel economy can often be considerably improved by attention to carburettor adjustment.

The fuel consumptions of tractors may vary considerably owing to differences in adjustment and maintenance; and it has been found that an increase of efficiency can be expected by adjusting the carburettor and governor, and servicing air cleaners, grinding in valves, and so on.

Diesel and High-Compression Petrol Engines. While the petrol-paraffin engine is the type most commonly employed on modern tractors, many manufacturers are not satisfied that it is the ideal, and engine design during the last few years has been proceeding along divergent lines. Some engines are of the full Diesel type, and these are popular in the large machines in this country owing to the high efficiency with which they burn a cheap fuel. In Germany the most important make of tractor has a heavy single-cylinder low-speed 2-stroke semi-Diesel or " hot-bulb " engine. But it is in the United States that the controversy concerning the best type of engine rages most fiercely, for in most States petrol which is used for agricultural purposes is free from tax, and high-compression engines are being used in some tractors. These engines burn " high-octane " petrol, and apart altogether from the fact that lubrication difficulties due to crank-case dilution are almost eliminated, some of the engines have established such high performance figures as to compel consideration. While the developments with respect to such engines can have little direct effect on tractors used in Britain owing to the high price of petrol compared with other fuels, American tractors have in the past occupied an important place in British farming, and the movement may have a considerable indirect influence in this country.

An interesting type of tractor engine that may have a big

FIG. 19.—SECTIONAL VIEW OF 2-STROKE DIESEL ENGINE INCORPORATING ROTARY BLOWER. (ALLIS CHALMERS.)

future is the " blown " 2-stroke Diesel. As it nears the bottom of the power stroke the piston uncovers intake ports in the cylinder wall, and large exhaust valves of the usual overhead poppet type in the cylinder head are opened. A powerful blower sends a blast of cool air in through the intake ports and out through the exhaust valves, with the result that the exhaust gases are swept out very effectively. The piston on the upward stroke then covers the intake ports, the exhaust valves close, and fresh air is compressed ready for another power stroke. The operation of this type of engine may be easily understood by reference to the sections of Appendix Six dealing with the 2-stroke cycle and Diesel engines.

The Air Cleaner. An efficient air cleaner is one of the most important desiderata in the tractor engine. Before air enters the carburettor it must be freed from grit, for in dusty conditions the life of an engine is short if grit is allowed to pass unimpeded into the cylinders. The air cleaner must be completely effective and easily cleaned. Grit may be removed from the air either by centrifugal action or filtering, or both. Filtering may take place through a dry material, but the more usual method is to filter through oil. A satisfactory type of cleaner in general use is the oil-saturated wire wool type. The inlet pipe is sometimes placed high above the tractor to prevent dust created by the outfit from being drawn into the engine.

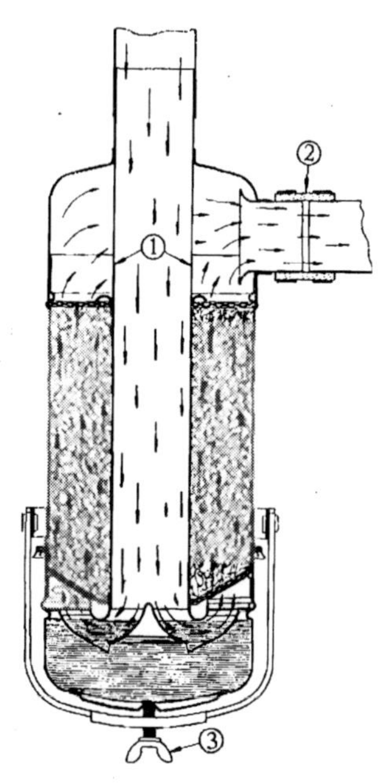

FIG. 20.—INTERIOR VIEW OF AIR CLEANER.

1, Air inlet; **2,** Air outlet; **3,** Sediment bowl. (Case).

The air first passes through a primary cleaner, so constructed that the induced air is given a swirl, thus flinging the heavier particles of grit through louvres around the edges, or into a glass jar at the side, which may be regularly emptied. The air then passes down into oil, and carries some of the latter up into the filtering element. Dust particles are removed by passage through the oil and the oil-saturated filtering element. The latter is kept in place by means of gratings, and washing out and refilling with a light engine oil or used crank-case oil is a simple operation. Washing out and cleaning should be carried out regularly according to the makers' instructions. Frequent changing of the oil is especially important in dusty conditions. The gratings and filtering element may become blocked in extremely bad conditions, and if this occurs they should be removed and washed in paraffin.

It is essential that the correct grade of oil be used in air cleaners. Use of too heavy an oil may result in bad running and excessive fuel consumption.

The Oil Filter. Grit and other foreign materials should be removed from the lubricating oil by means of an efficient oil filter. The oil pump situated in the crank-case forces oil through the filter before it is delivered again to the main bearings. Foreign matter remains on the filter, which should be so designed that cleaning is a simple operation. The filtering element sometimes consists of a cylinder of felt surrounding a central tube. Other types of filter are filled with fine brass wire which can be washed, or cotton waste, which is thrown away when dirty. Most " full-flow " filters have the oil pass from the inside of the element to the outside, but partial-flow filters often work in the reverse direction. Pressure-regulating and safety devices must be included in order that lubrication may not fail if the filtering element becomes clogged. A simple safety device consists of a ball valve which is normally held on to its seating by the force of a light spring, but rises from its seating and allows the oil to be short-circuited direct from the inlet to the outlet pipe when the pressure in the inlet pipe becomes much greater than that in the delivery pipe. With the full-flow type the filtering element itself is usually spring-mounted, and is automatically by-passed if excessive pressure develops.

The Governor is a device which automatically ensures a fairly constant engine speed under varying loads. The mechanism is so arranged that when the load is removed and the engine begins to go faster, the increased centrifugal force on the spring-loaded governor weights causes them to move farther from the axis of the governor shaft about which they rotate. This movement causes a collar to slide along the governor shaft and operate a linkage mechanism which closes the throttle and causes the engine to run more slowly. Conversely, if a heavy load slows down the engine, the movement of the governor mechanism automatically opens the throttle and tends to restore the engine speed to normal. Closing of the hand throttle lever, however, puts the governor out of action and allows the engine to " idle ".

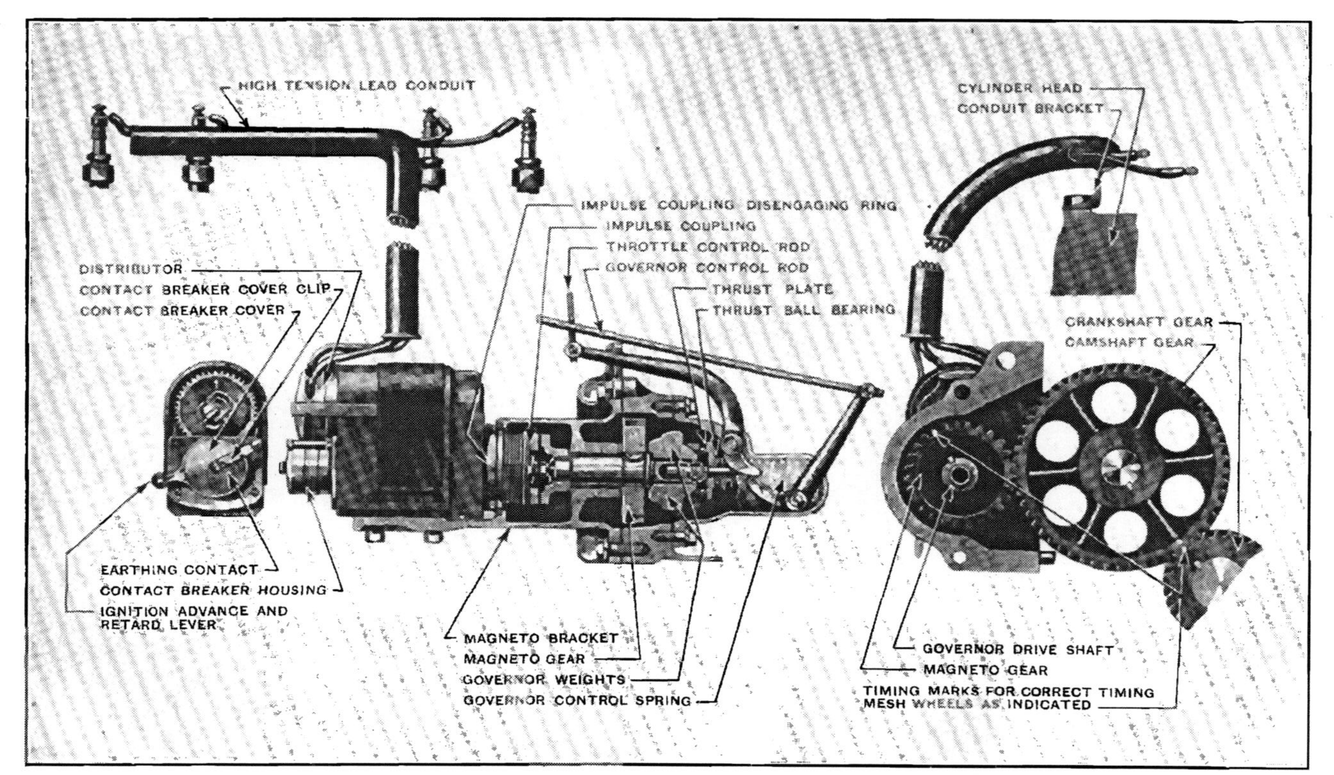

FIG. 21.—DIAGRAM SHOWING SECTION OF GOVERNOR, AND RELATIVE POSITIONS OF GOVERNOR, IMPULSE STARTER AND MAGNETO ON A TRACTOR. (FORDSON BEFORE 1952.)

Petrol-paraffin tractor engines usually run at 1,000-1,500 r.p.m. On many modern tractors it is possible to vary the engine speed between about 600 and 2,000 r.p.m. by adjustment of the governor. This adjustment is useful in operations such as threshing, when it is important to run the machine at the correct speed.

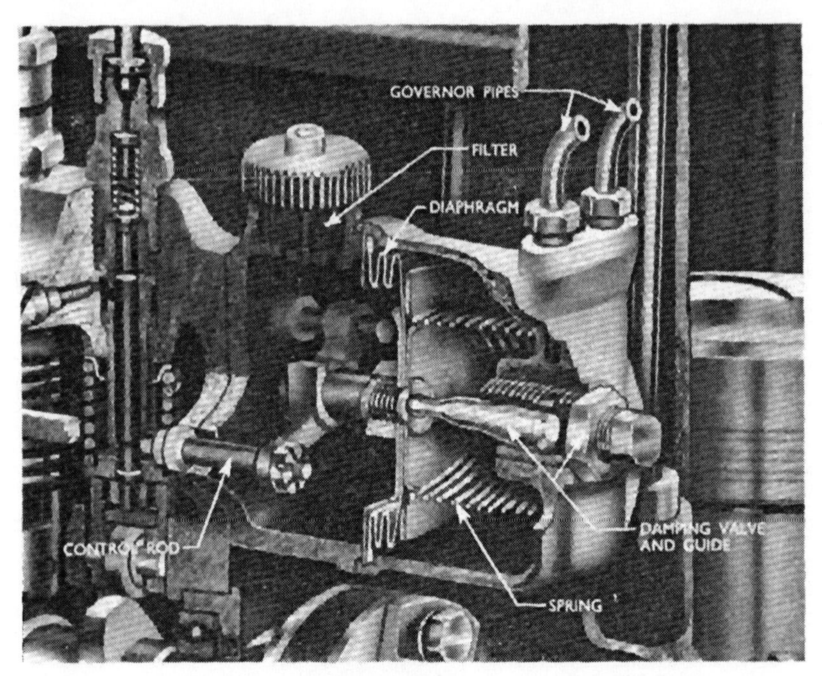

FIG. 22.—PNEUMATIC GOVERNOR USED TO CONTROL FUEL INJECTION PUMP ON DIESEL ENGINE. (FORDSON.)

Pneumatic governors are frequently fitted to diesel engines, and their mode of operation may be understood by reference to Fig. 22. Two pipes lead from the induction manifold to the governor casing, which has a diaphragm at one end. When the engine is idling, the throttle is almost closed, and the suction in the manifold is at a maximum. This negative pressure, transmitted by one of the pipes direct to the governor casing, is sufficient to draw the diaphragm to the right, against the force of the compression spring. This action pulls the control rod

3

to the minimum fuel position. As the throttle is opened to increase engine speed, the manifold depression decreases, the diaphragm moves to the left, and the delivery of fuel is progressively increased by corresponding movement of the control rod.

When the engine is at work and the throttle is held in a fixed position, engine speed does not remain exactly constant under varying load, but tends to increase if the load is lessened, and to fall if the load is increased. As soon as the engine starts to go appreciably faster, the manifold depression is increased, and the consequent movement of the governor control rod diminishes the fuel supply. Conversely, a fall in speed, resulting in reduced manifold depression for a given throttle setting, leads to an increased supply of fuel and substantial restoration of the speed to the governed setting.

The second suction pipe is connected to the inlet manifold at a point just outside the throttle ; so when the throttle is nearly closed, the air in this pipe is approximately at atmospheric pressure. At the governor end, the pipe leads to a damping valve which is connected to the diaphragm. Extreme movement of the diaphragm in response to a sudden depression of manifold pressure allows air to pass *via* the damping valve into the governor casing, thus reducing the suction in the chamber and preventing excessive fluctuation of the control rod, i.e. " hunting ".

Power Lifts. A power lift for the tools is now becoming a regular fitting, and this equipment must be considered one of the most important features of a modern all-purpose machine. The most common type of lift is a hydraulic device operated by the power drive shaft. Other types of lift include mechanical and pneumatic devices.

The hydraulic lift consists essentially of (1) a pump to provide oil under pressure ; (2) a control mechanism which regulates delivery of the oil ; and (3) a ram on which the oil pressure acts to lift the implement. The fundamental principles of the action of hydraulic lifts are briefly indicated in Appendix One (p. 552).

The ram is frequently built in at the rear of the tractor, and operates a horizontal cross-shaft to which are attached the lift arms (Fig. 24). Alternatively or additionally, oil may be fed by a flexible tube to an external lifting cylinder, which may be used for the operation of a middle or forward tool-bar, or a trailed implement (Fig. 27). In some circumstances, as when forward and rear tool-bars are being used simultaneously, it may be necessary to raise and lower them consecutively or independently, and some types of control mechanism are designed to permit this.

The main types of power lift pump are (1) the multi-cylinder reciprocating type, the plungers being driven by cams or eccentrics (e.g. the Ferguson) ; (2) a simple gear pump similar to that commonly used for engine lubrication, but much more accurately constructed (e.g. Fordson, David Brown, Nuffield, John Deere and International) ; and (3) various special types of high-efficiency rotary pumps. The pump must be capable of operating at a fairly high pressure, and must deliver a sufficient volume of oil to complete the lift in not more than 2 seconds.

The method of control naturally depends on the application required. If the only operation to be reckoned with is the raising and lowering of lift-arms by means of a built-in ram, a fairly simple form of control will suffice. This consists of a valve which in one position allows the oil to pass from the pump to the ram cylinder, and in the other position releases the oil from the cylinder and allows that coming from the pump to pass straight back to the reservoir.

Fig. 23 illustrates a control system of a type now commonly employed, which permits the implement to be fully raised, fully lowered, or held in any intermediate position. A single control lever can be put into any one of three positions—" up ", " down " and " neutral ", and there are two valves—one a by-pass and one a non-return valve. The by-pass valve, D, can pass oil from the pump directly back to the sump at all times except when it is held shut by plunger H, when the control lever is in the " up " position. The non-return valve, B, is operated by the oil pressure in the system at all times except when it is held open by plunger C, when the lever is in the " down " position.

The diagram shows the " up " position, valve D being held closed and oil passing through B to the ram. If the lever is moved to neutral, valve D is allowed to open, the pressure of oil in the ram cylinder closes valve B, and the implement is maintained in position. To lower the implement, valve B is held open by plunger C, while oil from the pump is allowed to by-pass through valve D.

The principle of operation of the Ferguson system of implement control may be briefly explained by reference to Figs. 24 to

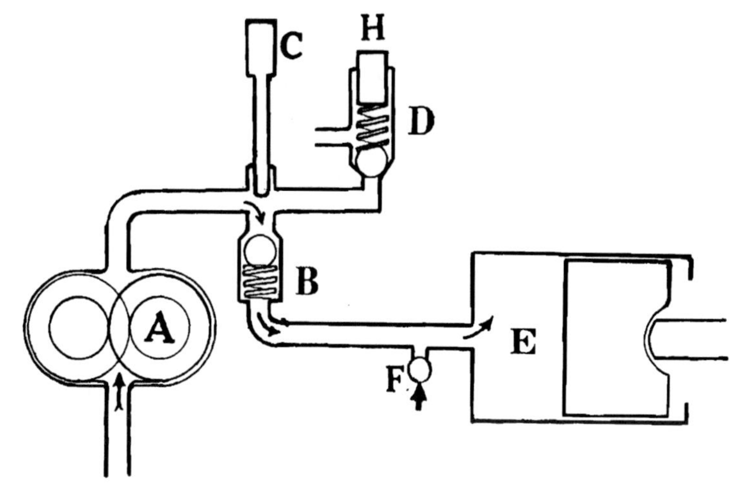

FIG. 23.—HYDRAULIC LIFT CONTROL SYSTEM.

A, pump ; **B,** non-return valve ; **C,** lowering plunger ; **D,** by-pass valve ; **E,** ram cylinder ; **F,** safety relief valve ; **H,** raising plunger.

26. A feature of this system is maintenance of the draught of the implement at a constant value which may be varied and pre-determined by the setting of the control lever in its quadrant.

An oil pump supplies oil to a ram cylinder which operates a transverse lifting shaft, in the way now usual with the 3-link system.

The hydraulic control linkage functions in three distinct ways : (*a*) hydraulic lift ; (*b*) hydraulic depth control ; (*c*) safety device to protect tractor and implement from hidden obstructions.

HYDRAULIC LIFT. Fig. 25 shows this in simplified form. When the control lever is moved forward, point A moves forward. The control fork pivots about point B and the control valve

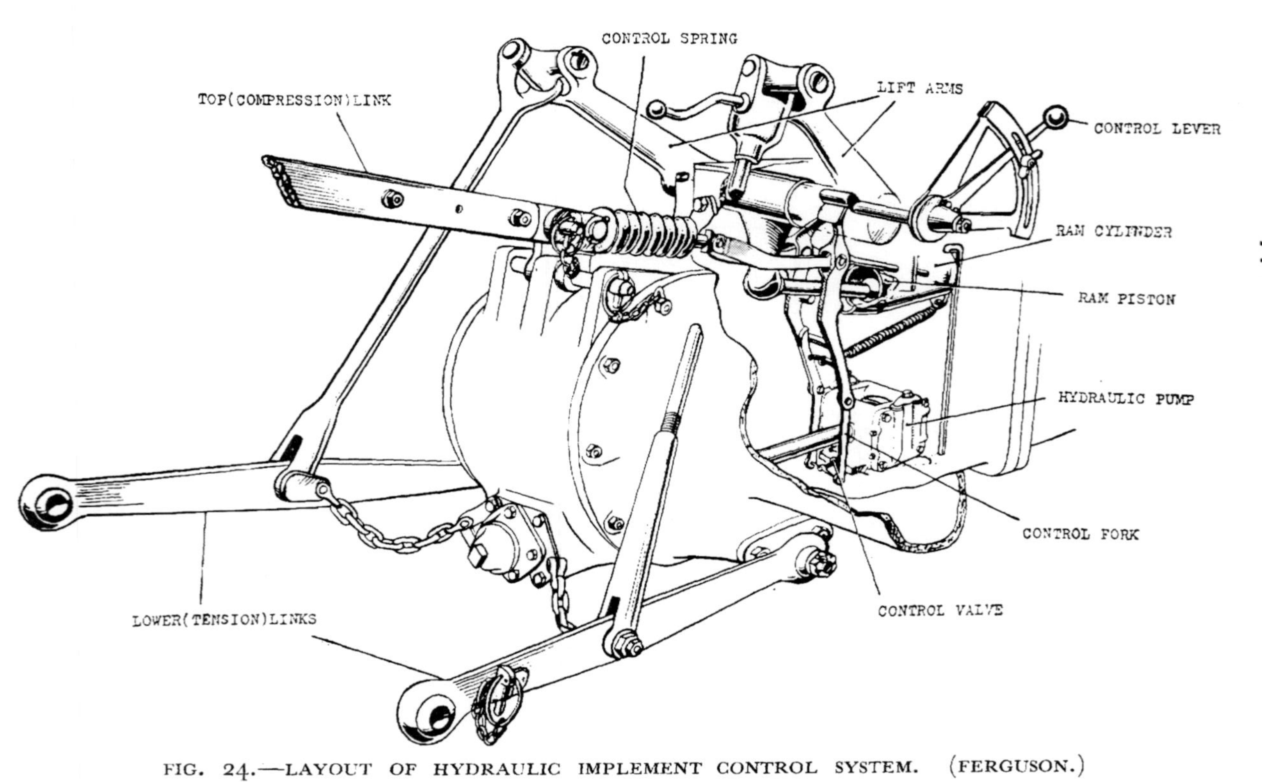

FIG. 24.—LAYOUT OF HYDRAULIC IMPLEMENT CONTROL SYSTEM. (FERGUSON.)

moves (1) to the rear, releasing oil from the ram cylinder, while cutting off the supply to the pump. The implement then lowers under its own weight. When the lever is raised, point A moves to the rear, the fork pivots about point B, and the control valve moves forward (2) retaining oil in the ram cylinder and at the

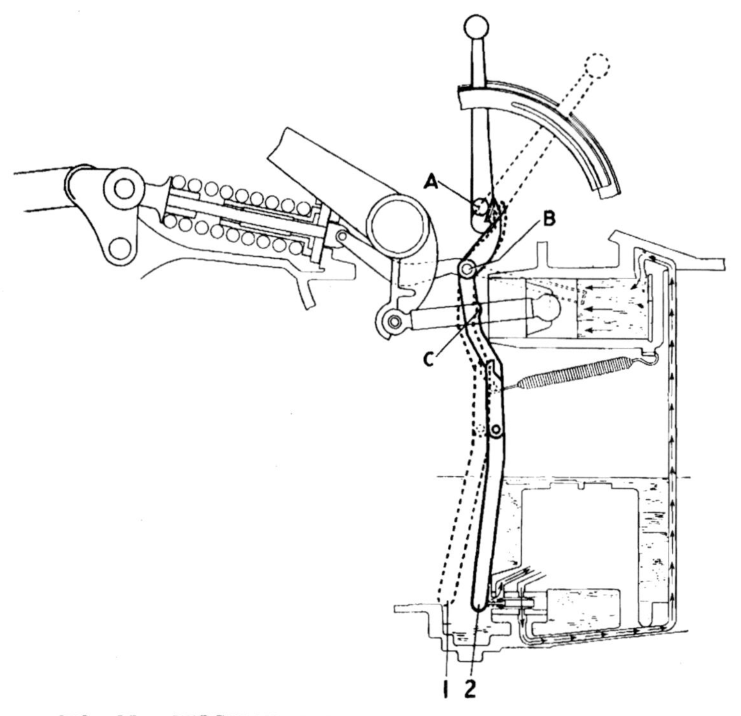

FIG. 25.—FERGUSON SYSTEM, SHOWING HYDRAULIC LIFT.

A, control lever boss ; **B,** control fork pivot ; **C,** lugs on control fork ; **1,** valve in lowering position ; **2,** valve in lifting position.

same time admitting oil to the hydraulic pump. This raises the implement.

When the implement is fully raised, the ram piston abuts against the lugs C and pushes the control valve to mid-position. In this position, the control valve prevents any oil being supplied to the pump, and also prevents the return of oil from the lifting cylinder to the sump.

DEPTH CONTROL. Depth control is achieved by utilizing the forces acting in the top link of the implement hitch. The greater the draught of the implement, the greater the force in the top link and the more the control spring is compressed.

The way in which the implement forces operate is by moving the position of the fulcrum about which the control fork rotates.

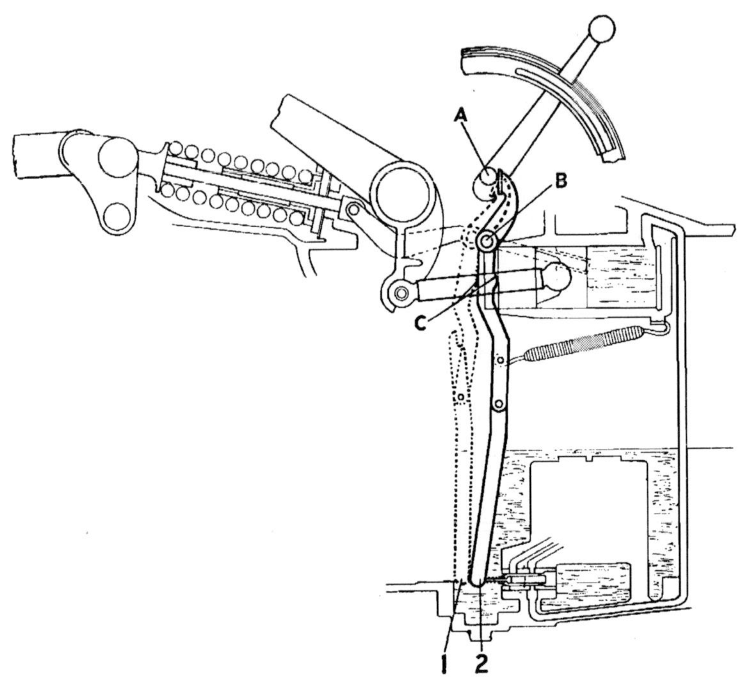

FIG. 26.—FERGUSON SYSTEM, SHOWING DEPTH CONTROL MECHANISM. **A,** control lever boss ; **B,** control fork pivot ; **C,** lugs on control fork ; **1,** valve in lowering position ; **2,** valve in lifting position.

(Fig. 26.) Suppose that the implement is at work and the draught becomes excessive. Fulcrum B is pushed forward (i.e. to the right in the diagram). This causes an anti-clockwise movement of the control fork about the fulcrum. The result of this is that the control valve is pushed forward, (2) oil is admitted to the pump, and the implement is raised, until the force on the control spring is no longer excessive, the control fork returns to its normal position, and the control valve reverts to mid-position. Conversely, a decrease in the force in the top link causes the implement to be lowered.

If the action of the automatic control device is excessive, its effects can be regulated by re-setting the control lever.

SAFETY DEVICE. If the implement strikes a serious obstruction, such that the control plunger is driven far forward, the lugs on control fork, C, strike the skirt of the lifting cylinder, and the control fork, pivoting about this point, is rotated in a clockwise direction, thus releasing the oil from the lifting cylinder. The effect of this is that the weight of the implement, and the downward soil forces on it, are suddenly dropped from the tractor drive wheels, and wheel spin occurs, thus saving the tractor and implement from excessive strain.

SLAVE CYLINDERS. Fig. 27 shows how the "delayed action" lift on the rear tool-bar of an International "Farmall" is secured. The control lever, C, is shown in the neutral position, when the oil from the pump is simply by-passed direct to the reservoir. To raise the tool-bars the lever is pulled back to D, where it is held by the projection on the underside of the locking-lever, X. The by-pass is then shut off by piston, Y, and the oil pressure begins to build up in the hoses connected to the lifting-cylinders.

The rear lifting-cylinder has a delayed-action valve, W, which is so adjusted that the pressure required to force oil through it is greater than that required in the front cylinders to lift the front tool-bar. When the front tool-bar is fully lifted, the oil pressure immediately begins to rise, and when it has reached a predetermined level the valve W permits oil to start passing into the rear cylinder. With the rear tool-bar in its turn fully lifted, the oil pressure rises still further, and acting on the spring-loaded cap on which the end of the locking-lever rests, it pushes up the end of the locking-lever, thus releasing the control lever and allowing it to slide back to the neutral position. The piston Y thus uncovers the by-pass again, and the pump idles. Meantime, the pressure in the two lifting-cylinders is maintained by the non-return ball valves.

To lower the front tool-bar, the control lever is pushed forward to B, causing the upper ball valve to be pushed off its seating and allowing the oil from the front cylinder to pass back to the reservoir. If now the control lever is pushed farther forward to A, the cam plate, Z, coming into contact with the pump body

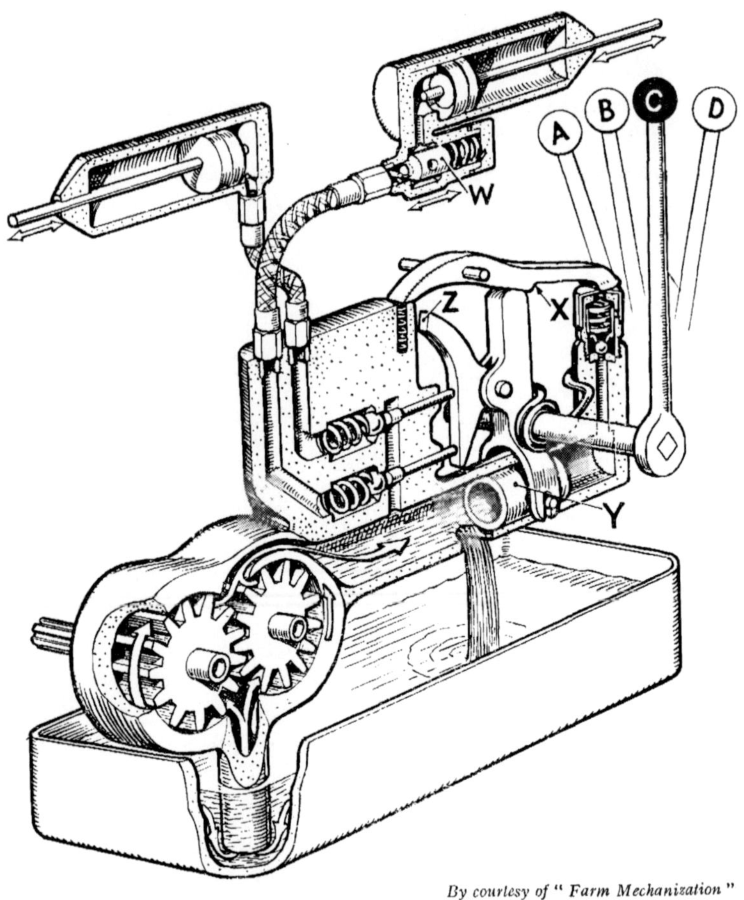

By courtesy of " Farm Mechanization "

FIG. 27.—DIAGRAM SHOWING DETAILS OF THE "FARMALL" LIFT FOR TWO TOOL-BARS. ONLY ONE OF THE TWO FRONT POWER CYLINDERS IS SHOWN.

at the top, pivots forward at the bottom and pushes the rear-cylinder non-return valve off its seating, thus releasing the oil and lowering the rear tool-bar.

3*

Points of importance in any hydraulic lift system are as follows :—

1. A pump of sufficient capacity to lift the required weight at the required speed, with a ram designed to suit.
2. A high standard of engineering precision throughout, to ensure efficient operation and long life.
3. Efficient filtration of the oil.
4. Pump should be under load only when actually lifting. It should be automatically relieved of load when implement is both fully raised and fully lowered.

FIG. 28.—MEDIUM-POWERED ALL-PURPOSE TRACTOR WITH 3-2 FURROW MOUNTED PLOUGH. (FORDSON.) NOTE WHEEL WEIGHTS ON TRACTOR, AND PLOUGH DEPTH CONTROL BY LAND WHEEL.

5. There must always be a safety relief valve to prevent the development of excessive oil pressures.
6. Ease and simplicity of control are essential.

The Power-Lift Linkage, and Implement Control. Applications of the hydraulic control system include raising and lowering of the following types of equipment :—

1. Front tool-bars, light bulldozer and front loaders. A hydraulic push-off device may be incorporated in manure loaders and hay stackers.

2. Under-belly or mid-mounted tool-bars of various types.
3. Rear tool-bars.
4. Mounted ploughs.
5. Mounted power-driven machines, e.g. potato diggers.
6. Trailed implements, tipping trailers, etc.

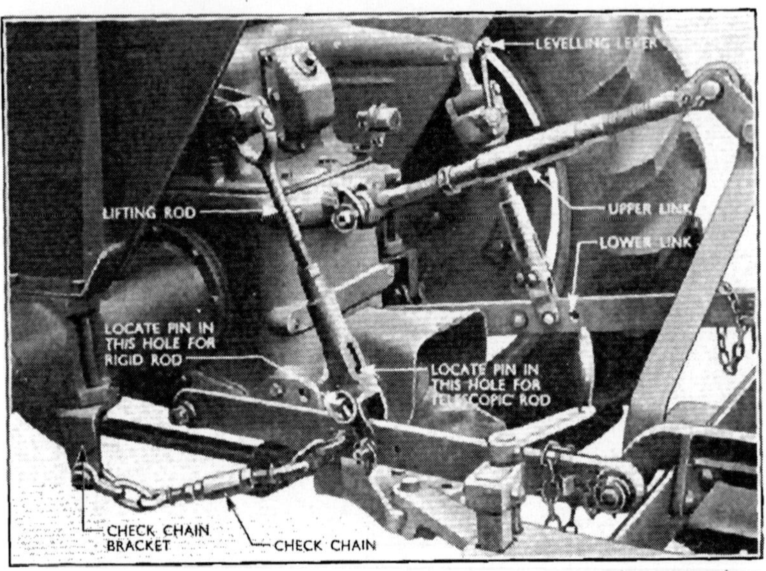

FIG. 29.—THREE-LINK HITCH FOR REAR TOOL-BAR. (FORDSON.)

The following types of linkage may be distinguished :—

(*a*) Tool-bar rigidly connected to the rear ends of a pair of long draw-links which pass beneath the rear axle and are pivotally connected at their forward ends to a bracket beneath the belly of the tractor. The pivot can usually be raised or lowered to control the pitch of the implement, and the lift is by means of chains or special links which allow the implement to " float ". Depth control is partially achieved by land wheels on the implement.

(*b*) Tool-bar connected to the tractor by a parallel-motion linkage, with depth control by wheels on the implement.

(*c*) Tool-bar connected to the tractor by a 3-link system (Fig. 29) which gives a " semi-parallel " motion, with a virtual hitch point near the front axle, and depth control by means of implement land wheels. Most tool-bars are allowed to float freely by using the lift rods in the position where telescopic action is provided.

(*d*) The Ferguson system, in which a 3-link hitch is employed, and the top link operates a spring-loaded control valve. For a given setting of the hand control lever the hydraulic system maintains a substantially constant draught on the implement. No implement depth-control wheels are used.

A preliminary study of systems (*a*), (*c*) and (*d*) relative to depth control with ploughs, published by the National Institute of Agricultural Engineering, shows that system (*a*) may result in the most satisfactory work over undulating or uneven land. On the other hand, it is not so attractive in many other ways as systems (*c*) and (*d*). The relative merits of depth control by hydraulic means or by ground wheels is still a subject of argument.

METHOD OF ATTACHING IMPLEMENTS TO 3-LINK SYSTEM. The attachment of heavy mounted implements having a depth wheel and cross-shaft control to tractors equipped with the 3-link system is an easy one-man operation provided that it is tackled in the correct sequence. This is as follows :

1. Back up squarely to the implement cross-shaft.
2. Let the top link settle as nearly as possible above the top-link bracket.
3. Stop the tractor, dismount, and adjust the length of the top link till the pin enters easily.
4. Attach the near-side link by adjusting the top link and the cross-shaft control.
5. Attach the off-side link by adjusting the top link, depth wheel, cross-shaft control and linkage levelling box.

There should be no need to move the implement during the hitching operation.

Tracklaying Tractors. Tracklayers are designed to secure good adhesion and transmit high drawbar pulls in difficult conditions, where wheels fail to secure an adequate grip on the soil. They are particularly suitable for use on steep hillsides, on heavy land, on fen land where particularly deep cultivations are required, and in all conditions where running wheels over the land may harm the soil structure. For one or more of these reasons tracklayers are preferred for many types of field work, but they suffer from the disadvantages of high first cost and high costs of overhaul when repairs or replacements are needed. A further disadvantage compared with a rubber-tyred tractor is unsuitability for running on metalled roads.

The chief users of large tracklayers (machines of 40 B.H.P. or more) are contractors who undertake heavy cultivations such as deep ploughing, subsoiling and mole draining. The larger sizes are also frequently used for bulldozing, etc. Such powerful machines are now almost invariably fitted with diesel engines and are extremely expensive. They naturally need heavy equipment to suit their power, and this also can be very costly.

An early tracklayer was built in California in 1904. Since that time, steady improvements in design and in the materials of construction have occurred and the modern tracklayer is a very efficient type of tractor with respect to tractive power. The track may be looked upon as a continuous rail made up of links and supported by the cleats. The tractor driving sprockets gear with the links of this rail, and the tractor rolls along it on the track rollers. The track is picked up after the tractor has passed over it and is relaid continuously. It provides a large area of contact with the ground and is eminently successful where adhesion is difficult or rolling resistance is high. A certain amount of power is, however, wasted in internal track friction. The weight of the tractor is carried on the track rollers and not on the driving sprocket, and the weight of the track itself is also partly removed from the driving sprocket by means of the upper track rollers.

The track itself generally consists of a series of links and ground-plates, coupled together by pin joints. It is flexible in both directions, so that obstacles are easily negotiated. Practically all heavy tracklayers are fitted with tracks of this type. The chief defect from an agricultural viewpoint is the wear that occurs at the pin joints, and the high initial cost.

A second type of track in fairly common use in Britain is the rubber-jointed type, where blocks of rubber are substituted for the pin joints. There is no pin movement with these tracks, internal deflection of the rubber blocks permitting the necessary

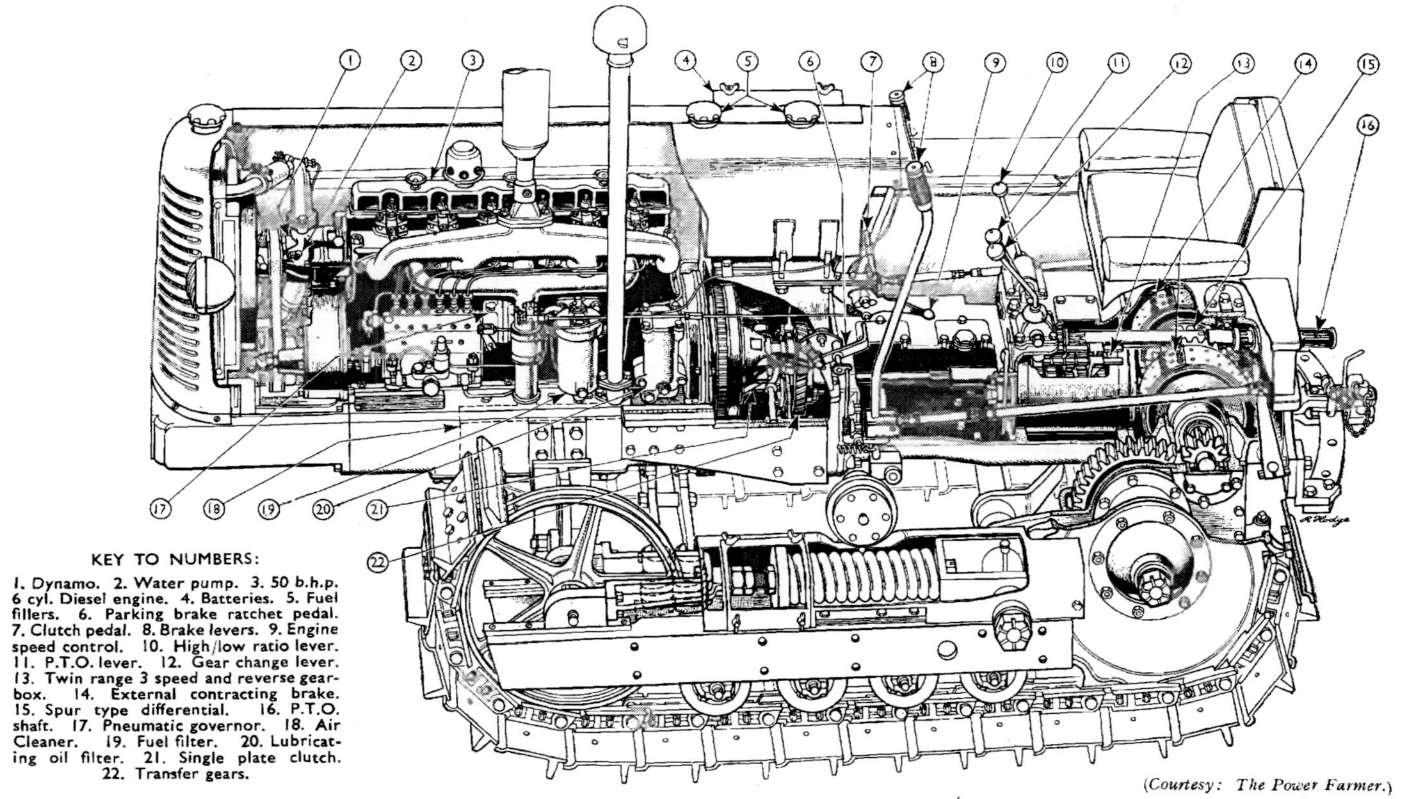

KEY TO NUMBERS:

1. Dynamo. 2. Water pump. 3. 50 b.h.p. 6 cyl. Diesel engine. 4. Batteries. 5. Fuel fillers. 6. Parking brake ratchet pedal. 7. Clutch pedal. 8. Brake levers. 9. Engine speed control. 10. High/low ratio lever. 11. P.T.O. lever. 12. Gear change lever. 13. Twin range 3 speed and reverse gearbox. 14. External contracting brake. 15. Spur type differential. 16. P.T.O. shaft. 17. Pneumatic governor. 18. Air Cleaner. 19. Fuel filter. 20. Lubricating oil filter. 21. Single plate clutch. 22. Transfer gears.

(Courtesy: *The Power Farmer.*)

FIG. 30.—SECTIONAL DRAWING OF MEDIUM-POWERED TRACKLAYER. (DAVID BROWN TRACKMASTER.)

flexibility. These tracks are not, however, flexible in both directions in the same way as the pin-jointed type. Indeed, they are built to a curve, and a considerable force is required to flatten the track. An advantage of this is that over rough ground the track rollers have a much smoother rail to run on than they do with the pin-jointed type. The disadvantage of these tracks is that rubber is not as strong as steel, and the tracks have to be made heavy. They are very satisfactory for small tractors (e.g. the Ransomes M.G. Cultivator, the Bristol, etc.), but are not a sound proposition for powerful machines of more than 6-7 tons or 50-60 b.h.p.

Experiments have been carried out over a long period with rubber tracks which have a steel-wire reinforcement. The advantages of this type are that adhesion is almost equal to that of a steel track in most agricultural conditions, and that the tractor can run along good roads without damage to either the road or the tractor. Disadvantages are a high manufacturing cost, and the fact that a scrubbing effect caused by turning in heavy-duty conditions causes excessive wear of the tread. Such tracks are seen to advantage when a very wide track is needed for traversing soft ground. For example, the Cuthbertson full-track " Water Buffalo ", or half-track outfits equipped with tracks 24 in. wide or more, will pull a drainage plough across bogs which are completely impassable to ordinary tractors or vehicles. In the Cuthbertson track, wire-reinforced rubber pads are squeezed between cleat plates on the outside and driving plates on the inside. The cleat plates may be of steel, rubber, or rubber bonded to a steel base.

Fig. 31 shows the method commonly employed for transmitting the power to and steering the tracks of a tracklayer. The tractor is turned by cutting off the power from one track by means of the clutches C and CI, which are operated by the steering levers, S and S-I. For example, in turning to the right, the operator pulls on the right-hand steering lever, thus disengaging the power from that track. The extent of the turn is determined by the length of time the steering clutch lever is held back. Sharp turns may be made by use of the individual brakes, B and B-I, but these need only be used when making short turns, when running light, or when preventing the tractor from rolling down a steep hill.

The steering of some machines is achieved by the use of brakes and a differential.

Some manufacturers employ "controlled differential" steering, operation of which may be illustrated by reference to Fig. 32. When the tractor is going straight ahead, the whole differential assembly driven by the bevel gear No. 7, rotates as a

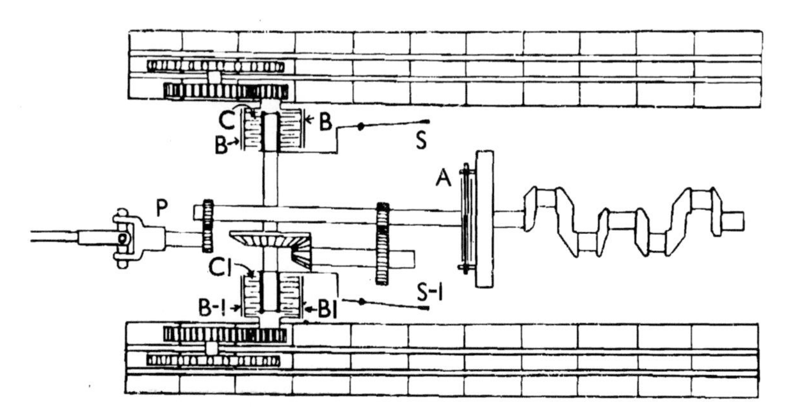

FIG. 31.—STEERING CLUTCHES AND BRAKES OF A TRACKLAYER (CATERPILLAR)

unit without any internal motion. To make a turn to the right the right control lever is pulled. This holds stationary the right steering drum No. 10, and the drum gear No. 6, which are fastened together and which rotate on a bronze bushing on their shaft. As a consequence, pinion gear No. 4 and differential gear No. 2, connected to bevel gear No. 7, by means of shafts, begin to turn forward. Final drive gear No. 8 is slowed down because No. 2 gear is turning and rotating in the same direction.

At the same time, differential gear No. 1, rotating around the final drive gear No. 9 in the opposite direction, is speeded up. This results in a positive drive to both tracks while turning. The disadvantage is that so long as the steering drum is held the urning circle is a fixed one, and to make a gradual turn the tractor has to do it in stages, going straight ahead for a time and then turning rather sharply. More perfect steering mechanisms have been developed for use in military tracklayers, but have not yet been adopted for farm tractors.

A slight disadvantage of tracklayers sometimes experienced in deep ploughing is that considerable side-draught may be unavoidable because it is necessary to run both tracks on the unploughed ground ; but against this must be set the advantage of no compression of the furrow bottom. On wet, heavy land, tracklayers, on account of their low ground-pressure, can often be used when other types poach the soil so much that they cannot be profitably employed.

FIG. 32.—DIFFERENTIAL GEAR FOR TRACKLAYER. (CLETRAC.)

Half-Track Equipment. The limitations of rubber-tyred wheels have caused a steady increase of interest in " half-track " assemblies which may be fitted in place of the rear wheels of general-purpose tractors. The object is to provide an outfit which is capable of overcoming difficult conditions and will provide a substantial drawbar pull in circumstances where any type of wheel would be unsatisfactory.

One successful half-track unit (Fig. 9) operates on a girder-track principle. Like the rubber-jointed track, this type is flexible in only one direction. There is no relative motion of pins and bushes at the point where maximum stress occurs, and this appears to result in reduced wear compared with the fully flexible type of track. This type of girder track gives the effect of a big wheel (Fig. 33).

As with full-track machines, half-track conversions should never be run with one track in the furrow when ploughing. On most types of soil such a practice will result in a track so treated having only half the life of one which is run on the land. When half-tracks are used for ploughing, suitable types of ploughs such as are designed for use with tracklayers should be chosen, and the swinging drawbar, which must pivot well forward, should be allowed to swing within reasonable limits.

Agricultural tracklayers and half-tracks are not designed for high-speed work, i.e. speeds of above 4 m.p.h. or so. One of

FIG. 33.—GIRDER TRACK DISCONNECTED, SHOWING THE "BIG-WHEEL" PRINCIPLE. (ROADLESS TRACTION.)

the advantages of the half-track is that when such high speeds are required it is possible to change back to pneumatic tyres. Many farmers find it convenient to fit half-tracks when starting autumn ploughing and to remove them when the heavy spring work is completed.

The girder type of half-track is available either with a standard track or with a skeleton track. The latter is intended, like skeleton wheels, for use on sticky clay soils or for work on seed-beds where the standard track causes excessive compaction of the soil surface. On sandy soils and light loams the skeleton track will dig in at high drawbar pulls, and the standard track is preferable for such conditions. In general, the skeleton track is preferable in conditions where skeleton type wheels give a better performance than standard types of steel wheels fitted with spade lugs.

It should be added that there are other types of half-track units in the development stage, at least one of which shows exceptional promise. It seems more than likely that more will be seen of half-track outfits of a roughly triangular shape, with the driving sprocket at the apex of the triangle and a simple bogie assembly carrying the weight of the tractor at the base.

Now that " all-purpose " tractors are regularly equipped with good steering brakes and a wide range of gearbox and engine

FIG. 34.—HALF-TRACK WITH WIDE CLEAT PLATES FOR USE ON BOGGY LAND. (CUTHBERTSON.)

speeds, the day of the half-track is due. It would not be surprising if simple half-track devices ultimately supersede full-track equipment for agricultural tractors up to about 40 b.h.p. Requirements of a fully successful half-track unit may be stated as follows :—

(i) The unit must be quickly and easily fitted or removed, so that changing from pneumatic tyres to half-track or the reverse may be undertaken as required.

(ii) The tractor must be as easily manœuvrable as the conventional full-track machine.

(iii) The outfit must be capable of operating efficiently on a wide range of surfaces and gradients.

(iv) The track must not pack the soil more than existing medium-powered full-track machines.

(v) Some adjustment of track width is desirable.

(vi) The track must be capable of transmitting fairly heavy drawbar pulls, but not necessarily to the point of stalling the engine at maximum rear axle torque.

It seems likely that all these requirements will be filled by outfits which are reasonable in both first cost and maintenance in comparison with conventional full-track machines.

Market-Garden Tractors. There are several quite distinct types of tractors designed for use in market gardens and for horticultural crops on general farms, and in order to avoid confusion it is best to distinguish the main groups at once.

The smallest may best be described as motor hoes. They are two-wheeled or single-wheeled tractors with engines of up to about 3 h.p., designed for hoeing, grass-cutting, spraying, etc. Some include a plough among their equipment, but in most instances these very small tractors are not really intended for serious ploughing. These machines are extremely versatile and useful, and the smaller ones find steadily increasing use in large private gardens, as well as on commercial holdings. The single-wheel type when equipped with hoes, cultivating tines, mowing cylinder or cutter bar is not difficult to balance, and has the advantage of needing no wheel adjustment when rows of differing widths are cultivated. When used for work like spraying or light transport, a 2-wheeled bogie may be attached. These small machines are easy to operate, and are powerful enough to cultivate, spray, mow and trim hedges quite effectively.

The next important group comprises the 2-wheeled tractors with larger engines, of 3 to 8 h.p., which can do quite satisfactory single-furrow ploughing as well as many other types of heavy cultivation including disc harrowing and rolling. Such tractors will also do multiple-row drilling and will pull a fairly heavy load on a trailer.

Out of these more powerful 2-wheelers have been developed a number of very small 3- or 4-wheeled tractors, some of which are much like a conventional 4-wheeled machine, while others have the engine and transmission compactly arranged over the rear driving axles and an open frame which is supported at the front by a single wheel or a pair of wheels, and carries the tools. In some cases the tractor becomes in effect a carrier for all types of mounted tools including hoes, drills and spraying equipment, and is frequently called a self-propelled tool chassis. With these machines much attention has been paid to points of design

FIG. 35.—LIGHT 2-WHEELED TRACTOR EQUIPPED WITH CULTIVATING TOOL-BAR. (BARFORD.)

which are important in securing really accurate work. Tractors of this type are now used on many of the larger market gardens, but are also being increasingly used as specialized drilling and hoeing units on general farms, especially where a big acreage of sugar beet is grown. One of the most widely used tractors of this general type is not designed for ploughing, but several manufacturers are aiming to develop machines which will operate a forward-mounted plough and will do deep cultivations, as well as inter-cultivations.

A type of market-garden tractor which is quite distinct is the baby tracklayer. One of these, the Ransomes M.G., has a single-cylinder 6 h.p. engine, and a maximum drawbar pull in normal working conditions of about 600-800 lb. It is particularly suitable for some kinds of light orchard work, being capable of working beneath very low trees. This little tractor can do light ploughing with its own single-furrow plough, disc harrowing, cultivating, seeding and inter-row hoeing.

FIG. 36.—DRILLING 6 ROWS OF VEGETABLE SEED WITH SELF-PROPELLED TOOL CHASSIS. (HUMBERSIDE ENG. CO.)

Another tracklayer (the Bristol), though considerably more powerful, is also of interest to many horticulturists on account of its small overall dimensions. This tractor has a 4-cylinder petrol/paraffin engine and is able to pull a 2-furrow plough in average working conditions.

Another important type of tractor now widely used on market gardens is the rotary cultivator or rotary hoe. Some of the smallest sizes are multi-purpose tractors which include a rotary hoe attachment along with a range of equipment suitable for use in large private gardens. This small size, as a rotary cultivator, is often considered too small for commercial holdings. The most important size on market gardens is the 2-wheeled special-purpose machine of about 6-8 h.p. (Fig. 86). Such rotary cultivators are capable of working to a depth of about 9 inches, and on some holdings they are invariably used for all seed-bed and plant-bed preparation, taking the place of the plough and the whole range of cultivation implements. Rotary cultivation is discussed in Chapter Seven.

FIG. 37.—SMALL TRACKLAYER WITH RUBBER-JOINTED TRACKS (BRISTOL.)

WALKING TRACTORS: CONSTRUCTIONAL FEATURES. Some of the points to study in choosing and using walking tractors are as follows :

For general work, the track width should be adjustable to permit working between rows, and there should be ample vertical clearance for straddling the rows of growing crops. Strong, well-designed tools are necessary, and lifting of the tools and control of the depth of working should be easy.

Methods of steering 2-wheeled tractors vary greatly. They include use of a differential gear and independent wheel brakes ; or dog clutches and overrunning ratchets. The necessity for power steering naturally depends on the size of the tractor. Some of the very light fractional h.p. tractors which are used only for the lightest hoeing and weeding tasks are easily steered with nothing but pawl and ratchet gears ; but the 3-8 h.p. machines used for ploughing and heavy cultivations require something better.

Gearboxes differ considerably, some having three forward gears and a reverse, others two forward and reverse, and others only one forward speed. Some makes employ an automatic centrifugal clutch which engages when the throttle is opened.

FIG. 38.—TWO-WHEELED TRACTOR AND PLOUGH. (TRUSTY.)

The usefulness of these walking tractors generally depends to a great extent on the range of equipment available. Most makes have the implements designed to form an integral part of the machine, on the " unit " principle. Attachment of the various implements should be a simple and quick job, and when the implements are attached and the tractor is at work the outfit should be well balanced, so that the wheels have maximum adhesion and the tools can easily be lifted out of work.

Ease of steering is one of the chief problems in the use of walking tractors for hoeing between rows of plants. When a two-wheeled vehicle turns, it pivots about a point in line with the axle, and if hoes are rigidly attached behind the machine, they first move in the direction opposite to that in which the tractor is steered. If the hoes are carried at the front, it is not easy for the operator to see them and, moreover, the torque reaction of the engine tends to lift them out of the ground. One satisfactory method of attaching hoes is by means of a pivotal attachment, the pivot being placed ahead of the main axle and the hoes running at the rear. A difficulty in this is that the hoes may swing sideways owing to a difference of resistance at the two sides, but this may be

FIG. 39.—SMALL RIDING TRACTOR WITH MID-MOUNTED TOOLS. (GARNER.)

controlled by exerting pressure on the hoe frame through a bar connected to the main steering handles.

For ordinary farm work, when cost and performance are considered, the two-wheeled tractor compares unfavourably with the small four-wheeled models now obtainable. Their use in market gardens may increase, but it is unlikely that the gap between the single horse and the "two-plough" tractor on ordinary farms will be filled by two-wheeled or front-wheel drive machines.

Much thought has been given by horticulturists to the development of a tractor which will do everything from deep ploughing and subsoiling to the finest inter-row hoeing. While the development of such a tractor is not impossible, it is too much to hope that such a universal machine could be produced at a price which would suit a grower with a 5-acre holding. The best compromise for many such growers seems to be to accept the fact that once in a while it will be necessary to engage a contractor for deep cultivations. For the rest of the work a small 4-wheeled riding tractor with rear-mounted engine and a range of forward-mounted tools, including a reversible single-furrow plough, seems likely to meet the needs of many growers.

Chapter I.—TRACTORS

GENERAL-PURPOSE TRACTORS

The general-purpose type of tractor, as its name implies, furnishes power for practically all farm work. Not only does it perform all the drawbar, belt, and power shaft jobs but, with the wide variety of integral equipment available for it, the general-purpose tractor puts speed and economy into many jobs for which the standard- or track-type tractor cannot be adapted, such as the cultivating of row-crops.

Figure 3—Cultivating corn with an integral two-row tractor cultivator.

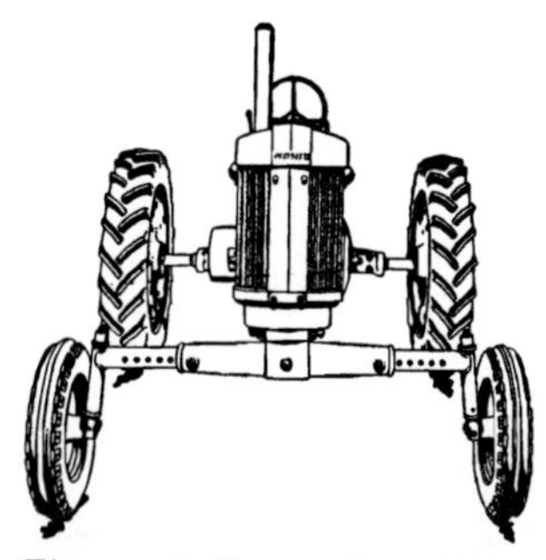

Figure 4—General-purpose tractor with adjustable front wheels; interchangeable with single front wheel, right.

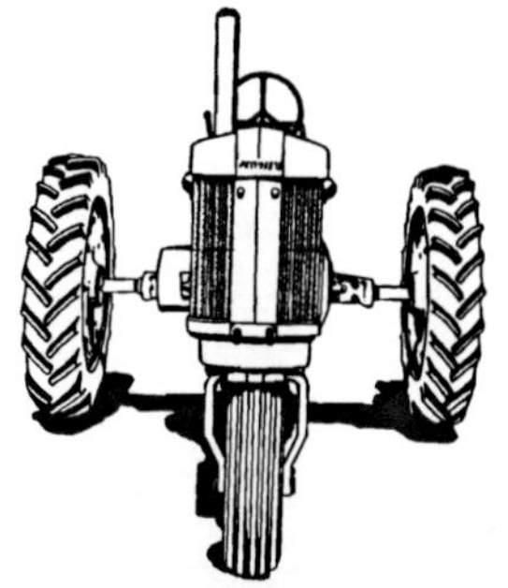

Figure 5—General-purpose tractor with single front wheel.

Taking into account the size of farms, the nature and relative importance of the various jobs to be done, manufacturers of present-day tractors aim to meet the need of every farm both in the matter of power required and type of equipment to be used for the crops to be grown. As a result, there is today a wide range of power sizes and types of general-purpose tractors to meet practically all requirements.

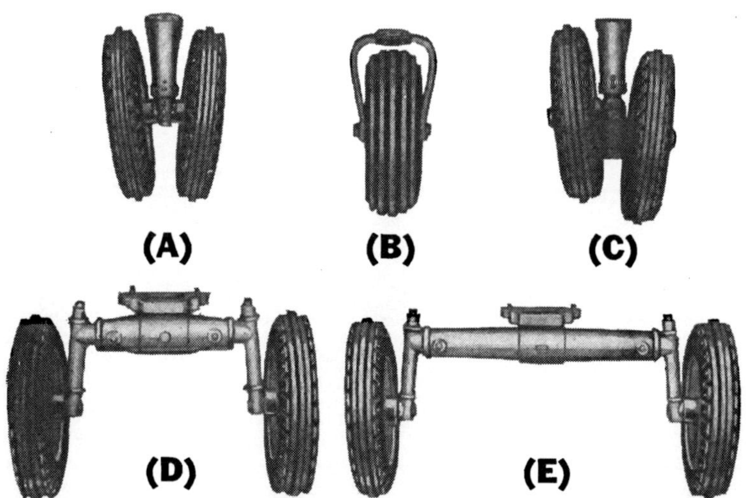

Figure 6—Some general-purpose tractors have two-piece front pedestals which permit the use of any of these front wheel assemblies: (A) dual front wheels; (B) single front wheel; (C) load equalizer wheels; (D) fixed 38-inch wheels; (E) adjustable front wheels.

GENERAL-PURPOSE TRACTORS

The conventional general-purpose tractor, Figs. 3 and 7, has two rear-drive wheels and a front steering member such as the regular dual front wheels or dual wheels equipped with the load equalizer. (See Fig. 10.) This basic design has many variations for specialized farm work.

For example, the tractor shown in Fig. 4, is equipped with the adjustable front axle for straddling wide beds, plowing, or working in extremely soft ground conditions. Wheels can be set in different positions to meet varying conditions. Oftentimes this front axle can be interchanged with a single front wheel (Fig. 5) which is essential for good work in narrow-spaced row crops.

Some conventional general-purpose tractors have special 2-piece front pedestals which permit interchanging the dual front wheels, used in general row-crop work, with any of the assemblies shown in Fig. 6. These include the load equalizer wheels; single front wheel; the fixed 38-inch tread front end, designed for bedder equipment; and the adjustable front axle, mentioned previously.

Figure 7—The small tricycle-type, general-purpose tractor is ideal for cultivating.

Sizes of general-purpose tractors vary from small tricycle type or four-wheel type, as shown in Figs 7 and 8, to big-capacity, 3-plow tractors. The small horsepower models are designed to bring all the advantages of power farming to the small-acreage farmer or to serve as auxiliary or "helper" power on the larger farms. A complete line of integral equipment, easily attached and detached, and controlled by the hydraulic power control system, is generally available for tractors of this type.

Figure 8—The small general-purpose tractor of the four-wheel type.

In heavier soils and on large-acreage farms where row crops are raised, the general-purpose tractor of three-plow power meets the power requirements of most farmers. It is fully adaptable to the varied farm operations including

Figure 9—Cultivating corn four rows at a time with an adjustable-tread tractor.

plowing, planting, cultivating, and harvesting, plus work requiring belt power.

Weight Is Factor. A general-purpose tractor must be heavy enough to give good traction efficiency in plowing and similar heavy work, yet no heavier than needed, because a larger part of its work is on mellow soil. Weight must be properly distributed to gain efficient traction and to maintain stability. The engine must have enough power for the heavier drawbar jobs, yet be efficient at lighter loads. The clearance of all parts that pass above cultivated plants must be sufficient to allow the tractor to pass over them without harming them, yet the machine must not be top-heavy.

Figure 10—The front wheel load equalizer is shown here in X-ray view to show differential construction.

A typical general-purpose tractor, with adjustable rear-wheel tread, which can be equipped for a wide variety of uses in almost any row crop, is illustrated in Fig. 21. Two- and four-row planters, two- and four-row cultivators for corn, cotton, and other crops, multi-row cultivators for special row crops, and two-, three-, and four-row bedders for cotton are some of the equipment that can be used with this tractor. For such jobs as plowing, the rear wheels can be set in 56-inch tread, which largely overcomes side draft. Fig. 9 shows the four-row cultivating unit attached to the adjustable tread tractor.

Some modern tractors have means for quickly changing the rear-wheel tread. The general-purpose tractor mentioned previously has wheels which are adjustable from 56 to 88 inches. (See Figs. 11 and 12.) The wheel is jacked up three clamp screws which hold the wheel in position are loosened, two jack screws are tightened to free the wheel on the axle, and the adjusting nut is turned to move the wheel in or out. The operation is reversed to relock the wheel in any spacing setting desired.

Manufacturers of tractors and farm equipment now provide a wide variety of equipment for their tractors, making it possible to grow and harvest practically any crop, using tractor power exclusively. The attachments and machines available are so numerous as to make impractical a complete consideration of them in this text. The implement dealer's store provides the best place to see and study the various equipment available for each community.

Clearance Important. The general-purpose type tractor must be so constructed as to allow all necessary clearance above the growing crops. Ample clearance is gained in the

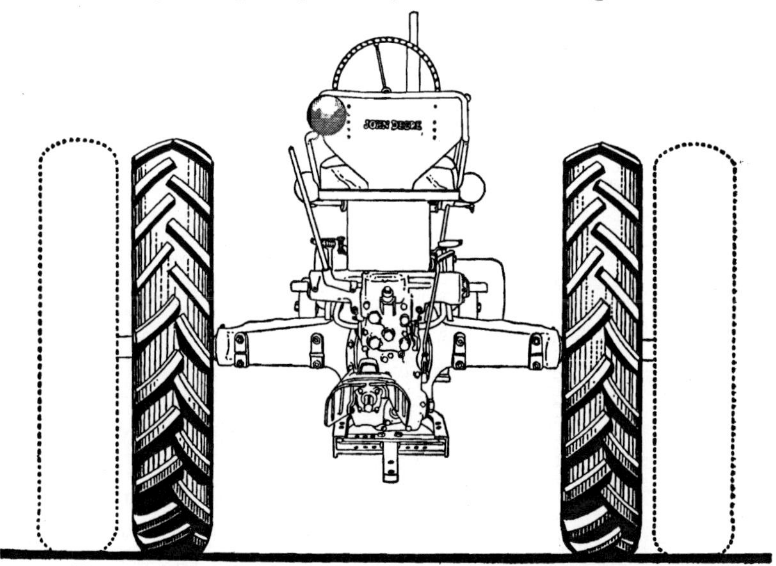

Figure 11—Diagram showing maximum variation in rear wheel tread.

adjustable-tread, general-purpose tractor by several important features of construction. By mounting the front of the tractor on a single support and extending rear wheel tread to straddle two rows, the engine is placed between the rows. In addition, the high drive wheels, in combination with the properly designed rear axle housing, provide ample clearance for cultivating practically all row crops.

Figure 12—Quick change of the tread of the rear wheel is a modern convenience row-crop farmers need.

It is highly desirable in planting and cultivating to turn completely around without stopping and be in position to continue back on the next set of rows. To make this possible, there is a separate brake for each rear wheel on the general-purpose tractor shown. Pressing the brake pedal for the inner wheel holds the wheel back and aids the front wheels in swinging the tractor around sharply.

High-Crop or High-Clearance Tractors. In some sections of the country, extremely tall, bushy, and, sometimes, fragile crops are grown. These include such specialized crops as tomatoes and sugar cane which require maximum tractor clearance.

Figure 13—The general-purpose tractor of the high-crop or high-clearance type.

The answer to this need for extra clearance is found in the high-crop or high-clearance tractor. (See Fig. 13.) Normally this tractor is the same design as the regular general-purpose tractor except for the added clearance. Thus, growers of tall crops can get extra, damage-free cultivations that help produce better yields and bigger profits.

Forward Speeds. In a general-purpose tractor, flexibility of speed has much to do with capacity and efficiency. In cultivating or transplanting, especially, there are times when it is necessary to go very slow. At other times, both speed and effectiveness are gained by traveling fast and throwing the soil briskly.

To meet this wide range of speed requirements, most general-purpose tractors have several forward speeds in the transmission gears, providing extremely slow speeds for certain field jobs as mentioned previously and higher speeds for transporting equipment on highways from barnyard to fields and return. These several forward speeds are provided so that the tractor can be operated at full throttle at all times, thereby assuring maximum engine efficiency.

Figure 14—Cultivating and fertilizing with a general-purpose tractor.

The power take-off device, which supplies power directly by shaft to machines being pulled by the tractor, has found wide application and great usefulness on tractors of both the standard and the general-purpose types. It is discussed in detail on page 47.

Care Important. It is very important that the tractor be given proper care. If the owner is dependent upon his tractor for all farm jobs, delays are costly. Careful handling, strict attention to oiling, adjusting and repairing the tractor and the equipment that is used with it, as directed by the manufacturer in the Operator's Manual, will result in greater satisfaction and greater net profits.

STANDARD-TYPE TRACTORS

While the general-purpose tractor, described in preceding pages, meets the needs of the row-crop farmer in plowing, planting, cultivating, and harvesting his crops, the particular power requirements of the small-grain grower and the orchardist are best met by tractors of standard design, especially adapted to the work at hand. On larger farms, where row crops are grown, standard-type tractors are often used to supplement the general-purpose tractors in preparing seedbeds and harvesting the crops.

What has been said about the design and care of the general-purpose tractor applies so generally to the standard types that a further discussion is unnecessary.

The standard-type tractor, furnishing power at three outlets, the drawbar, belt, and power take-off, is used for practically all power requirements except planting and cultivating. Many standard-type tractors are equipped with hydraulic systems for raising, lowering, or adjusting drawn equipment through a remote cylinder.

STANDARD-TYPE TRACTORS

The tractor, shown in Fig. 15, is a typical two-three-plow tractor of this type. For larger farms, standard-type tractors having three-four-plow or more power are also in use.

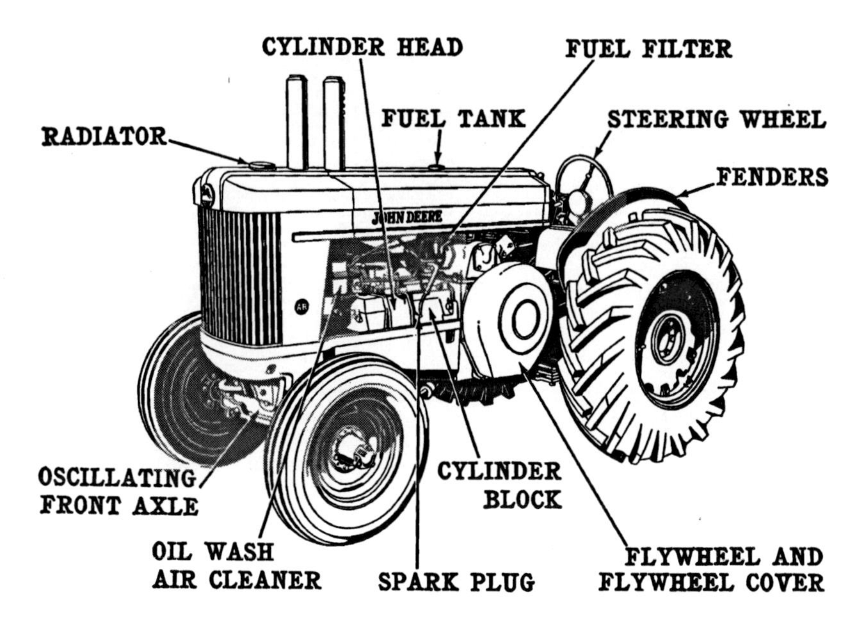

Figure 15—Standard-type tractor of two-three-plow power.

A further variation of this type is the orchard tractor. It is built low and compact with wheels and pulley shielded for working close to trees and under low-hanging limbs in orchards or groves (see Fig. 16).

On large-acreage farms where greater power is required for heavy-duty plowing, disking, seeding, and harvesting operations, the wheel-type Diesel tractor has increased in popularity during recent years. (See Fig. 17.) Fuel oils for Diesel tractors usually cost less, making Diesels far more economical to use on big-capacity farming jobs.

Figure 16—An orchard-type tractor working in a California orchard.

Track-Type Tractors. In extreme farming conditions, track-type tractors offer several definite advantages. These tractors have the flotation necessary for working in light soils, loose soils, rough terrain, in woodlands, etc., and they have the stability that is essential for working on extreme hillsides.

Track-type tractors vary in size, depending on the farming requirements. Typical of the smaller tractors of this type is that shown in Fig. 18. The traction mechanism of this tractor is essentially two endless, metal-linked belts or chains known as tracks. Each runs on two steel wheels, one of which is a sprocket wheel and acts as the drive; the other serves as an idler. Steering is accomplished through the tracks themselves by reducing the movement of one track below that of the other. Track rollers on the underside of the track frame act as supports for the machine and that part of the track in contact with the ground.

Figure 17—The wheel-type Diesel tractor of four-five-plow power.

Track-type tractors are used extensively in orchard work, for farming operations on extremely hilly sections and in light soils, for terracing, land clearing, and, particularly, for earth-moving and leveling operations in irrigated sections.

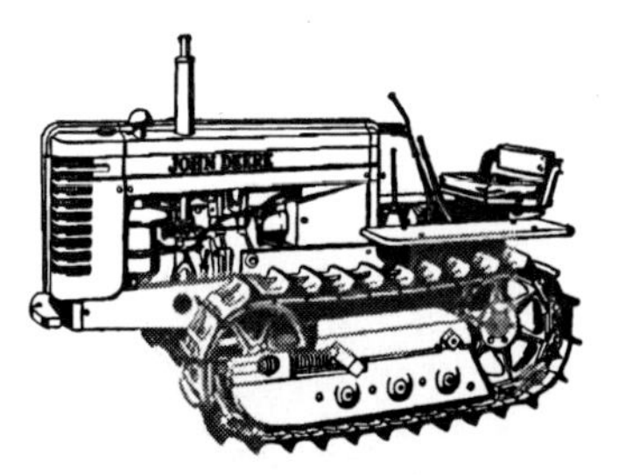

Figure 18—The track-type tractor of two-three-plow power.

Questions

1. *Name and describe the different types of tractors.*
2. *What are some of the advantages of the general-purpose tractor?*
3. *Why is weight an important factor in the general-purpose type tractor?*
4. *What is meant by a high-clearance tractor and where is it used?*

5. *Whdt is the main difference in construction between the general-purpose tractor and the standard-type tractor?*

6. *What are the important advantages of the standard-type tractor?*

7. *Name the different types of standard-type tractors and describe their particular uses.*

8. *What is a track-type tractor and where is it used?*

9. *How is a track-type tractor controlled?*

10. *What types of tractors are best suited for the farming in your community?*

TRACTOR FUNDAMENTALS

All farm tractors, regardless of their size, type (general-purpose, standard, or crawler types), or kind of engine (Spark ignition or Diesel), are made up of four basic and fundamental units.

The heavy-duty Diesel tractor pulling a big-capacity offset disk harrow.

These include (1) the engine which is the source of power; (2) the transmission which makes this power available at the drawbar, power take-off, power lift, belt pulley, and which provides means for varying the forward speed to meet the job at hand and the condition encountered; (3) final drive (including the differential) which delivers the power from engine through transmission to the rear wheels; (4) clutch which acts as a coupling to connect the engine to the transmission and belt pulley. The efficiency of a tractor depends upon the proper functioning of all units—a condition which exists only when all units are properly adjusted and maintained.

Before attempting to operate a tractor or engine, the operator should make a careful and thorough study of the instruction book furnished by the manufacturer.

Although general engine principles are the same, each make of tractor or engine will have different operating procedures. The basic principles, common to practically

Figure 19—Keeping a tractor in good working condition makes the job easier for the tractor operator.

all internal-combustion engines used on the farm, will be discussed in the following pages.

Internal-Combustion Engines. An internal-combustion engine is an engine in which the heat or pressure energy necessary to produce motion is developed in the engine cylinder, as by the burning of a gas, and not in a separate chamber as in a steam engine boiler.

The fuel, mixed with air, ignites, burns rapidly, expands inside the cylinder, pushes the piston back, turns the crankshaft, and so develops power. The power generated can be applied to the operation of machines through the belt pulley in the case of the engine, and through the belt pulley, drawbar, power take-off and, in many cases, through the hydraulic control system of the tractor.

The modern hydraulic control system supplies power to raise, lower, or set equipment at any in-between working position desired. On general-purpose tractors, integral equipment is controlled through the tractor rockshafts; drawn equipment is controlled through the remote cylinder which attaches to the implement and is connected to the tractor by flexible oil lines. On standard-type tractors, hydraulic power can be used to raise, lower, or adjust drawn equipment through the remote cylinder.

There are two general types of internal-combustion engines—two-stroke cycle and four-stroke cycle. The two-stroke cycle engine has a power impulse or working stroke every revolution. The four-stroke cycle engine burns its fuel charge every second revolution. There are four strokes of the piston from one power impulse to the next. Practically all farm engines and tractors are of the four-cycle type, either spark ignition or Diesel.

The strokes of the spark ignition engine are:

First: suction or intake. Here, the piston draws a charge of fuel and air into the cylinder through the inlet valve.

Second: compression. The piston, on its return, compresses the fuel and air mixture into the end of the cylinder called the combustion chamber. Full power is secured only with good compression.

Third: expansion or power stroke. At a point slightly in advance of full compression, an electric spark, produced by a magneto or battery, ignites the fuel. This causes a sudden high expansion pressure to act on the piston, pushing it back so that work is performed.

Fourth: exhaust. On its return from the power stroke, the piston pushes the burned gases out of the cylinder, through the open exhaust valve, and then through the exhaust manifold.

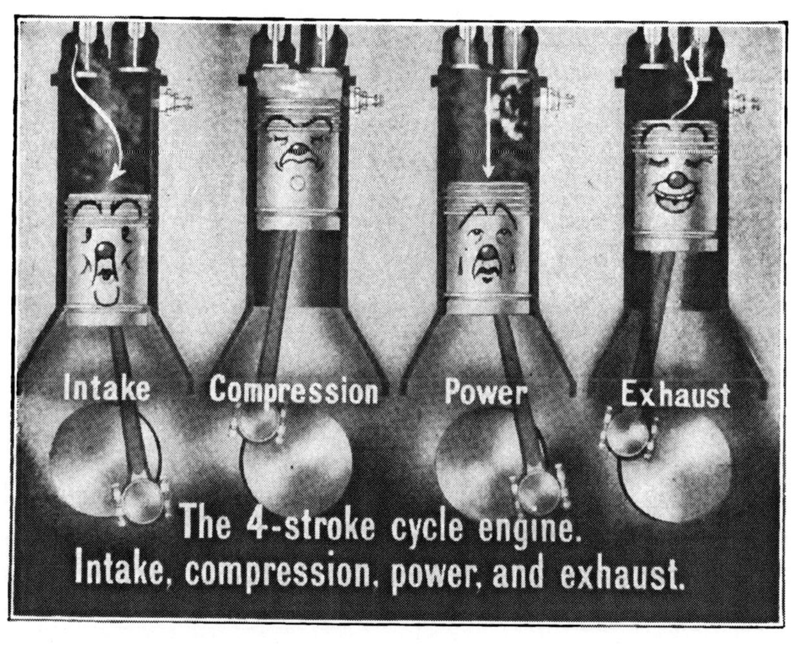

Figure 20—Illustrating the four strokes of the four-cycle engine.

It is well known that air or gas under pressure produces heat. The Diesel engine utilizes this principle to burn the fuel. That is, air is compressed until the resulting high temperature is sufficient to ignite the fuel injected after compression is practically completed. Since the temperature of compression is well above the minimum fuel combustion temperature, no ignition devices or spark plugs are required.

The strokes of the Diesel are similar to those of the spark-ignition engine except in the third, or power stroke, at which point the fuel is burned. Just before the piston reaches the end of the compression stroke, fuel is injected into the combustion chamber and ignition, induced by the heat of compression, begins as the piston approaches the top of the stroke. If the fuel were drawn in with the air, under most conditions it would ignite before the end of the compression stroke, making the engine buck or attempt to run backwards. Therefore, fuel must be injected at or slightly before the end of the compression stroke. The expanding gases, released by the combustion

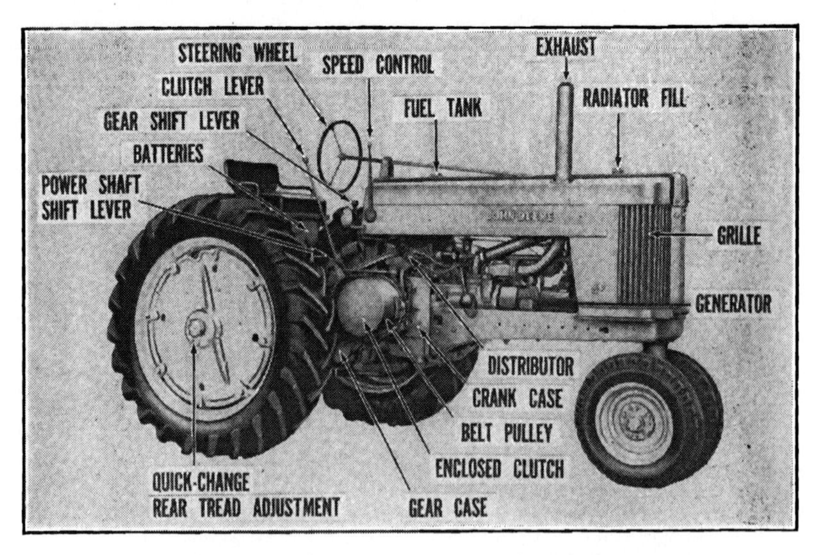

Figure 21—Adjustable rear-wheel tread, general-purpose tractor.

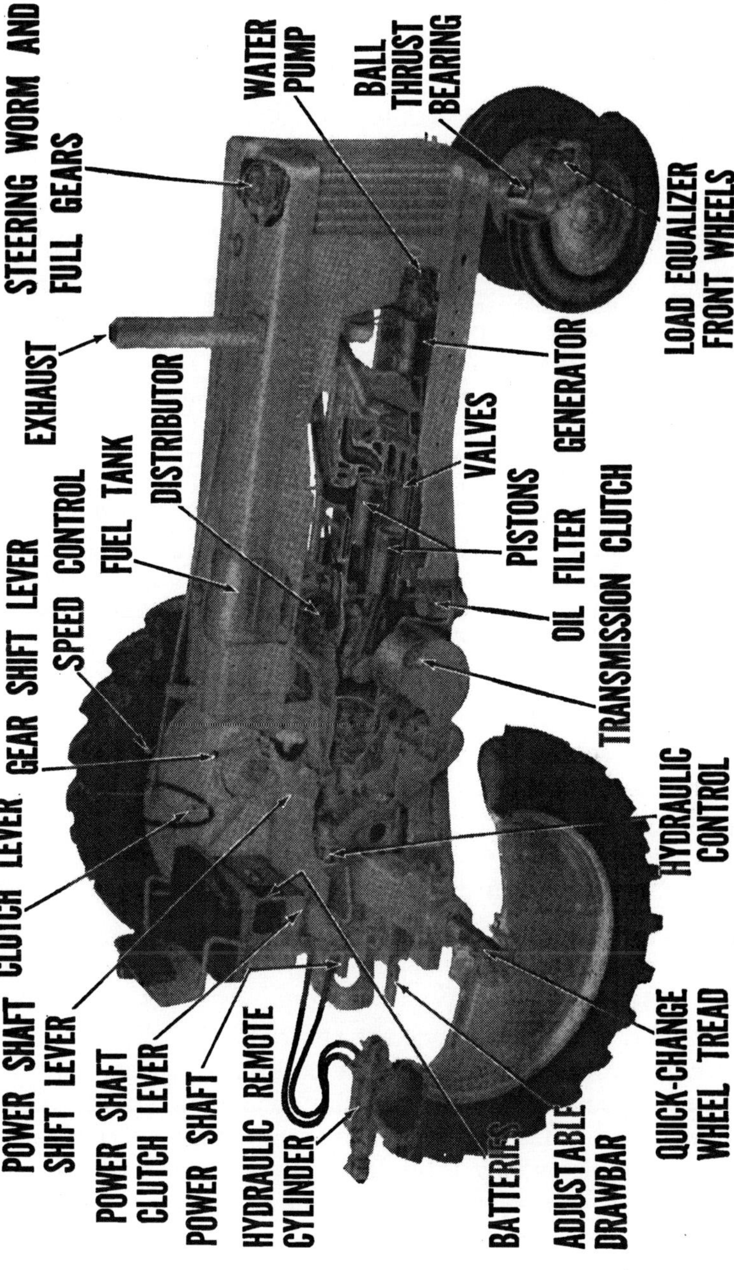

Figure 22—Cross-section view of the general-purpose type tractor.

of the fuel, force the piston back to exert power on the crankshaft.

Thus, on all types of four-cycle engines these events—*suction* or *intake, compression, expansion* or *power,* and *exhaust*—make the complete cycle.

Spark-Ignition Engine. For the discussion of this part of the chapter, the tractor shown in Figs. 21 and 22 will be used as a basis. It is a typical general-purpose farm tractor having a two-cylinder, horizontal engine available in two types: with high compression ratio, to burn gasoline, and, with lower compression ratio, to burn the heavier fuels such as distillate and tractor fuel, and gasoline also.

When a new tractor is delivered, it is ready to give efficient service for a long time, under normal conditions, without a great amount of adjustment. The operator's chief responsibility is in correct lubrication and proper care. However, when trouble arises, the operator should be capable of analyzing his tractor and making the day-to-day adjustments that fall within the range of his skill and the equipment of his farm shop.

As the source of tractor power, the engine of the tractor is in operation during every minute the tractor is at work whether on drawbar, belt, or power take-off. For this reason, it is well to gain a thorough understanding of the essential requirements for most efficient engine operation.

Air-Fuel System. It is wise to consider fuel and air at the same time, for the successful operation of the engine depends upon a correctly-proportioned mixture of fuel with air. The proportions are controlled by adjustment of needle valves on the carburetor. When the mixture is correct, the engine runs smoothly, delivering its maximum power; too much fuel in the mixture, called a "rich" mixture, is indicated by a black, smoky exhaust and irregular running of the engine; too little fuel, called a "lean" mixture, is indicated by a "popping back" through the carburetor, misfiring of the engine, or by a high-pitched "ping" referred to as a pre-ignition knock.

The fuel system of the gasoline tractor, which will be discussed more completely in this chapter, consists of the fuel tank, carburetor, and manifold. The all-fuel tractor, on the other hand, has a fuel system consisting of the fuel tank, a small tank for gasoline used in starting, a three-way valve which permits the flow of fuel from either tank, and a carburetor and manifold.

Gasoline is drawn into the carburetor which serves to atomize the fuel in air, producing a highly combustible gas. This gas is drawn into the combustion chambers, placed under pressure by the piston on its compression stroke and ignited or burned by a spark, timed to fire at the proper instant to deliver full power of the burning fuel to the piston.

The Carburetor. Fuel must stay liquid while in the carburetor until discharged in the form of a fine spray through a nozzle. The carburetor has its own local fuel supply (float bowl) carried at a constant level by means of a float. The nozzle or jet forces the liquid fuel from this bowl into the air stream. Nozzles are located centrally in what is called a venturi tube or barrel which is a constricted section of the intake manifold designed to increase greatly the velocity of air at this point.

Figure 23—The duplex carburetor meters identical amounts of fuel and air to each cylinder.

The tractor shown in Fig. 21 is equipped with duplex carburetion. (Fig. 23.) This carburetor has two venturi which meter fuel in identical amounts to each cylinder, thus providing quicker response and smoother engine performance.

Increasing the air velocity increases the ability of the air stream to pick up the fuel from the nozzles and atomize it. The throttle, which controls the amount of fuel to reach the engine, is located beyond the jet nozzles and venturi tubes. In most tractors the throttle valve is controlled by a governor mechanism.

The manifold heat control valve, located in the engine manifold, improves engine performance and efficiency in hot and cold weather. (See Fig. 24.) Turning the valve to the "hot" position causes the exhaust gases to warm the incoming fuel and air mixture before it enters the combustion chamber of the engine. This position is used in cold weather, below 32° F.

Figure 24—Two-position manifold valve, set at the hot position, *right*, diverts exhaust gases to pre-heat incoming fuel for cold weather operation.

Cold setting, *left*, expels burned gases directly into exhaust, keeping the incoming fuel mixture cooler in hot weather.

Turning the valve to the "cold" position causes the hot gases to leave the engine without warming the incoming fuel and air mixture. This position is for use in weather above 32° F.

Checking the Fuel System. When any one or more of the three essentials, fuel, air, or spark, is deficient or lacking, the engine will operate poorly or fail entirely to operate. It is wise, therefore, to check the "important three" for lack of efficiency in the engine, difficult starting, or entire failure to start. In checking the tractor, start with first essentials first and check through each possible source of trouble.

This procedure eliminates guesswork and removes the need for "back tracking" in making the service checkup.

The most important point in caring for and adjusting the fuel system is keeping out dust and dirt. Fuel should always be stored in clean containers, protected from dust, dirt, and water; it should always be strained when filling the tanks. When engine trouble occurs, it is well to look for the cause of the trouble in the fuel system first.

First, shut off the fuel at the fuel filter by turning the shut-off valve. Then, remove the glass bowl and clean it thoroughly. After the bowl is removed, turn the shut-off valve to see if fuel flows readily from the tank. If not, the tank must be cleaned. Replace the fuel bowl, being sure the gasket fits properly and the screen is clean and in good condition. All fuel lines should be checked.

The screen of the gasoline carburetor can be cleaned easily. Flush the sump by removing the sump plug and turning on the fuel. Then, remove the strainer plug with attached screen. Clean the screen and replace screen and plug.

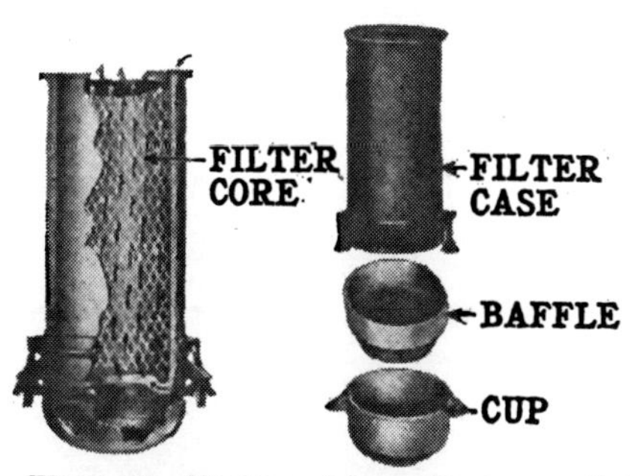

Figure 25—Air cleaner shown in cross-section disassembled for servicing.

Checking Air Cleaner. When we consider that every gallon of liquid fuel consumed by the engine must be mixed with nine thousand gallons of air, the importance of the air cleaner (Figs. 25 and 26) becomes apparent. The sole function of the air cleaner is to provide a continuous flow of clean air to the carburetor where it is mixed with the fuel and drawn into the combustion chamber. The air cleaner removes from incoming air, dust and grit particles that would injure the cylinders and working parts if drawn into the combus-

Figure 26—Air cleaner should be checked every ten hours.

tion chamber. As the air is drawn into the cleaner, the dust and heavier particles of dirt are caught and retained in a mist of oil created by the draft of air drawn through the cleaner.

Before each day's work, the oil sediment cup at base of filter should be detached. If the oil is thick with suspended dirt or if there is more than 1/2-inch of dirt in the bottom, the air cleaner should be serviced. It should be serviced at least every 30 hours. The dirt-filled oil should be removed, and the entire cup washed in kerosene to remove all the sediment. (See Fig. 26.) The cup should then be refilled to bead mark with new engine oil and replaced. When engine difficulties occur, the air cleaner should be given a routine check since a clogged air cleaner or a badly dented or crimped air intake may so constrict the passage of air as to make engine operation impossible.

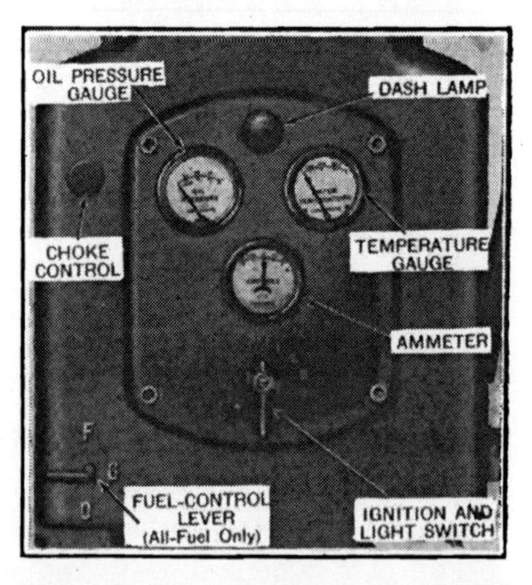

Figure 27—Most tractors are equipped with instrument panels similar to this shown, including the oil-pressure gauge, dash lamp, water temperature gauge, ammeter, choke control, and ignition and light switch. Fuel control lever used on all-fuel tractors.

Ignition System. In the spark-ignition type tractor engines, the air-fuel charge is ignited by means of a high-voltage electrical spark jumping a spark plug gap. There are two common types of ignition systems: (1) the battery distributor and (2) magneto. The principal difference in the two systems is the primary source of electrical current.

Figure 28—A tractor should have all operating controls convenient to the tractor operator, including the throttle or speed control lever, clutch lever, gear shift lever, power shaft shift lever, brake pedals, power shaft clutch lever, and hydraulic remote or rockshaft lever.

In the battery system the necessary current is produced by chemical action within the battery, while in the magneto system the armature of the magneto is rotated to produce current. The battery system will be discussed more thoroughly in this text.

The battery-distributor-type ignition system consists of a battery, a coil which transforms low voltage current to high voltage, a distributor which carries the high voltage current to the proper plug, and the spark plugs which release the spark in the combustion chamber. In this type of system, a generator is required to keep the battery charged and there must also be a switch to break the circuit so that the battery will not discharge when the engine is not in operation.

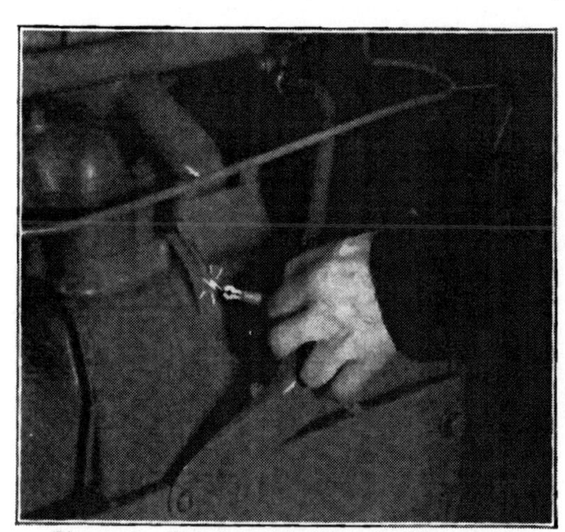

Figure 29—Simple method for checking spark at spark plug.

Checking the Ignition System. With fuel and air flow established, check the ignition system to make certain that a good "hot" spark reaches the compressed gas. To check for spark at the combustion chamber, remove spark plug wire, leaving spark plug in position, and hold end 1/4-inch from the engine as shown in Fig. 29. Turn flywheel, or crank the engine. If sparks jump from end of wire to the engine, the ignition system and spark plug wires are in good condition; spark plugs are at fault.

If the plug is dirty, it should be cleaned. If the porcelain insulator is cracked, the plug must be replaced. The electrode gap should be spaced .030-inch; always use a spark plug gauge for correct setting. (See Fig. 30.)

If no spark occurs upon checking at the terminal, check the source of spark, the distributor. Remove the switch-to-distributor lead from the distributor circuit. Place the opposite end of the lead 1/4-inch from some metal part of the tractor. If no spark occurs, the trouble is in the battery-distributor circuit.

If spark does occur here, the distributor is operating satisfactorily; it is evident then that spark-plug wires are broken or badly frayed, or that the insulation has deteriorated with age, causing a short circuit which "grounds" the spark before it reaches the plug. Replace spark plug wires. If no sparks occur upon checking in this manner, remove distributor cap and check points for proper clearance as given in your tractor instruction book or service

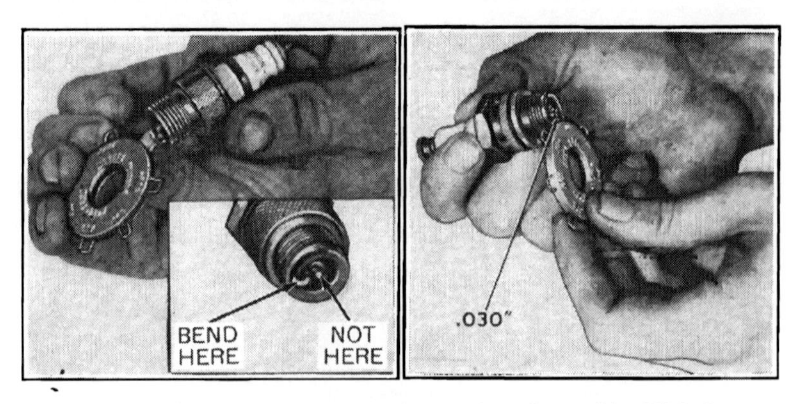

Figure 30—Plugs should be cleaned frequently and spaced to .030-inch.

manual. If points are badly pitted, burned, or dirty, they should be replaced. Otherwise carefully remove them, clean and dress each point to a smooth flat surface, using a fine hone. Do not file points. It is not necessary to remove pits. Just be sure the surfaces are clean and flat.

Figure 31—Adjusting the distributor breaker point gap, set at .021-inch

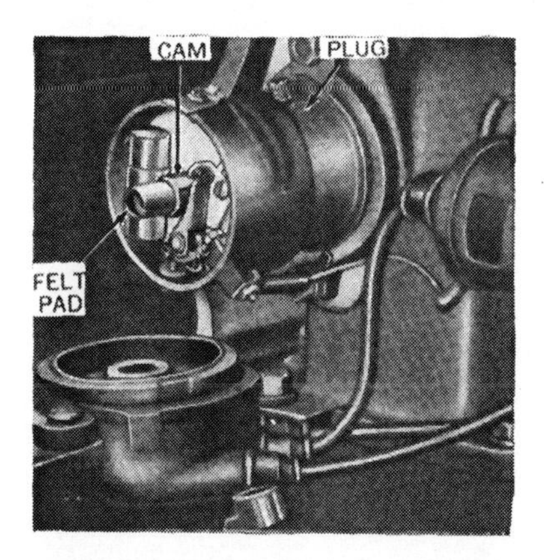

Figure 32—Servicing a distributor is simple; grease breaker point cam and oil felt pad on end of cam as per manufacturer's directions

With points replaced and properly gapped or spaced, check again. If no sparks occur, remove the unit and take it to your tractor service dealer for repair. Ignition repairs call for a fine degree of skill and the use of special shop equipment; for this reason, it is unwise to attempt repairs in the farm shop. When replacing the ignition parts, follow the manufacturer's instruction book for correct procedure to insure proper "timing" or firing order of the cylinders.

The magneto-type ignition system can be checked in a similar manner. Complete instructions will be given in your tractor operator's manual.

The Battery and Its Care. The battery stores chemical energy which can be converted into electrical energy whenever you need it, for starting and operating the engine, lights, etc.

Many present-day tractors are equipped with 12-volt systems to provide power for lights, starters, etc.

The tractor under discussion (Fig. 21) is equipped with such a system. It includes a heavy-duty battery with a rubber case and anti-leak battery caps. No other battery should be used as replacement; be sure only a battery recommended by the manufacturer is used as a replacement on your tractor.

Figure 33—Battery should be checked at frequent intervals. Here the specific gravity is being checked with a hydrometer.

At least every 30 days or every 120 hours the battery should be wiped off with a damp cloth. Loosen any corrosion around terminal connections and apply a solution of 1/4 pound of soda to one quart of water. Flush the outside of the battery with clear water. Make sure the vent holes are open in each cap. Battery connections should be clean and tight; a coating of vaseline on each terminal connection will retard the accumulation of corrosion.

Check the electrolyte (acid and water level) in the battery every 120 hours or 30 days for proper level—more frequently in hot weather. Add distilled or soft water (if not available, use drinking water) until the recommended level is reached. Never permit the electrolyte level to go lower than the top of the cell plates. Never add water in freezing weather until after the engine has started; water will not mix with the electrolyte until the generator passes a charging current into the battery.

The specific gravity of the electrolyte should be checked with an accurate hydrometer before adding water. (See Fig. 33.) Specific gravity should not go below 1.225 which is half charge. When fully charged the reading will be approximately 1.240 to 1.255, depending on make of battery.

Cold weather affects the battery adversely in several ways. It reduces the output of the battery even if fully charged. Freezing weather may damage a battery beyond repair. Therefore, never allow a battery to stand long in the winter time without checking its conditions and recharging if necessary.

If you are having trouble keeping the battery charged during cold weather, moving the wire from one terminal to another converts the current voltage regulator into a straight voltage regulator and permits a higher charging rate and consequently a faster build-up of the battery. See your operator's manual for complete instructions before making this change.

Always follow the manufacturer's suggestions in storing a battery or installing the battery in the tractor.

Starting and Lighting Equipment. This equipment, more and more widely used on modern tractors, requires but a small amount of attention. However, it is highly important that what little servicing is required should be done at periodic intervals so that the system will function properly and dependably.

Starting and lighting equipment includes three important units: the storage battery, already discussed; the generator, which develops the energy to be stored in the battery; and the starting motor, which is called upon to "turn over" or crank the tractor engine.

The generator converts mechanical energy into electrical energy. The charging rate, set at maximum inside the generator, is controlled by the voltage regulator used to prevent overcharging of the battery. Generator output, or charging rate, should be regulated to the demand made upon the battery for starting and lighting. If you operate your tractor without the use of lights or use them only occasionally, the generator charging rate should be reduced. If, during busy seasons, you operate all night, the charge rate must be increased to maintain the battery at full charge. To change charging rate, remove the generator dust band and loosen the brush-holder screw. Move the brush in direction armature rotates to increase, and move in opposite direction to decrease. After obtaining desired charging rate, tighten the brush-holder screw. Replace the cover band, making sure it is tight and covers the openings properly.

The generator should be lubricated sparingly, only a few drops of oil at a time. The drive belt should be adjusted for proper tension. See your operator's manual for complete instructions.

All connections in the entire system should be kept clean and snug-fitting. When your new tractor is delivered with starting and lighting equipment, it is advisable to check carefully the instructions covering this equipment to familiarize yourself with the servicing required.

Checking Compression. As mentioned previously, efficient engine operation depends upon proper mixture of fuel and air, proper compression in the combustion chamber, and spark properly timed to ignite the mixture.

We have discussed the method of checking the fuel system, the air flow to the carburetor, and the ignition system; compression should be the next point to be given consideration.

Proper compression requires a perfect seal of the combustion chamber which will insure full power from every charge of fuel entering the chamber. Two factors are involved in gaining the perfect seal necessary for efficient operation: proper seating of the valves (which permit intake of fuel and passage of burned gases and which are "timed" to close at the time of the power stroke to seal the combustion chamber), and proper fit of pistons and piston rings which seal the chamber to prevent passage of gas and power between piston and cylinder wall.

As the piston moves forward in the cylinder, it compresses

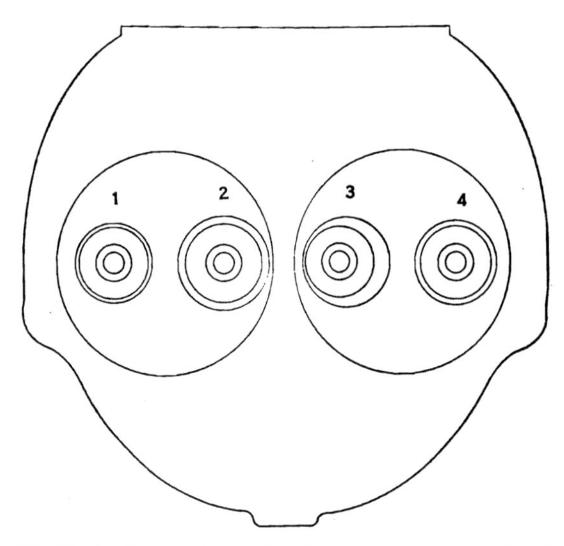

Figure 34—The ideal valve seat is one that forms a perfect seal of uniform width over the complete circumference of the valve-port. A seat too narrow (1) tends to cut or pound a groove in the valve, resulting in lost compression. With the seat too wide (2), it is almost impossible to get a perfect seal and, here again, loss of compression results.

Where the seat has been worn off center (3) due to worn valve guide or any other cause, it is impossible to obtain the proper seal between valve and seat which results in lost compression and uneven operation of the engine.

The properly seated valve (4) forms the perfect compression seal, and insures proper engine performance for the longest period. Width of seat varies with power and type of engine. See manufacturer's instruction book or consult your dealer for exact width for your engine.

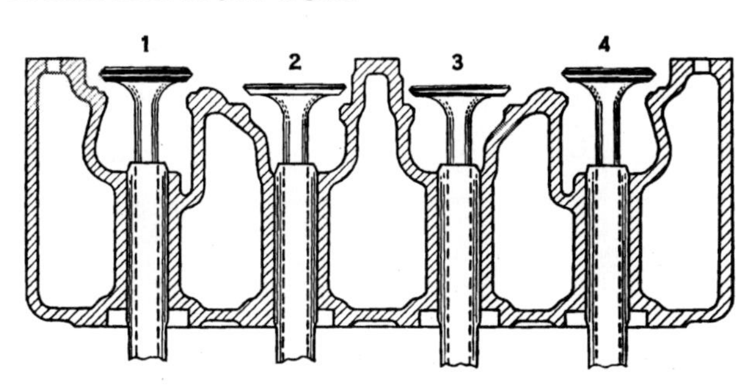

the air in the combustion chamber creating a "cushion" of air. When cranking the engine, this cushion will be noted as each piston approaches "dead center," or the end of its stroke. If no definite cushion or compression is noted at uniform intervals as the engine is cranked, weak compression is indicated.

Energy required to turn the engine should be alike for all cylinders indicating that compression is equal in all

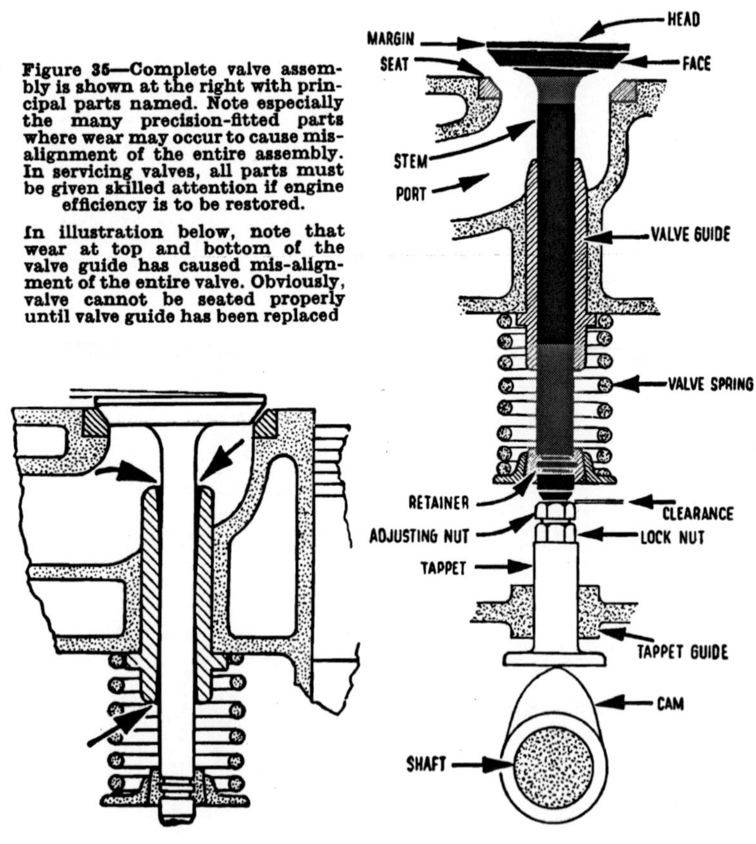

Figure 35—Complete valve assembly is shown at the right with principal parts named. Note especially the many precision-fitted parts where wear may occur to cause misalignment of the entire assembly. In servicing valves, all parts must be given skilled attention if engine efficiency is to be restored.

In illustration below, note that wear at top and bottom of the valve guide has caused mis-alignment of the entire valve. Obviously, valve cannot be seated properly until valve guide has been replaced

Illustration showing how worn valve guide affects proper seating.

Cross-sectional view of complete valve assembly with parts named.

cylinders. If energy required to bring one piston over center, or "compression," is greater than that required for another, or if a "hissing" sound indicating the escape of air is heard, compression is weak, with resultant loss of power under operating conditions.

The cause of poor compression may be found in poorly adjusted valves, improperly seated valves (see Fig. 34), worn valve guides (Fig. 35), worn piston rings or cylinder walls, or improperly sealed cylinder head gaskets which permit passage of air or water under pressure. Before attempting extensive repairs, it is well to check the valves for proper tappet clearance and adjust tappets to tolerances recommended in the instruction book or service manual for your particular tractor. Thickness gauges should be used in determining proper clearance (see Fig. 36).

If restoring proper tappet clearance fails to restore compression, it is well to enlist the aid of a skilled tractor serviceman to restore the engine to full efficiency. While the actual principles of engine construction and function are simple, the servicing of such parts as valves, pistons, and piston rings of the modern precision-built tractor involves the use of highly specialized equipment which the average farmer would find unprofitable to install in his home shop. The use of special steels in valves dictates the use of special stones and facing tools in their servicing (see Fig. 37); likewise, the highly precise job of

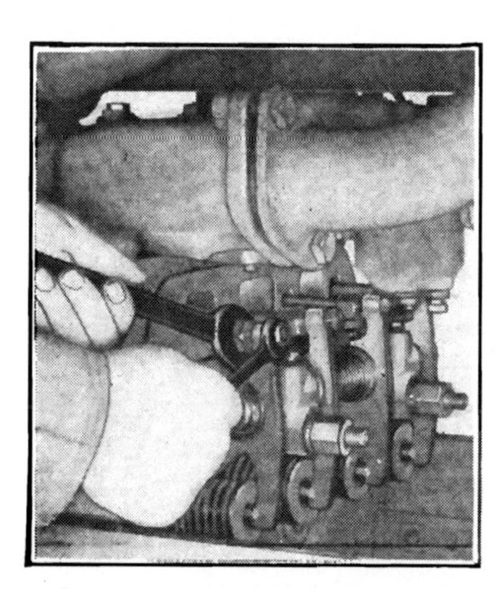

Figure 36—Thickness gauges should be used when adjusting valve tappet clearance.

replacing pistons, piston pins, and piston rings which involves removing and, in many cases, adjusting pressure lubricated bearings, is one for which special tools have been developed.

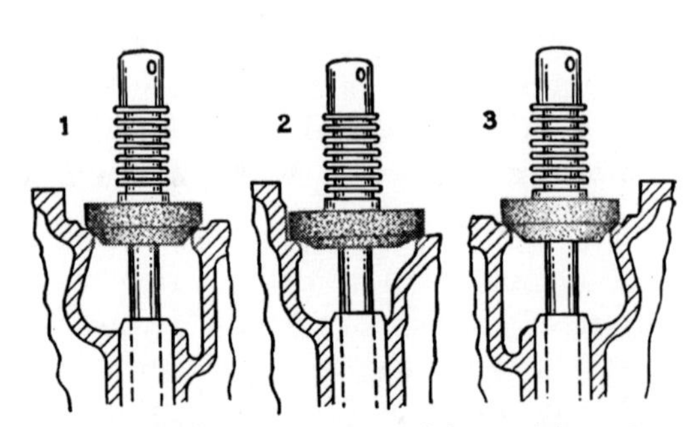

Figure 37—Special stones used in servicing valves. (1) Rough-cutting. (2) Rough stone used for narrowing the seat. (3) Fine stone for finishing.

Diesel Engines. While the Diesel engine is more costly to build, it has certain economy characteristics that enable it to convert fuel into horsepower at extremely low cost. It is for this reason that Diesel power is so effective in reducing costs on the heavier jobs where periods of operation are prolonged. It follows then, that Diesel power is used widely in the larger tractors, as shown in Figs. 17 and 38, in many stationary jobs such as hydraulic pumps and electric generators, and in railroads, trucks, and ships.

As mentioned on page 18, air only is drawn into the cylinder of the Diesel engine on the intake stroke. The engine speed is controlled by the amount of fuel injected. In order to assure satisfactory performance it is essential that only clean fuel be used in the Diesel engine. Every precaution is taken in the fuel system to keep the fuel clean.

The fuel system of the Diesel engine consists of the fuel tank, transfer pump, fuel filters, injection pump, and injectors. On the Diesel tractor shown, the incoming fuel flows by gravity from the tank into the glass sediment bowl where moisture and dirt settle out. (See Fig. 39.) The fuel flows on into the transfer pump which, in turn, forces the fuel through two micronic paper filters. These filters remove dust, dirt and abrasive materials which

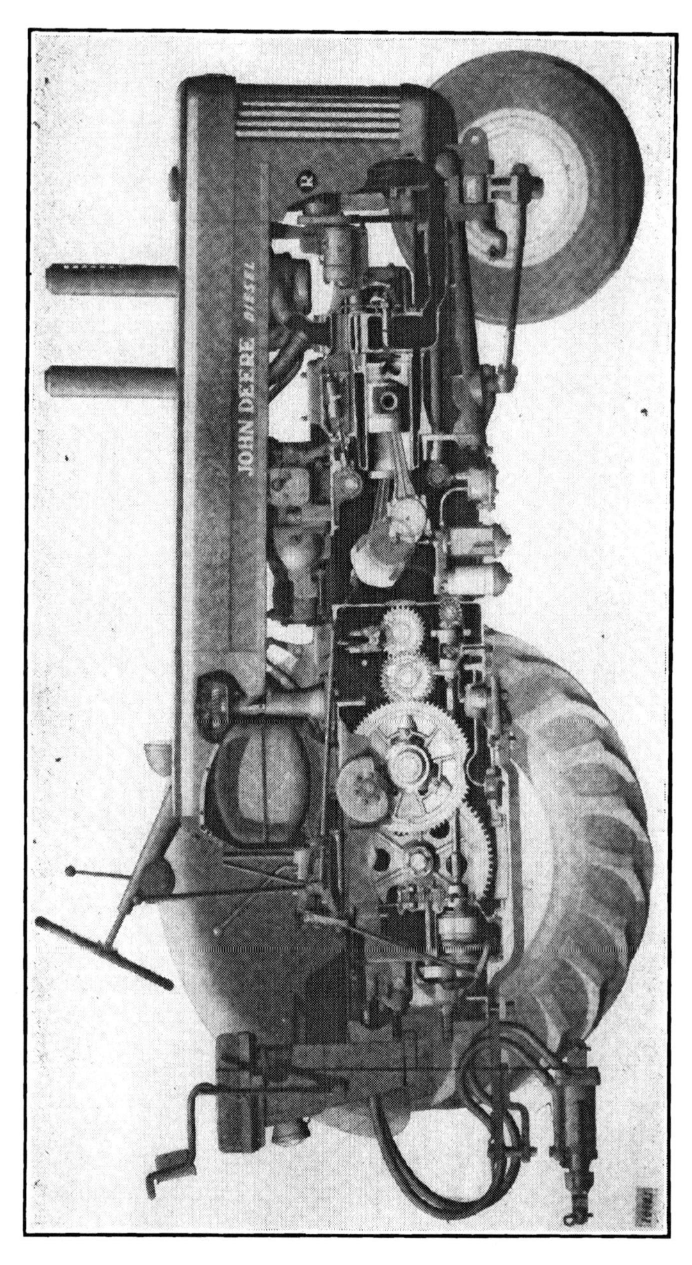

Figure 38—Cross-section view of a Diesel tractor.

might injure the closely fitted parts of the injector pump and injectors.

This system of filters is intended only to remove the dirt and water which may normally enter the fuel tank during tractor operation in the field and the condensation which may form in the tank. Under normal conditions, these filters should not require service too frequently. Follow the recommendations of the manufacturer set forth in the operator's manual.

Figure 39—The three-stage fuel filtering system used on one type of Diesel tractor.

The clean fuel now flows to the injection pumps which force the fuel under high pressure into injection nozzles. The fuel is then broken into a fine mist and sprayed directly into the combustion chamber.

Repairs or adjustments on the injection pump should be made by a competent mechanic in your dealer's shop. On the other hand, injector nozzles can be checked for proper operation, as recommended in the operator's manual.

Diesel Starting Engine. One of the problems in design of the Diesel engine, regardless of size, is the amount of energy required to crank the engine. Obviously very high compression pressure is the reason. Hand cranking is usually out of the question and, in most cases, a mechanical starter is used. In the case of the tractor being discussed, a small, high-speed, two-cylinder engine equipped with electric starter furnishes power for starting the Diesel engine. This small engine is identical in principal to the standard gasoline engine discussed in previous pages.

In starting the Diesel engine, the compression pressure is released. As soon as the Diesel engine is turning over properly the compression release lever is disengaged, the fuel-injector pumps are engaged and firing begins. The auxiliary engine is then disengaged and stopped.

The Cooling System. The purpose of the cooling system, regardless of whether Diesel or spark ignition type of engine, is to dissipate the heat of combustion and friction and to maintain proper engine temperature for most efficient engine performance. When we consider, for example, that the temperature in the combustion chamber of a spark-

Cross-sectional view of the thermo-siphon cooling system.

ignited engine under load may run as high as 1250° F, the need for an adequate, efficient cooling system, properly maintained, is apparent.

The spark-ignition tractor shown in Fig. 21, is cooled by water forced through the system by a water pump. Water is cooled through the radiator by a blast of air drawn through the radiator by the fan. Proper operating temperature is maintained by means of a thermostatically controlled radiator shutter. The water pump provides a constant forced circulation of the water.

The cooling system of the Diesel tractor, discussed previously, and that of many general-purpose tractors is through the thermo-siphon or temperature-controlled system. When the cylinders warm up after starting the engine, the warmed water rises and is displaced by cooler water. The constantly rising warm water from around the cylinders causes a circulation through the radiator where the water is cooled.

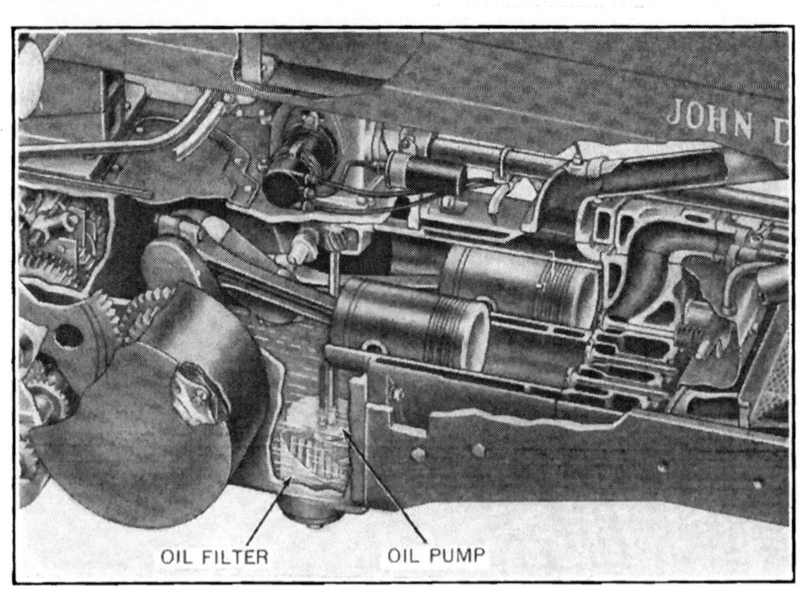

Figure 40—Horizontal cross-section showing oiling system.

The radiator consists mainly of a core of vertical tubes attached to which are fins that form extra cooling area. As the fan draws a steady current of air through the radiator, the water is cooled as it flows downward or is forced through.

While the manufacturers of most tractors provide screens to prevent foreign matter from entering and clogging the radiator tubes, it is the operator's responsibility to see that this screen is kept clean. Make certain that only clean water is placed in the radiator and that the water level is always above the radiator tubes. Extremes of temperature may cause serious damage to the tractor. Water should never be poured into an empty cooling system when the engine is hot, nor should cold water be poured into a hot

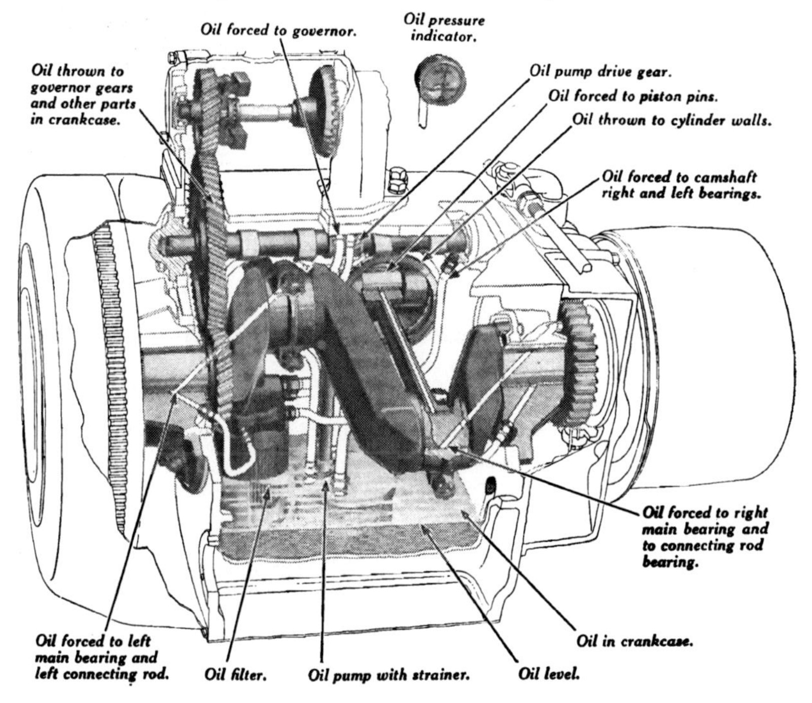

Figure 41—Cross-section of the oiling system showing how all working parts are automatically oiled.

engine, or hot water poured into a cold engine. Many cooling systems today are designed to operate under pressure. Since a sudden release of pressure may result in scalding of the operator, it is recommended, as a safety procedure, that the tractor be permitted to cool before the radiator filler cap is removed.

The Oiling System. Of all farm machines, the tractor requires the most careful oiling. Due to the nature of its work and the large amount of friction surface in its bearings and cylinders, the tractor must be properly lubricated with good oil and grease if it is to develop its maximum efficiency and last a normal length of time. No other factor affects the life of a tractor so greatly as does oiling.

Engine lubrication is of special importance, for engine parts are built to fine precision, many parts being held to tolerances as close as one-thousandth of an inch. Close tolerances generate considerable heat of friction which must be dissipated or conducted away. Add to the heat of friction, heat generated by combustion and we have a three-fold reason why engine lubrication should be given exacting consideration.

The engine of the general-purpose tractor shown, is provided with a positive-driven, full-force-feed pressure lubricating system. A gear-type oil pump forces oil under pressure through the replaceable oil filter element, into the main bearings and through the drilled crankshaft to the connecting rod bearings, then through holes in connecting rods to piston pins (see Fig. 41).

When the engine is started, the oil indicator (see Fig. 41) will show pressure if the oiling system is working properly. If it does not show pressure, the operator should check the supply of oil in the crankcase. If the oil level is correct, the trouble may be in the oil strainer screen or in the pressure relief valve. Another possibility is that the oil gauge itself may not be functioning properly, in which case the gauge should be checked with a master gauge by your implement dealer. To insure lubrication, the indicator must show pressure when engine is running.

After every ten hours of operation, the level of oil in the crankcase should be checked and fresh oil added if necessary. After every 120 hours of operation, the crankcase should be drained completely and refilled with fresh oil of proper weight or viscosity for the temperature range in which the tractor will be called upon to operate (see Fig. 42). Oil of the wrong weight can result in loss of power, excessive fuel consumption, and undue wear on moving parts.

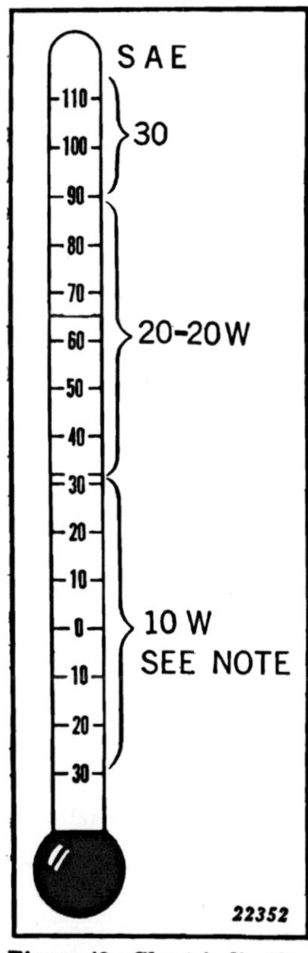

Figure 42—Chart indicating correct weights of oil to use at various outdoor temperatures.

At the time the crankcase oil is replaced or changed, the replaceable filter element should be removed and a new element installed. (See Fig. 43). The importance of the filter cannot be overestimated. The modern tractor engine, built to close tolerances as mentioned previously, can be seriously damaged by grit particles as small as one-thousandth of an inch. For this reason, the filter element is designed to remove these tiny particles. However, when the filter element becomes clogged with grit, it cannot permit the further passage of oil. Grit-laden oil, then, is by-passed through the pressure relief valve to the bearings, pistons, rings, and other precision parts.

A mistaken idea, prevalent among some tractor operators, that *clear* oil is *clean* oil has resulted in serious damage to many a fine tractor engine. Here is a test that you can make in the classroom or at home: fill a test tube with clear oil; add a teaspoonful of clean sand; shake the tube. The oil remains clear, the grit will settle, but the clear oil is laden with abrasive particles which, if placed into a crankcase, would bring destruction to bearings in a short time. The same test proves the fallacy of the ideá that if oil "feels" good, it is safe oil for the crankcase, for when the engine is stopped, the heavy grit particles settle to the bottom far from the check cock and out of reach of the dip stick.

Clear oil is not always *clean* oil; likewise *dark* or *black* oil is not necessarily *dirty* oil. When oil becomes dark or discolored in service, the discoloration is due primarily to the entry of soft carbon which, in itself, is a lubricant.

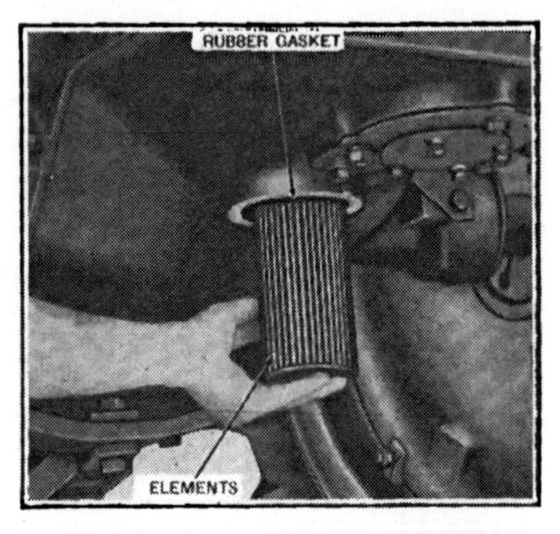

Figure 43—Oil filter should be changed at least every 120 hours of operation.

The present-day heavy-duty oils are proof of the point established in the preceding paragraph. The modern heavy-duty oils enter the crankcase just as clear as the regular type oils, yet, upon draining, they will be discolored, in many cases, actually black. These oils have the faculty for carrying particles in suspension to be drained out with the used oil rather than to drop into the crankcase to foul the clean oil replacing it.

From the foregoing paragraphs, it is obvious that we cannot depend upon our eyes or our fingers to judge the condition of the oil; therefore, it is the part of true wisdom to follow the manufacturer's instructions for periodic changes of oil and filter elements. When replacement of filter element is indicated, it is of greatest importance that the replacement unit be of the size and type recommended by the manufacturer of your particular tractor.

The Ventilating System. The tractors discussed on previous pages are equipped with automatic crankcase ventilation systems. This means a constant circulation of clean air is drawn through the air cleaner and through the crankcase.

The illustration, Fig. 44, shows how this automatic ventilation system works on the general-purpose tractors. Air that has been cleaned by passing through the air cleaner is drawn through a pipe, tapped in the air inlet elbow, and forced into the crankcase by a rotor pump. The air circulates throughout the crankcase, collecting harmful vapors and is then expelled out through a vent ahead of the tappet case cover. This ventilating system, as mentioned, is entirely automatic.

The air cleaner system and its care were discussed thoroughly on page 24.

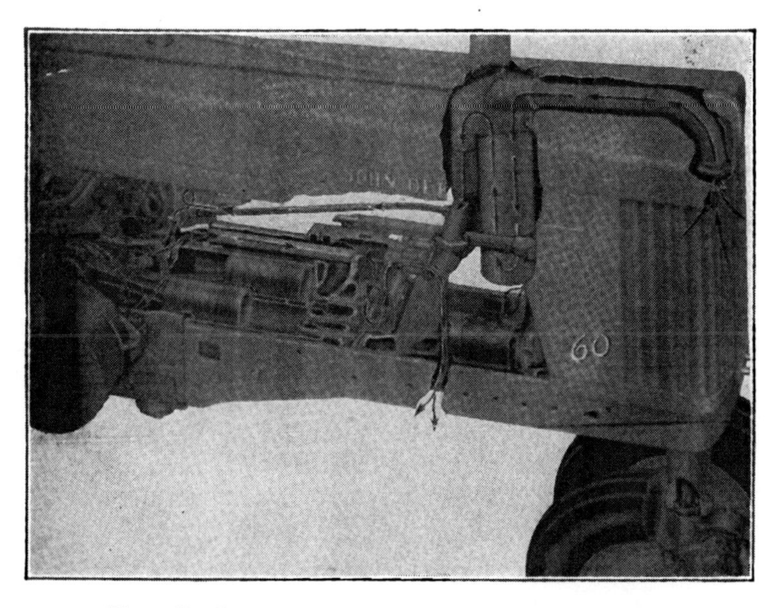

Figure 44—Cross-section of the crankcase ventilation system.

Transmission System. The transmission system, as its name implies, transmits or delivers the power from the engine to the drive wheels where it is used to pull loads; to the power take-off where it is available to operate equipment requiring power in addition to that necessary for forward travel; and to the power lift, where it is used in raising and lowering integral equipment and, with remote cylinder, for raising, lowering, and adjusting drawn equipment. Power is transmitted to the belt pulley direct.

Power is transmitted from the engine to the drawbar through the clutch and transmission gears and through the final drive or differential to the axles. Sliding pinions of varying sizes in the transmission are shifted to mesh with corresponding gears to provide for various tractor speeds forward and one speed in reverse as mentioned.

A differential is a special arrangement of gears which permits each drive wheel to turn independently as the tractor turns a corner. It permits one drive wheel to rotate slower or faster than the other when turning or when working in rough conditions.

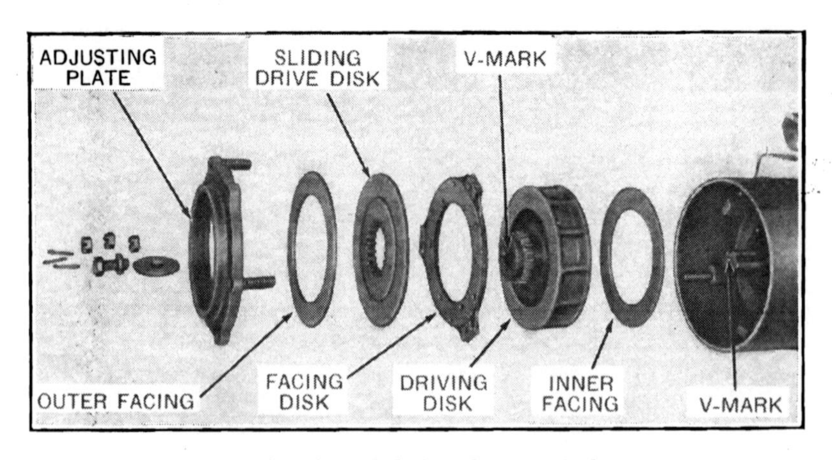

Figure 45—Clutch "exploded" to show principal parts.

The sole purpose of the clutch (Fig. 45) is to link or connect the power of the engine to the load. Without such a link, no provision could be made for transmission of power to the various power outlets or for changing speeds or direction of travel. To function properly, the clutch should engage smoothly, picking up the load gradually rather than with a jerk.

The clutch on the tractor used for this text, is properly adjusted when the nuts are drawn up exactly to the same tension, and the clutch operates with a snap, requiring some pressure to lock it. If it is necessary to tighten the clutch, each nut must be turned down to the same tension, disregarding the number of exposed threads (see Fig. 46).

To replace clutch facings, remove dust cover and clutch adjusting plate (see Fig. 45).

When installing new clutch facings, make certain that the inside or first clutch facing is in proper position while clutch drive disk is being replaced. Install second clutch facing in clutch adjusting plate, making sure to have the

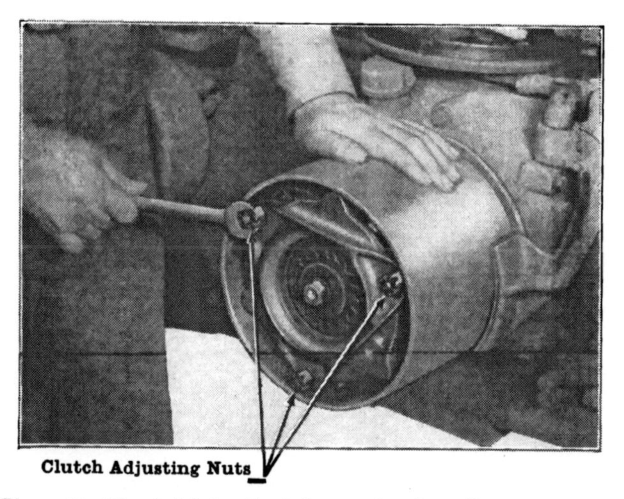

Figure 46—The clutch is adjusted properly when all nuts are drawn up to even tension.

three short springs in place. Adjust clutch as described. Always refer to the operator's manual for proper directions in servicing the clutch on your particular tractor.

Lubrication of Transmission. The transmission and differential units require little attention other than efficient lubrication which is highly important.

In the enclosed transmission and differential, gears, shafts and bearings operate in a bath of heavy oil. For proper lubrication of these parts, the correct oil level should be maintained. Periodic check, following the manufacturer's instructions, will determine when oil should be added. Seasonal change in weight of oil used in transmission and differential, in line with the manufacturer's instructions, is necessary for efficient operation and long life.

The heavy gears in transmission and differential depend upon a cushion of oil to relieve the tremendous shock of starting and the constant pressure of working under load. Too thin or light an oil will be "squeezed" out by the teeth; too heavy an oil will not be carried around by the gears. A simple test to prove this point is to run a cold gear over a heavy grease—note that the gear simply makes a print or track in the grease without picking it up.

In addition to the engine, transmission, and differential units, there are several oil fittings that require regular attention with a high-pressure grease gun. A thoroughly lubricated tractor will last longer and give better service than one that is given only ordinary care.

Power Shaft or Take-Off. A power shaft or take-off is a third means of taking power from the tractor engine to drive such machinery as the combine, baler, and corn picker. It drives the entire mechanism of this type of machine, the wheels acting merely as supporting or carrying members.

The power take-off consists mainly of a shaft, extending back from the tractor, which is driven by the regular transmission. Its operation is controlled by a shift lever usually placed for convenience of the tractor operator.

Hitch and power take-off locations on all general-purpose tractors are now standardized so that tractor-drawn equipment, powered through the take-off, is readily interchangeable, thereby saving the operator considerable time and labor in shifting from one machine to another. It is wise, therefore, when buying new equipment for use with older tractors which do not have standardized hitch and power take-off, to convert the tractor to standard rather than to buy the new equipment with parts to adapt it to the old tractor. In this way, your power equipment will be adapted to your present tractor and to the new tractor you may acquire later. Your implement dealer will offer further information at the time you purchase equipment.

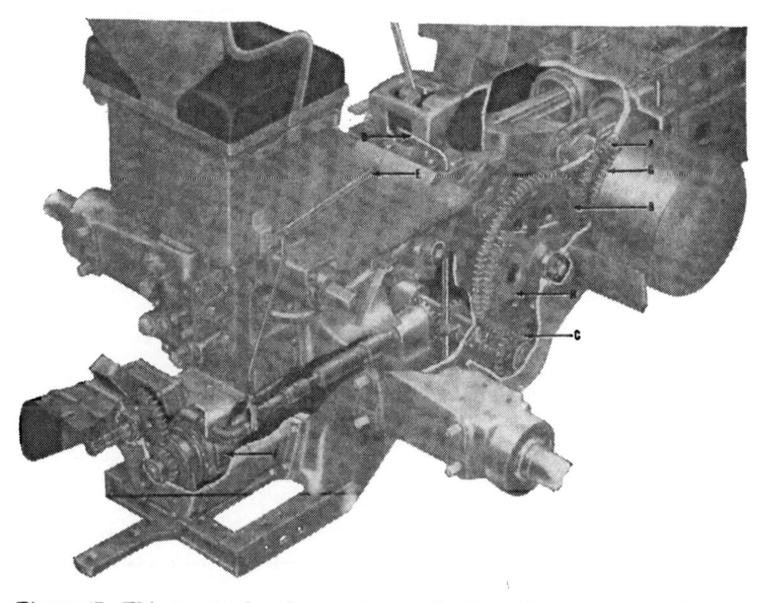

Figure 47—This tractor has been cut away to show the power take-off power train. Sliding gears and drive shafts operate in oil. (A) power shaft idler drive gear; (B) power shaft idler gear; (C) sliding gear and shaft; (D) shifting lever for sliding gear and shaft; (E) power shaft clutch lever; (F) power shaft clutch; (G) belt pulley gear and (H) first reduction gear.

"Live" Power Shaft. The general-purpose tractor described on previous pages is equipped with a "live" power shaft. (See Fig. 47.) It provides continuous running power for operating power take-off machines as long as the tractor engine is running. Its power is completely independent of the transmission clutch. The operator can stop the forward travel of the tractor at any time and start it up again while the power-driven machine continues to operate at full speed without interruption. This is especially advantageous in harvesting heavy crops with a minimum of loss.

On the tractor shown in Fig. 47, power is delivered directly from a drive gear on the crankshaft to an idler gear. Both gears operate whenever the engine is running. A sliding gear and shaft transmit power through the power train to a spring-loaded, over-center, wet-disk type clutch. Engaging the clutch delivers power to the splined PTO shaft which drives the equipment.

This clutch enables the operator to bring the power-driven implement up to full speed before starting the forward movement of the tractor. It completely protects the PTO shaft and cushions shock on the implement. Even when with the engine at wide open throttle, the clutch can be engaged instantly without damage. The power shaft clutch has an independent oil reservoir and lubrication system; it operates in a constant bath of oil supplied by a separate pump.

Operator's manual should be consulted for proper servicing.

Differential Brakes. The foot-operated or rear wheel differential brakes, one for each drive wheel, are provided to facilitate short turning either right or left when working in row crops, to stop the tractor, and to hold the tractor stationary on belt work.

In operation several cautions should be used in applying differential brakes: when traveling at high speed as on highways or in going to and from fields, brakes should be applied gradually and uniformly to both wheels to prevent skidding of the tractor and, in extreme cases, upsetting. Never brake one wheel when turning at high speed. Observe, always, that brakes are provided solely for the purposes outlined above.

Differential brakes require only occasional adjustment which should be made in accordance with instructions given in manufacturer's service manual or instruction book. Whenever a brake adjustment will not correct or improve defective brake action, the brakes should be relined.

Figure 48—Front wheel bearings should be checked for proper adjustment.

Lubrication of Front Wheel Bearings. Lubrication and adjustment of front wheel bearings should be given special attention. While the front wheels on most tractors are lubricated by means of a high-pressure fitting, it is advisable to remove the wheels for occasional servicing.

To service front wheel, remove the wheel, take out all old grease, examine the bearings (Figs. 48 and 49), clean the two felt washers in gasoline, resoaking in transmission oil before replacing. If these washers are worn thin, replace with new ones. Pack hub and bearing with wheel bearing grease, replace wheel on spindle, and adjust as follows: relieve bearings of all weight by running one wheel up on a block or plank to raise the other. Turn adjusting nut tight; then, back off the adjusting nut 1/3 to 1/2 turn. Adjust opposite wheel in same manner. Wheels should rotate freely but without end-play. Lock adjusting nut at proper point.

Front wheels of many general-purpose tractors may be reversed as shown in Fig. 50, to provide easier steering control, especially in listed crop territories where it is necessary to keep the front wheels on the ridges. This extra clearance is an advantage in exceptionally muddy conditions, since mud will not accumulate under the frame. In normal conditions, steering is easiest with the wheels in narrow setting.

Another important aid to easy steering, operator comfort, and increased tire life is the load equalizer (Fig. 10), which equalizes the front end load over both wheels, permitting the tractor to "walk over" surface irregularities and to conform with ridges in the field. Riding is greatly improved because up and down movement of the front end is cut exactly in half as the tractor travels over rough ground.

Hydraulic Power Control. This control supplies power for raising and lowering integral equipment through the tractor rockshafts and for making field adjustments of drawn equipment through a remote cylinder attached to the implement and connected to the tractor by flexible oil lines. Both rockshafts and remote cylinder are actuated by one lever.

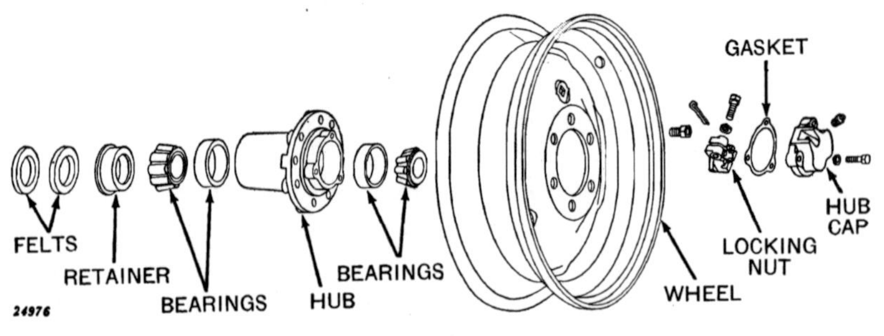

Figure 49—"Exploded" view of front wheel bearings and related parts.

The tractor shown in Fig. 21 can be equipped with "live" high-pressure hydraulic control which is direct engine-driven and operates completely independently of both the transmission clutch and the power take-off. Power is taken from the crankshaft gear and transmitted to the hydraulic control pump through the cam gear and idler gear. (See Fig. 51.) Whenever the engine is running the hydraulic power is available to the operator.

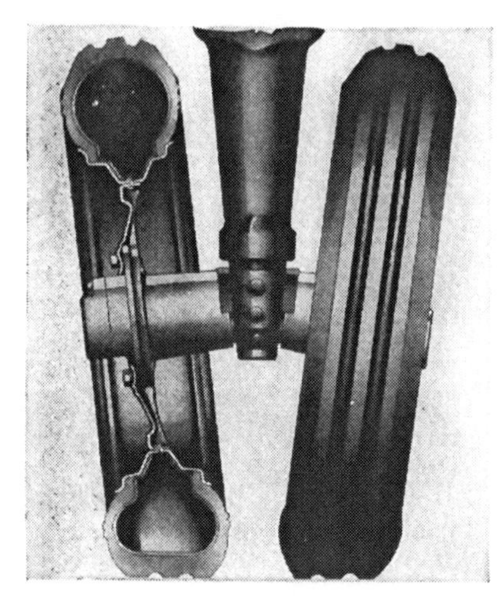

Figure 50—Front wheels set in normal position (top) and in wide setting as used on listed ridges or for easier steering in muddy conditions.

When the hydraulic power is used for remote cylinder operation (see Figs. 52 and 53), the power can be used to regulate and adjust drawn equipment. With this valuable extension of power, the height of the combine platform, for example, may be raised or lowered quickly, easily, and exactly to meet varying conditions; the plow may be raised just enough to pass over a bad spot in the field; the disk harrow may be straightened to cross a grassed waterway and angled quickly returning to work at the pre-set working angle.

Any degree of variation within the extreme limits set by the operator may be gained, simply by moving the control lever in the selective range, yet, when the lever is moved quickly to either extreme, the maximum lift or drop is obtained promptly.

Figure 51—Hydraulic power pump control is engaged through the control leader. Operating lever has five operating positions; neutral, slow raise, fast raise, slow drop, and fast drop.

In making the simple change from rockshaft operation to remote cylinder control of drawn equipment, several basic considerations should be remembered. While the piston in the remote cylinder operates on a reciprocal motion, powered on both the outward and inward strokes, the cylinder should be placed so that the lifting load is always applied on the outward stroke, thereby taking advantage of the full area of the piston for the heavier load.

In coupling the oil lines to the tractor, knurled coupling adapter rings should be drawn up snugly by hand only—never with a wrench.

Since valve openings, or apertures, are extremely small—many of them smaller than the lead in a fine-line pencil—it is of utmost importance that dirt and other foreign

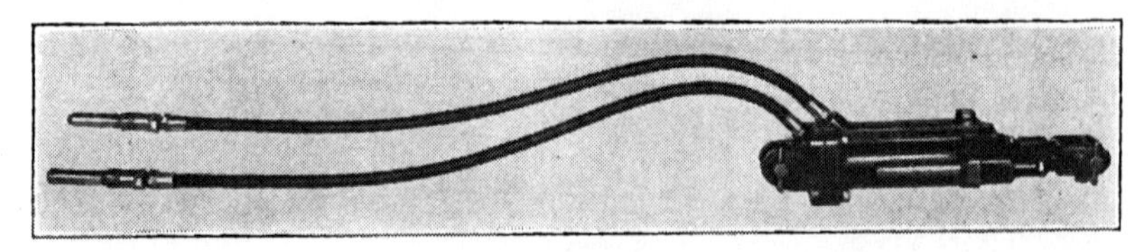

Figure 52—Remote cylinder complete with hydraulic hose. Fittings are capped to protect from dust and dirt.

matter are kept out of the system. The manufacturer has provided every protection possible, yet it is the operator's responsibility to use extreme care in keeping the system clean.

The hydraulic unit requires very little servicing other than a periodic check for proper oil level. Change oil twice a year, preferably before spring and winter seasons. The entire unit should be drained and refilled to proper level with the correct grade of oil. When storing the remote cylinder, the piston should be pushed all the way into the cylinder to protect the polished shaft from exposure to dust and moisture.

Hour Meter. The engine hour meter, available in various types, enables you to take the "guesswork" out of servicing your tractor (see Fig. 54). It keeps an accurate account

Figure 53—Power lift equipped for remote cylinder operation with hydraulic cylinder connections in position.

of the engine operating time. You protect vital parts of your tractor by lubricating "on schedule" instead of by guesses of hours or days of work. You can change oil exactly at 120 hours as recommended, replace the oil filter, service air cleaner, hydraulic system, transmission and steering at the proper time, giving full preventive maintenance care to your tractor.

Figure 54—The engine hour meter.

Maintenance of Rubber Tires. The broad use of rubber tires on farm tractors and machinery has resulted in a great saving both in time and operating cost. There are, however, certain basic fundamentals in the care of tires that should be followed carefully if the owner is to derive maximum benefit from his investment.

First and most important is to maintain proper pressure for the work at hand. Your best guide to proper inflation is the instruction book covering the particular tractor or implement under consideration. Read your instruction book or consult your dealer concerning proper inflation, and *check air pressure regularly*. Underinflated tires suffer from rim bruises, sidewall snagging, and carcass failure. Overinflation increases tread wear (on tractors and ground-driven implements) and because of reduced traction, weakens the carcass, and hastens weather checking. An

Figure 55—Proper inflation is of utmost importance in the life of rubber tires.

air pressure gauge and a good tire pump are essential in maintaining proper inflation. Proper inflation is especially important where fluid weight is used since the air space is greatly reduced (see Figs. 55 and 56). A special air-water gauge should be used for testing tires carrying fluid weight.

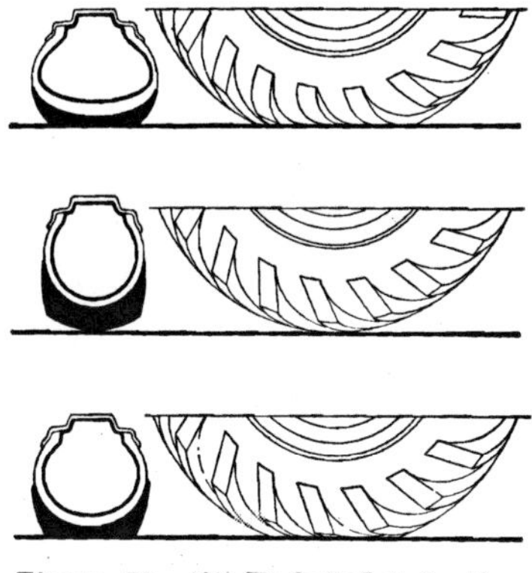

Figure 56—(A) Underinflated tire, (B) overinflated tire, and (C) properly inflated tire.

Grease and oil are natural enemies of rubber. Protect tires from oil and grease as much as possible. Should tires become spattered with oil or grease, wipe them off with a rag dampened with gasoline—but do this job *outside* the implement shed to reduce fire hazard. Never allow tires to stand in barnyard acids. If spray chemical gets on the tires, wash it off.

Inspect tires periodically for carcass breaks and cuts and have them repaired immediately. No cut is too small to require attention, for if it is not repaired, further damage will result.

Use tractor wheel weights (according to manufacturer's instructions) to secure maximum traction and minimum slippage.

Avoid high transporting speeds. Implement tires, unless otherwise specified, are not designed for speeds exceeding fifteen miles an hour. Take added precautions as tires age.

Don't overload. This applies particularly to combine grain tank extensions. Reduce speed and load, if possible, on rough ground.

Protect the tires of idle implements from sunlight.

When a rubber-tired implement is to be idle for a considerable time, block up the axles to take the weight off the tires, but leave the tires inflated.

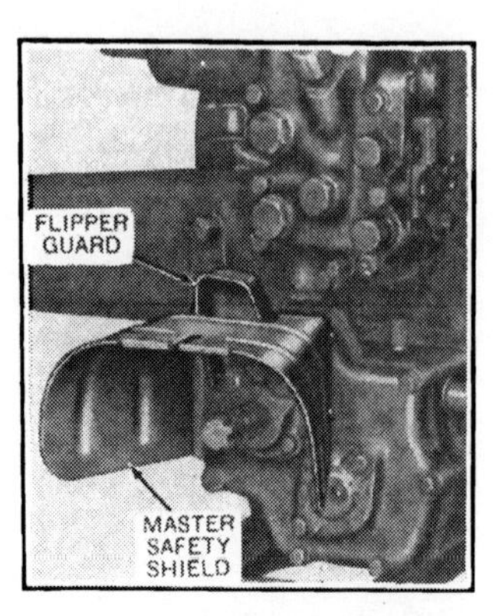

Figure 57—Shield should remain over the power shaft at all times.

Safety Precautions. Every year farmers are killed by accidents, many of which could have been avoided by using care in working with and around machinery. These are but a few of the many safety suggestions which should prove practical when working around the farm.

Only one person—the operator—should be permitted on the tractor platform when tractor is in operation.

Refuel your tractor only when the engine has been shut off. Do not smoke or use an oil lantern while refueling.

Be sure power take-off shields and guards are in place and in good order before starting field work.

Tractor brakes should be adjusted properly.

Do not oil, grease, or adjust a farm machine that is in motion.

Clothing worn by tractor or machine operator should be fairly tight and belted. Loose jackets, skirts, shirts, or sleeves should not be permitted because of the danger of getting into moving parts.

Never drive a tractor too close to the edge of a ditch or creek.

When your tractor is hitched to a heavy load, always hitch to drawbar and never take up the slack in chain with a jerk.

Always keep tractor in gear when going down steep grades.

Drive at speeds slow enough to insure your safety. Reduce to low speed before turning quickly or applying individual brakes. Drive slowly over rough ground.

Keep a firm grip on the steering wheel at all times when the speed is increased.

When hitching a drawn implement to the tractor, back the tractor past the clevis. Then move it slowly forward so that, in making the connection, the tractor will be moving away from you.

Provide a first-aid kit at the house for use in case of accident, and use the proper antiseptics on scratches, cuts, etc., without delay to prevent the possibility of blood poisoning.

Finally, remember this: An accident is usually caused by someone's carelessness, neglect, or oversight. The life you save may be your own.

Questions

1. *What is an internal-combustion engine?*

2. *What is the difference between a two-stroke cycle engine and a four-stroke cycle engine? What are the four strokes of the latter? What is the chief difference between the spark-ignition and Diesel engines?*

3. *What three elements are required for efficient engine operation?*

4. *What parts make up the fuel system; how would you check each part?*

5. *What is the importance of the air cleaner; how is it serviced?*

6. *Describe the function of the distributor. How would you check it for operating efficiency?*

7. *Describe a step by step check of the ignition system.*

8. *How does the oiling system of the tractor illustrated operate? Why is an oil filter of special importance in the modern tractor?*

9. *What is the function of the valves?*

10. *Why is the cooling system necessary? What attention does it require?*

11. *Describe the transmission of power from engine to drive wheels; belt pulley; power take-off; and power lift.*

12. *What is a differential unit? Describe a clutch.*

13. *What are the advantages of a power-shaft attachment?*

14. *How does the remote cylinder increase the usefulness of hydraulic power?*

15. *How should a battery be cared for?*

16. *Mention several of the important points in proper care of pneumatic tires.*

17. *Name five safety precautions everyone should know.*

WORKING PRINCIPLES OF TRACTOR ENGINES

ALL tractors have internal combustion engines, the operation of which depends upon the expansion that takes place when a compressed mixture of air and petrol, or paraffin or vaporizing oil or fuel oil, is fired. Part of the required rise in temperature comes from the compression itself. Indeed, in Diesel heavy-oil engines no external means of firing the mixture of fuel and air is provided. Air alone is drawn into the engine and compressed. Towards the end of the compression stroke, the required quantity of fuel is injected as a fine spray into the cylinder. The temperature of the compressed air is high enough to fire the fuel as it enters, and the combustion of the air and fuel mixture produces the power.

Since the fuel fires as it enters the combustion chamber, and since the spray of fuel continues during about 20° of rotation of the crankshaft, the whole of the charge does not explode instantaneously as it does in a spark ignition engine. The impulse to the piston is, therefore, sustained for a longer period than it is in a spark ignition engine. The impulse is, indeed, rather a push than a sharp blow, and this is one reason why these compression ignition engines, or Diesel engines, can be run at high compression ratios which would cause 'knocking' in a spark ignition engine.

The expansion caused by the explosion of the mixture of fuel and air is converted into useful power by the movement of the piston, which is forced down the cylinder away from the chamber where the gas has exploded. The piston is connected by a rod to the crankshaft. At the piston end of the connecting rod is a gudgeon-pin bearing. At the crankshaft end is a metal-lined bearing called the big-end bearing. The upward journey of the piston compresses the air or fuel and air.

Big-end bearings have a white metal lining in the housing. This white metal, composed of tin, antimony and copper, melts at a fairly low temperature. Therefore, if the lubrication of the engine should fail, and the bearings become overheated, the white metal runs before any great damage is done to the machined surface of the crankshaft. The lining and housing are in halves. The halves of the steel housing are separated by a series of several thin packing washers called shims. When the bearing has worn slack it can be tightened by removing some of these shims.

The little-end bearings on some makes of engine have a gudgeon pin which is held fast in the piston. The connecting rod has a bushed bearing which rotates or, more truly, swivels on the pin. In other

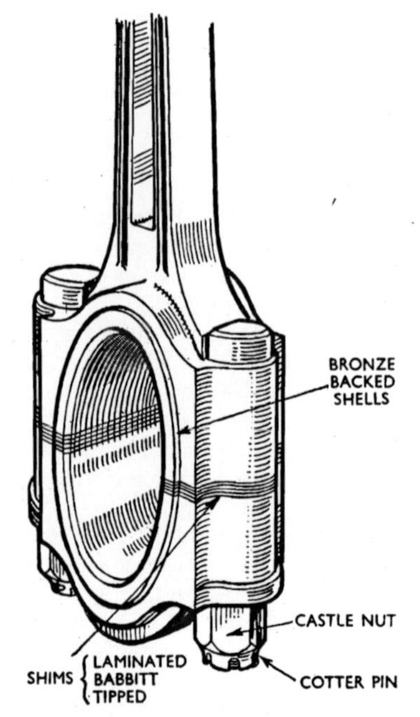

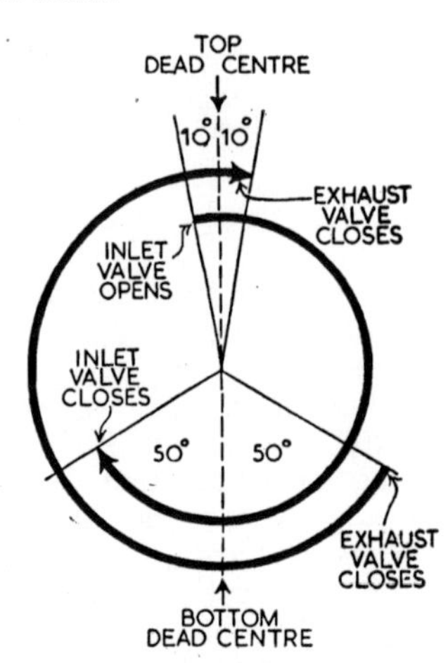

(Left) The parts of a big-end bearing. (Above) During the exhaust and induction strokes of the cycle in a four-stroke engine, the inlet valve does not shut completely before the exhaust valve begins to open, and the exhaust valve does not shut before the inlet valve opens.

makes of engine the pin is not held fast in the piston but is able to turn in a bush on the piston as well as in a bush on the connecting rod. This latter arrangement forms what is called a fully-floating little-end bearing.

FOUR-STROKE ENGINES

In a four-stroke engine, the piston makes two other journeys in each cycle of operation of the engine, a downward suction stroke for drawing the mixture, or air, into the cylinder before the compression and power strokes take place: and an upward scavenging stroke to drive out the burnt gases from the cylinder.

It can be seen that the passage through which the mixture, or air, enters the cylinder, and the passage through which the burnt gases pass out of the cylinder into the exhaust pipe, must not remain open all through the cycle. The inlet passage must be open when the piston is travelling down its suction stroke, and the exhaust passage must be open for its scavenging stroke. During the compression and power strokes, both passages must be closed.

The passages are opened and closed by mushroom-like valves, which are lifted from their seatings by cams on a shaft driven by the engine. The shape and size and position of these cams are fixed carefully so that the valves shall open and close at the right times in the cycle of the engine's revolutions. The inlet valve does not shut completely before the exhaust valve begins to open. There is some 'overlap' when both valves are open. Nor does the exhaust valve close before the inlet opens, though in this case the overlap is generally smaller. In both cases, the object is to allow time for all the exhaust gas to have left the cylinder before the new gas enters. The size of the overlaps, however, and the timing of the opening and shutting of the valves in relation to the top dead centre of the travel of the piston, vary with the design of the engine. Alternatively, the position of the opening and closing of each valve may be described as a given number of degrees of angle of rotation after top dead centre or bottom dead centre.

On some engines the valve ports are at the side of the combustion chamber of the cylinder. In other engines the valves are in the top of the combustion chamber, above the piston, and these engines are called overhead-valve engines.

TWO-STROKE ENGINES

In a two-stroke engine there are no valves and no separate suction and scavenging strokes. The compressing of the mixture, or air, is done in the crankcase, and the expelling of the exhausted gases is taken care of by the deflector-shaped dome of the piston, which ejects them through a port in the side of the cylinder — a port that is uncovered when the piston is at the bottom of its stroke.

The mixture, or air, for the engine is admitted not directly into the cylinder, but first into the crankcase, or sump, of the engine. When the piston ascends it increases the volume of the crankcase compartment. This causes a partial vacuum and draws the mixture from the carburettor, or air, into the crankcase. When the piston comes down, it reduces the volume again, and drives the mixture, or air, through a passage into the cylinder. The port of this passage is an opening in the cylinder wall which is uncovered by the piston at the right part of the stroke.

DIESEL ENGINES

In some Diesel engines the fuel is injected directly into the cylinder. In others it is injected into a separate chamber.

In direct injection engines the combustion chamber is usually a recess in the piston crown. In indirect injection engines the piston nearly fills the cylinder at the end of the compression stroke and the combustion air is displaced into a separate spherical chamber, connected with the cylinder by a passage. During compression and combustion heat is lost from the gases to the cylinder and combustion chamber walls.

The combustion chamber of the direct injection engine of a given cylinder capacity has a smaller surface area than that in an engine with separate chamber, so the heat losses are smaller. The direct injection engine usually starts more easily than the indirect injection engine and has a lower fuel consumption. However, for efficient combustion there must be intimate mixing between the air in the combustion

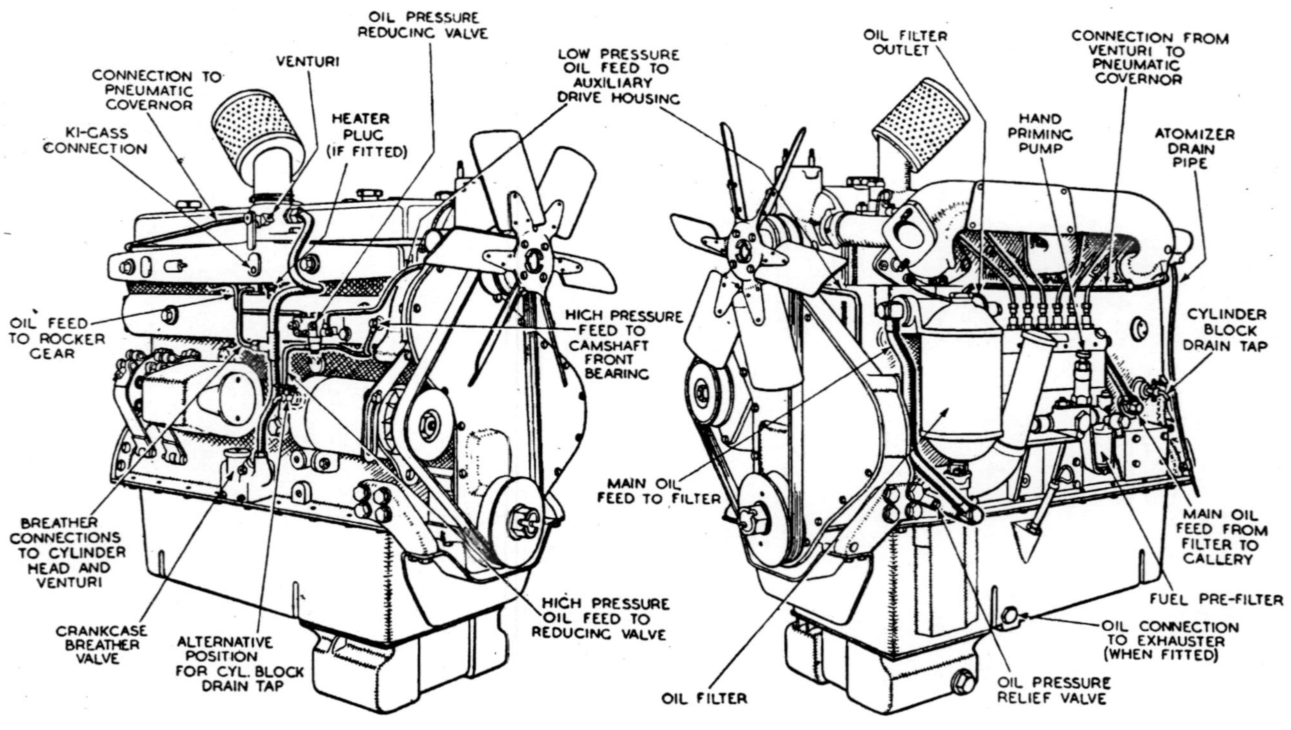

The Perkins P6 Diesel engine.

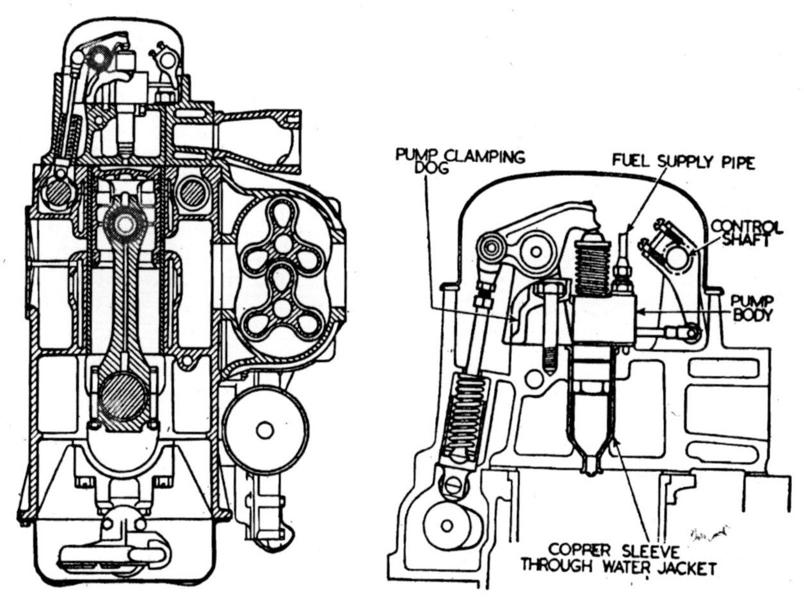

(Left) Cylinder of a General Motors two-stroke Diesel engine as fitted to some Allis-Chalmers tractors. The lobes of the blower which forces air into the cylinders are shown. Unlike the cylinders of the usual transfer-port two-stroke engine, each cylinder has exhaust valves through which the burnt gases are pushed by the incoming air from the blower. In the head can be seen the combined fuel pump and injector. (Right) Cylinder head of General Motors two-stroke Diesel engine. This shows how the plunger of the combined pump and injector is operated through a push rod and rocker from the engine camshaft.

chamber and the injected fuel and the more the air in the combustion chamber is moving about, the easier it is to get this intimate mixing. The amount of air movement which can be induced in the combustion chamber of the direct injection engine is smaller than in that of the indirect injection engine. Therefore, the direct injection engine relies to a greater extent on the way in which the fuel is injected into the cylinder. Usually in the direct injection engine the fuel is sprayed through four or more fine holes into the relatively still air in the open chamber. In the indirect injection engine the fuel is usually sprayed through one single large hole into the rapidly moving air in the separate chamber. The indirect injection engine is less sensitive to the condition of the injection equipment and to the quality of the fuel used.

Low-speed, low-compression, two-stroke heavy oil engines, requiring a hot pin to aid ignition, can burn inferior fuels satisfactorily. Fuel consumption of these semi-Diesel engines is higher than that of a full Diesel.

The progressive push mentioned earlier as a characteristic of the

Diesel engine is more manifest in engines in which there is a pre-combustion chamber. An engine which has direct injection into the main combustion chamber has a sharper combustion; indeed the fuel is able in some conditions to ignite even more quickly than the fuel in a petrol engine. In the pre-combustion type of engine, however, the combining of the fuel with the air takes place gradually, that is to say all the fuel does not come into contact with all the air instantaneously. The pre-combustion chamber contains insufficient air for the fuel injected to be burnt out completely. It should be noted that the pre-combustion chamber is not actually cut off by any airtight door from the main combustion chamber, but is connected with it through only a restricted passage and it takes an appreciable time for the fuel left unburnt by the spray to be driven into the main combustion chamber for the remainder of the combustion and expansion to take place.

A two-stroke engine can give more power for a given size than a four-stroke, but to do so it must have a supercharger. In supercharged two-stroke Diesel engines the air is pumped direct into the cylinder. It does not enter the crankcase. Some two-stroke Diesel engines employ only crankcase compression for scavenging, making the engine simpler, but reducing the power to about the same as a four-stroke of the same dimensions.

Injection Pumps

The purpose of the injection pump is to deliver fuel to the cylinder under high pressure and at a predetermined point in the cycle. The quantity of fuel varies according to engine load and speed. The usual injection pump has an enclosed camshaft and is of the constant-stroke type; the amount of fuel injected is controlled by rotating the plungers in their barrels so that the effective stroke is changed.

Driven at half engine speed by the timing gears at the front of the engine, the four cams on the camshaft operate pumping elements consisting of a plunger and barrel and a valve known as a delivery valve for each cylinder. The barrel has two diametrically opposed holes or ports near the top and the plunger has a vertical groove running downward from the top face, connecting with a 45° slot in the plunger side.

The lift pump supplies fuel to a gallery running along the top of the injection pump housing which connects with the ports in the pumping element barrel. When the camshaft is in such a position that the plunger is at the bottom of its stroke, fuel under slight pressure enters the barrel through the ports. It then fills the groove and slot in the plunger and the space immediately above the plunger top. As the camshaft revolves the plunger rises in the barrel, closing the two ports and cutting off the fuel escape. Fuel above the plunger is now subjected to the pressure of the rising plunger, causing the delivery valve at the top of the pumping element to open. Fuel under high pressure then

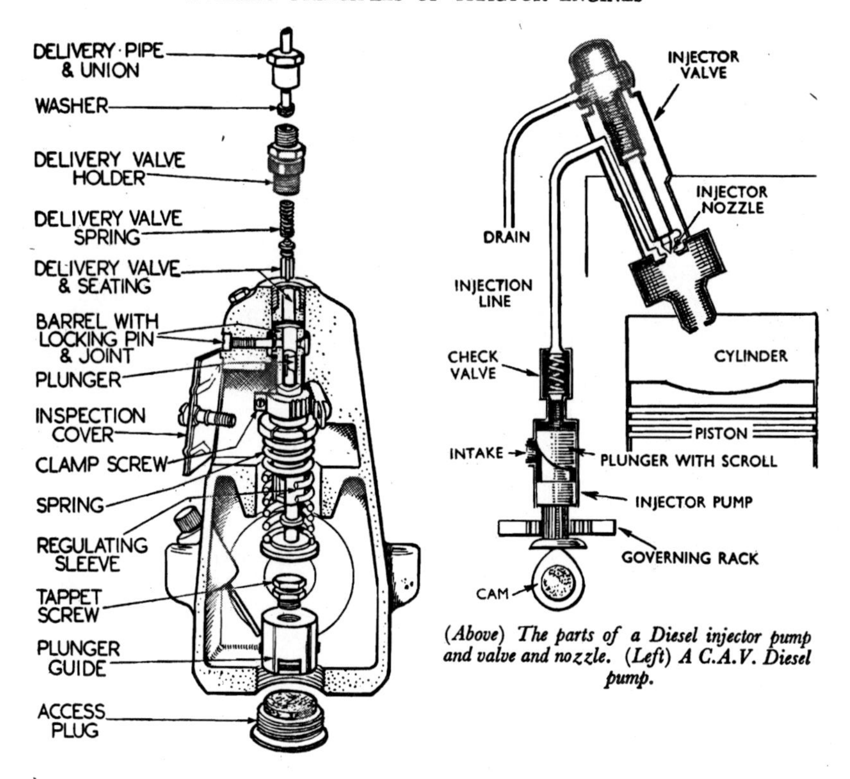

(Above) The parts of a Diesel injector pump and valve and nozzle. (Left) A C.A.V. Diesel pump.

passes through the open delivery valve to the injector. Injection ceases when the upper edge of the 45° slot in the plunger begins to uncover one of the ports, thus allowing fuel above the plunger head to escape down the vertical groove and 45° slot and then out through the uncovered port.

Engine speed being controlled solely by the amount of fuel injected, delivery has to be varied according to engine requirements. At the lower end of each plunger is a small arm which engages with a fork fitted to a rod known as the control rod. If this rod is moved to the left or right it causes the plungers, through their arms, to be revolved within the barrels, altering the position of the 45° slot in relation to the port. This will allow the fuel to escape from the top of the plungers either earlier or later according to the direction in which the control rod is moved. The point at which injection commences is always constant, this being determined by the point at which the plunger top cuts off the fuel escape through the ports. The period of injection, which directly controls the amount of fuel injected, is varied by the position of the plunger arm and the 45° slot relative to the escape port. Moving the control rod to the right reduces the fuel delivery and to the left increases it.

To stop the engine, fuel delivery is completely cut off by moving the plunger arm until the groove in the plunger lines up with the escape port, thus preventing any pressure from building up in the pump element.

A delivery valve is fitted above each plunger and barrel, so that at the end of the fuel delivery period a needle valve in the injector nozzle is shut rapidly, preventing any tendency for the fuel to dribble. If there is any dribbling, poor combustion and nozzle tip carboning result.

The delivery valve consists of a conical seat with a lower parallel portion having a sealing face which is a lapped fit in the bore of a guide. Before the fuel can flow through the delivery valve, the valve must travel upwards against the pressure of a spring until the sealing face is clear of the guide bore, so allowing the fuel to pass over the flutes and round the valve into the pipe. At the end of the fuel delivery period the fuel pressure is reduced in the barrel when the slot registers with the port as described earlier. The delivery valve spring, assisted by the high pressure still existing in the pipeline, causes the valve to close rapidly. During its closing the parallel portion of the valve, which acts as a small piston, increases the volume in the pipeline before the valve reaches its seat. Thus the pressure in the pipe is suddenly reduced, permitting the needle valve in the injector nozzle to snap shut on its seating, giving a clean cut-off to the fuel spray.

In a multi-cylinder pump, the fuel enters the main fuel gallery through the pre-filter on the left of the pump. This filter is fitted mainly to collect any scale or dirt originating in the pipe between the main fuel filter and the injection pump.

The plungers are operated by tappets above the camshaft, springs returning the plungers on the down stroke.

To start the engine it is necessary for the pump to supply the normal running delivery of fuel. Under cold conditions it may be necessary to inject excess fuel over and above the running fuel delivery to ensure initial combustion. This is achieved by an excess fuel starting device.

Each cylinder has an injector fitted to it, capable of spraying the fuel, as it is delivered from the injection pump, into the high-temperature air charge within the cylinder.

Fuel Injectors

The injector consists of the nozzle holder and the nozzle with needle valve.

Fuel is forced from the injection pump to the injector inlet adaptor. Incorporated in this adaptor is a clearance filter protecting the injector nozzle from any scale or dirt originating in the pipe between the injection pump and the injector. The fuel, under pressure, passes down the hole in the body to the annular groove in the face of the nozzle, thence through passages to the nozzle valve seat in the lower part of

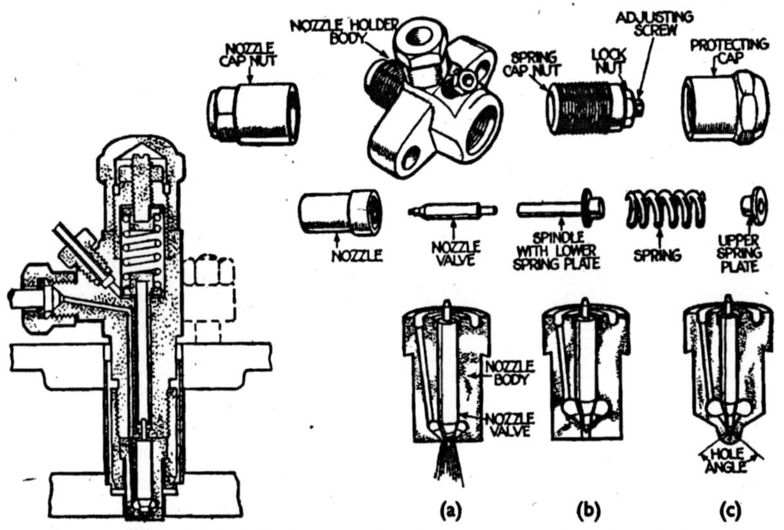

One type of C.A.V. Diesel fuel injector (left) and its parts. Three different kinds of nozzle are shown, for use in different designs of combustion chamber; (a) *pintle type,* (b) *single hole type, and* (c) *multi-hole type. The nozzle holes may be as small as 0.2 mm in diameter and it is essential that their dimension shall remain constant.*

the nozzle. The pressure of fuel at this point lifts the needle valve against the pressure of the spring and the fuel is then injected into the cylinder in a highly atomized state, through the four small holes in the tip of the nozzle.

A small quantity of fuel leaks up between the stem of the needle valve and its guide, providing lubrication. The excess is led away through the leak-off pipe connected to the top of the injector cap nut and returned to the fuel system.

SPARK IGNITION ENGINES

The charge of air and petrol or vaporizing oil for a spark ignition engine is mixed before it enters the cylinders. Air is drawn through a choke tube in the carburettor past the tip of a jet through which fuel is drawn by the reduction in pressure caused around the jet orifice by the flow of air past the constriction. The fuel leaves the jet as a fine spray which mixes intimately with the air. The proportion of fuel to air and the quantity of mixture drawn into the cylinders at each charge are controlled by the size of the jet orifice, and by the throttle valve and the choke valve.

The level of fuel in the jet is maintained by a float-operated valve in the pipe supplying the chamber of the carburettor with fuel from the

B

tank. The float is connected through linkage to a valve and maintains a constant level of fuel in the bowl. In this way it is possible to maintain a uniform and carefully regulated flow of liquid fuel to the jet or jets.

In the simplest form of carburettor a single jet meters the liquid fuel from the bowl into the air stream. The jet is located centrally in the venturi tube, which is the name given to the choke or constricted section of the intake manifold. Because the venturi has a reduced cross-sectional area, the velocity of the air travel at this point is increased. This greater velocity increases the ability of the air stream to pick up fuel from the jet and to atomize it.

At the beginning of the intake manifold, just beyond the jet nozzle and venturi tube, is a butterfly valve. The degree of opening or closing of this valve controls the amount of carburetted fuel charge to reach the engine, and it is called the throttle valve.

For ease in starting, a damper, or choke valve, is fitted in the air intake opening to the carburettor. In ordinary running this valve is kept wide open, but when it is necessary to increase the richness of the fuel mixture sufficiently to enable a cold engine to be started, the choke valve is used to restrict the flow of air entering the carburettor. This restriction increases the suction on the carburettor jet and causes a greater proportion of fuel to be drawn in relation to the air. This enriches the mixture.

Speed and load conditions affect the mixture proportion needs of an engine. To accommodate these variable needs, the carburettor has to be made a little complicated. In carburettors with a single jet and with a single place for air entrance, the mixture strength cannot be maintained uniformly constant at different engine speeds. As the engine speed increases, the flow of fuel in response to the suction increases faster than the flow of air, and the mixture becomes too rich. Several different methods are used to control the mixtures at various speeds and at various engine loads: an auxiliary air valve, an auxiliary or compensating jet, an economizer valve and power jet, a plain tube and air bleed valve, or float chamber suction control.

The auxiliary or compensating jet carburettor has two jets or nozzles. One jet works as a simple jet and delivers a mixture which increases in richness as the air speed increases. The other jet is connected to a well which is always at atmospheric pressure, and the flow of fuel into it is always at a constant rate. It delivers a mixture which decreases in richness as the air speed increases, because more air is then drawn in with the fuel.

Full power needs a richer mixture than part throttle operation. One way to provide this additional richness is by a piston and valve called an economizer. Under part throttle operation, the vacuum above the throttle is higher than it is when the throttle is fully open. This vacuum holds the economizer piston up and the check valve is then open and the economizer valve closed. This shuts off additional fuel from the power jet. When the throttle is opened the vacuum falls, and the piston

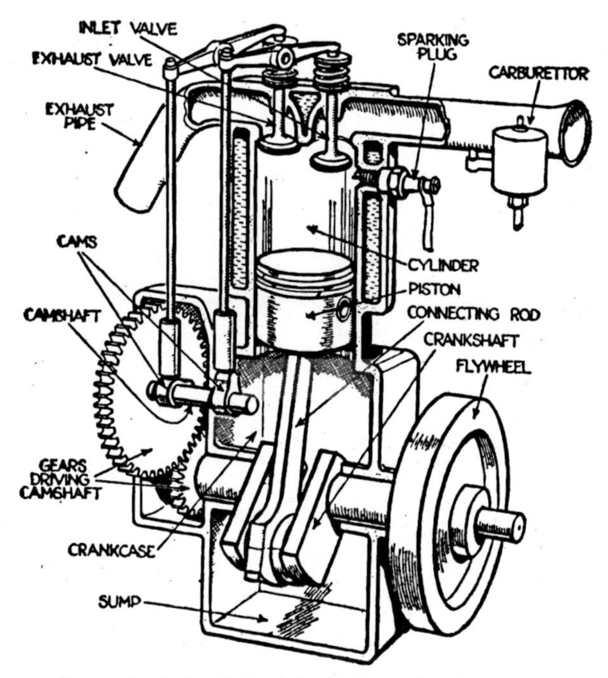

Layout of a single-cylinder overhead-valve spark ignition engine.

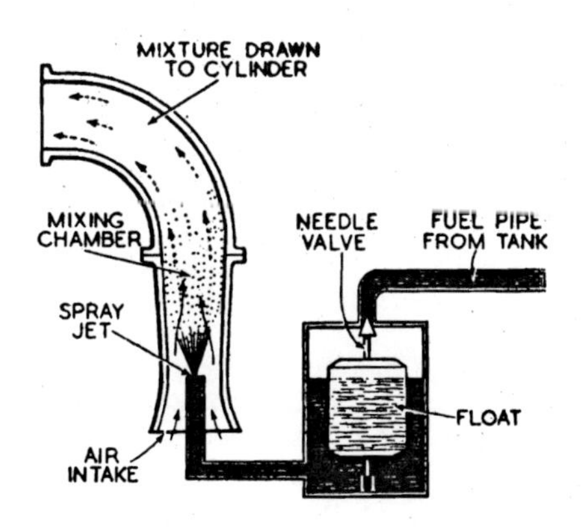

Main components of a carburettor.

drops. This causes the check valve to seat, and prevents a flow of fuel back into the bowl. The economizer valve is pushed open. The fuel displaced by the falling piston is forced out through the power jet, and as long as the throttle valve is held open the piston will remain at the bottom holding the economizer valve open, permitting an additional flow of fuel and enriching the mixture. The power jet measures only enough additional fuel to develop full power. When the throttle is partly closed the vacuum increases above it, the piston is drawn up to the top and the economizer valve closes. This shuts off the additional flow of fuel and consequently a more economical mixture goes to the engine.

To reduce fuel consumption under certain load conditions, some American tractor engine carburettors are designed to carry a slight vacuum in the float chamber. This vacuum is maintained only during operation of the engine loads at over about a quarter capacity and less than about three-quarters capacity. During such operating conditions the slight vacuum maintained decreases the amount of fuel drawn through the main jet and so makes the mixture weaker. For heavy load conditions, when the throttle is opened wide, the vacuum in the float chamber is released, and this permits a free flow of fuel through the jet and enriches the mixture.

In some carburettors the main jet and a compensating jet discharge into the air stream through a number of small openings extending across the diameter of the intake throat with the object of securing better atomization of the heavier types of fuel.

At usual idling speeds there is insufficient air velocity at the ends of the ordinary fuel jet or jets to raise the necessary fuel for proper operation. To provide this fuel in carburettors of other than the auxiliary air valve type, an auxiliary jet called an idling jet is added. This works when the throttle valve is almost closed and it conducts fuel around the throttle valve into the intake manifold.

A spray of petrol becomes, even at low temperatures, so finely divided that it is virtually a vapour. Paraffin, however, does not change to a mist so easily, and it is necessary to heat the mixture of air and paraffin before it enters the cylinder. The heat breaks up the large globules which would be too big to explode and would hang in the cylinders and even seep down into the sump and dilute the lubricating oil.

Oil Dilution

Even in a hot engine paraffin or vaporizing oil does not volatilize quite completely, and there is always more dilution of the lubricating oil in a paraffin engine than in a petrol engine or diesel engine. Tests on the dilution of lubricating oil in engines running on paraffin have been carried out in 24-hour runs at about three-quarters rated load. Each full run was divided into three eight-hour periods with a two-hour

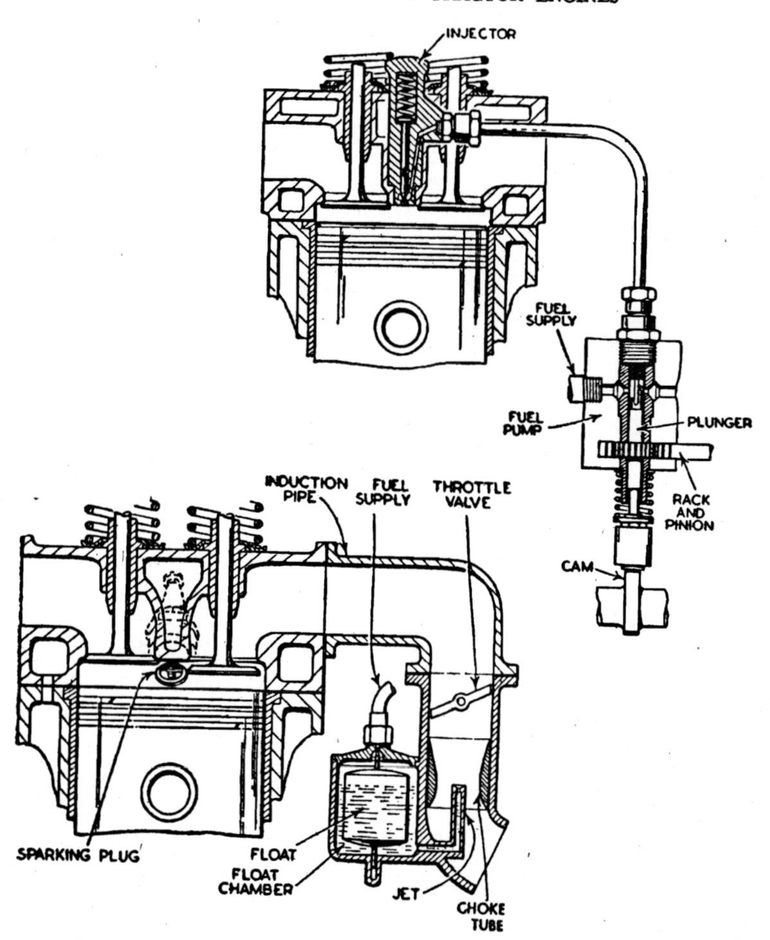

Differences between the components needed for a compression-ignition Diesel engine, and those needed for a petrol or paraffin spark-ignition engine. Top, cam-operated fuel pump and injector; bottom, carburettor and sparking plug.

interval between. In one set of tests, the temperature of the cooling water was not controlled, so that it averaged about 140°F over the whole test, and after each re-start the tractor was allowed to run on petrol for 15 minutes before changing over to paraffin fuel. In another set of tests the water temperature was controlled by means of a radiator blind so that it averaged 189°F and the changeover from petrol to paraffin was made on each occasion as soon as the water temperature had reached 194°F. At intervals during each test samples of oil were taken for dilution analysis. At the end of 24 hours' running without temperature control the oil in the sump contained one part of vaporizing oil diluent to three parts of lubricating oil and its viscosity had fallen from 225 seconds to 86 seconds on the Redwood scale.

In the second case, after 24 hours' running with the temperature controlled at about 194°F, the sump contained only one part of diluent to nine parts of oil and the viscosity had fallen only to 154 seconds Redwood. If the tests had been run under field conditions, with the tractor continually starting and stopping and with cold winds blowing around the vaporizer, the dilution in the cold test would probably have been greater. Measurements which are taken in the field from time to time indicate that it is not uncommon to find farm tractors working with radiator temperatures as low as 112°F.

The percentage of dilution does not go on increasing with the number of hours the tractor has run since the oil was newly put in. There comes a time when an equilibrium is reached between the quantity of paraffin or vaporizing oil that is being added to the oil and the quantity of vaporizing oil which is being driven away from the lubricating oil. The point at which the equilibrium occurs depends on the temperature of the lubricating oil.

Fuel Vaporization

The mixture of air and paraffin or vaporizing oil on its way to the cylinders is caused to pass over baffle plates in a vaporizer. The baffle plates are heated by the exhaust gases.

Paraffin is not sufficiently volatile at ordinary temperatures for a cold engine to be started on it. Therefore a paraffin engine must be started on petrol, and the fuel supply tap must not be switched over to paraffin until the vaporizer has become hot.

In a petrol engine no special arrangements for heating the induction system need be made. Indeed, too hot an induction pipe can reduce the power output of the engine by expanding the mixture of fuel and air so that its density decreases, and a full charge of the mixture cannot be drawn into the cylinder. Sufficient vaporization is achieved with only a small area of hotspot: little or no jacketing of the inlet manifold is required to maintain the fuel mixture temperature.

Vaporizing oil has a higher boiling point than petrol, and engines designed to use it must have some system of vaporization by exhaust heat. There are two general methods. In the first, a normal petrol-type carburettor is used, and the induction pipe is jacketed by an exhaust gas pipe, or a tubed heat exchanger is incorporated inside the induction system. In the second method the fuel is pre-heated by being made to pass through narrow channels in a plate bolted on to the outer surface of the exhaust manifold. Thereafter the gaseous fuel and cool air are carburetted and pass into a normal induction system. This latter system is unusual but it has the advantage of drawing a fuller charge of mixture into the engine. In the system in which the mixture of air and fuel is heated, the density of the charge whch the engine draws in is reduced by expansion by heat. Thus the maximum possible power output is lower in the engine with this kind of vaporizer if everything else in the design is kept the same.

The expansion of the charge due to pre-heating affects power output but does not necessarily affect fuel efficiency. Efficiency is, however, affected by compression ratio, and is greatest when this is high. The limit to the compression pressure depends upon the anti-knock value of the fuel; the anti-knock rating of the lowest grade of petrol is about 70 octane while that of vaporizing oil is about 50 octane. When a fuel of low octane rating is to be used, the engine has to be designed with a low compression ratio if destructive detonation is to be avoided.

The onset of detonation does not, however, depend only on the octane number of the fuel employed. The shape of combustion chamber, the heat capacity of the sparking plug, and the timing of the spark also have great influence. In engines designed for high-compression operation on low-octane fuel the combustion head is shaped so as to achieve a complete scavenging of the burnt fuels and to direct the gas flow over the valve seats and sparking plug housing in such a way as to help to cool them. The engine can, however, use low-octane fuel only while the combustion chamber is free from carbon. As soon as any thickness of carbon forms, hotspots develop and detonation takes place unless the ignition timing is retarded. A clean engine with 4·5 to 1 compression ratio will run smoothly on low-grade paraffin even with the ignition in the fully advanced adjustment. When the engine becomes carboned it will knock if it is run on low-grade paraffin, but it will continue to run satisfactorily on ordinary commercial vaporizing oil with an octane rating of 50.

On some vaporizers the induction manifold is completely surrounded by the exhaust chamber, both being cast in one unit. The exhaust gases in the chamber are controlled by means of the adjustable exhaust shutter which is provided with set positions to regulate the volume of the gases passing round the induction manifold, thus varying the temperature of the induction manifold heating chamber to suit the operating conditions. The exhaust shutter control lever is usually marked 'hot' and 'cold'. A screw and locking nut is located at the end of the lever to secure the shutter at one of the several indentations in the side of the vaporizer body.

With the lever set at the 'hot' position, the shutter directs the gases around the back of the induction manifold and out through the exhaust pipe, thus causing the manifold to heat more rapidly and to a greater extent than when in the other positions. This is the most suitable arrangement for operating under light and average loads.

When the lever is adjusted so that the shutter is slightly open, a small amount of the gases is allowed to escape under the shutter and the remainder passes round the induction manifold, giving a slightly reduced heating effect. This is the most satisfactory in this country when operating under average and medium heavy load conditions.

With the lever set at a position halfway between the second and the fully open setting a greater distribution of the gases under the shutter and around the induction manifold and a corresponding reduction in the temperature of the manifold are permitted. In very warm weather this position is suitable for light and average loads, also for maximum loads such as continuous belt work in cooler weather.

In the 'cold' position the shutter is fully open and permits the gases to pass out through the exhaust chamber into the silencer without any restriction, thus keeping the manifold at the minimum temperature, and this is the most efficient position under maximum load conditions in warm weather.

FUEL TRANSFER PUMPS

On most tractors the fuel tank is situated too low to ensure proper fuel supply to the carburettor or injection pump by gravity. This is overcome by fitting a lift pump capable of supplying fuel under slight pressure from the tank. This transfer lift pump is usually operated by an eccentric on the engine camshaft, and the pumping action is obtained through the upward and downward movement of a flexible diaphragm, consisting of several sheets of fabric specially treated to render them impervious to fuel oil.

As the camshaft revolves, the fuel pump eccentric causes the rocker lever in the pump to pivot, moving the diaphragm downwards. The spring beneath the diaphragm will, as the camshaft revolves further, cause the diaphragm to be pushed upwards. Whenever the engine is running, then, the cam and the spring combined give a continual flexing up and down movement.

As the fuel lift pump is always capable of supplying more fuel than the engine uses and as the requirements vary considerably with engine speed and load, some wastage would result unless the flow of fuel from the pump under pressure were controlled. When the fuel filter and the fuel injection pump are full the lift pump is forcing fuel into the connecting pipe, where it builds up until it forms a back pressure which is sufficient to hold the pump diaphragm down against its spring. It also holds the connecting link away from the rocker arm in such a way that it rocks idly on the cam without actually working the lift pump diaphragm. This continues until the injection pump, requiring more fuel, causes the pressure above the diaphragm to be reduced; the spring then forces the diaphragm upwards and the normal pumping action is resumed.

Fuel is forced under slight pressure through a filter from the fuel lift pump and flows down into the body of the filter around the outside of the filter element. As this fuel is under slight pressure it is forced through the element, leaving any particles of foreign matter behind. It then passes up the centre of the element and through the connection to the injection pump. A filter is fitted in the injector to prevent damage by dirt from the injector pipe lines.

ELECTRICITY

The spark to ignite the compressed mixture in a petrol or paraffin engine is produced by the jumping of a high-tension current of electricity across a gap. It will be well therefore for us to consider the nature of electricity before we enquire into the electrical components of the tractor engine.

Some substances contain free electricity in the form of electrons (particles of negative electricity) which have escaped from their atomic system; these substances are called conductors. By the application of a potential difference, these free electrons can be made to drift through the conductor and create in it a current of electricity. The current direction is opposite to the direction of electron drift.

If a cell or battery is connected to the ends of a copper wire, the free electrons flow along it and a current of electricity is set up, just as when we start a water pump, a current of water is set up in the pipes to which it is connected. The battery does not make the electricity any more than the pump makes the water; both cell and pump are merely agents whereby something which already exists is set in motion. It can be said that it is the pressure difference which the pump creates that causes the motion; this comparison is valuable because whatever engine we use to move the water, the pressure difference is ultimately the physical condition causing the motion and we can therefore say that pressure difference is essential to a flow of water. Similarly, electrical pressure difference is the physical essential for the flow of electrons or electricity, and any agency which produces such a pressure will, if applied to a conductor, set up a flow of electrons known as an electric current.

A battery is such an agency, and so is a dynamo or generator. These devices do not create electricity, but generate an electrical pressure difference, or as we have called it, potential difference (P.D.). We also speak of cells and dynamos as possessing an electromotive force (E.M.F.) for they can set electricity in motion if a complete circuit exists. The term E.M.F. refers to the electrical pressure generated in the interior of any source, which results in the production of a P.D. between its terminals. P.D. always implies a comparison between two points.

Solid conductors are not the only things which contain electric particles capable of being set in motion; fluids and gases can be made to conduct quite well. The current jumping the sparking plug points is conducted through air or through the fuel mixture.

It is best, when practical problems have to be tackled, to think of electricity as a positive flow down the slope of potential, like water flowing downhill, rather than as electrons drifting uphill.

Electricity can be provided by a cell, or battery of cells, or it can be generated by a dynamo. On many tractors we have both a battery and a dynamo and we will describe first the action of the battery.

Primary cells, as in pocket lamp batteries, which actually make the current by chemical decomposition of the materials in their construction, are not used on tractors. Tractor batteries are re-chargeable, and usually they are supplied with electricity by a dynamo driven from the engine. They are called secondary cells or accumulators, and generally they are of a type employing lead and an acid electrolyte.

The acid electrolyte is dilute sulphuric acid and the plates are lead grids packed with a grey paste of lead oxides. When the battery is being charged, the paste in the plate where the current enters is gradually oxidized. It changes colour from grey, and becomes dark brown owing to the formation of lead peroxide. Meanwhile, the other plate is losing oxygen and the lead oxide paste becomes finely divided lead. The brown plate becomes the positive plate when the cell is used for providing current, and as the cell discharges, the paste in the positive plates loses oxygen and reverts from lead peroxide to lead oxide, and the metallic lead in the negative plate acquires oxygen until it also becomes lead oxide.

Tractor engine batteries have several positive and several negative plates in each cell. The plates are arranged alternately positive and negative. They are placed close together and are separated by porous sheets of insulating material of a kind unaffected by acid. The two outer plates are always negative. Alternate plates are joined together so as to form only two terminals for the whole cell.

Each cell gives about 2·2 volts and the batteries are usually made up with 3 or 6 cells to provide current for a 6- or 12-volt equipment of accessories.

Magnetic lines of force can be caused to generate an electric current, and it is true also that an electric current can be made to produce magnetic lines of force. An electric current flowing in a wire produces a magnetic field around the wire. If the wire is wound round into a coil the current produces a magnetic field whose lines of force form closed loops linking the turns of the coil. If the current in such a coil is suddenly stopped these lines of force collapse and in so doing cut the wire forming that coil, or any other coil wound over the first. A current is caused to flow in the coil whose conductors are cut by the lines of force. The greater the number of turns in a coil the greater will be the induced voltage set up by the collapsing magnetic field. In this manner very high voltages can be produced; and this is the basic principle of the ignition coil and of the magneto.

Both of these changes, magnetism into electricity, and electricity into magnetism, are used in tractor equipment. Dynamos usually have electromagnets, not permanent magnets, so that electricity is providing the magnetism and the magnetism is generating the electric current. The agent providing the energy is the engine driving the dynamo.

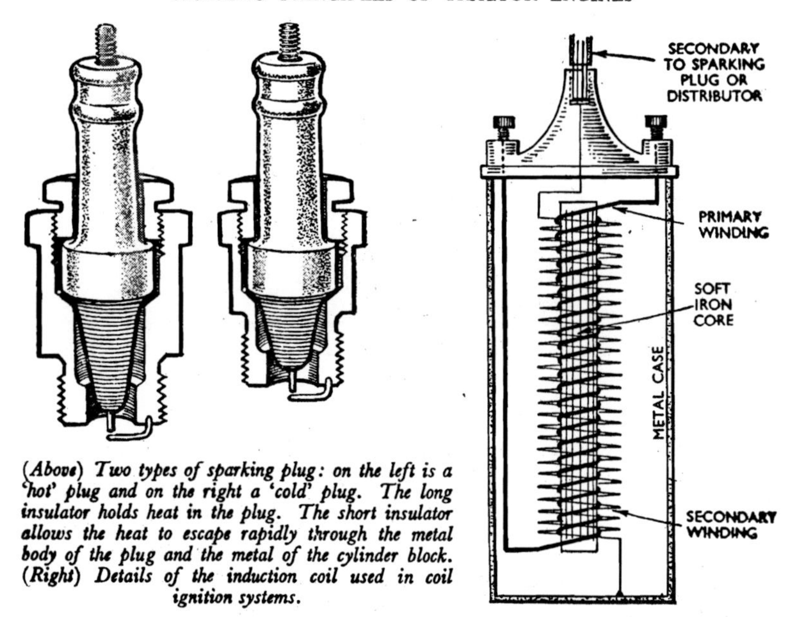

(Above) Two types of sparking plug: on the left is a 'hot' plug and on the right a 'cold' plug. The long insulator holds heat in the plug. The short insulator allows the heat to escape rapidly through the metal body of the plug and the metal of the cylinder block. (Right) Details of the induction coil used in coil ignition systems.

The spark for ignition in a petrol or paraffin engine is produced at the gap between the electrodes of a sparking plug. These electrodes are separated by an annular insulator made of porcelain or mica or some other reconstituted material. The high-tension current is provided by an induction coil and battery and dynamo, or by a magneto.

THE INDUCTION COIL

An induction coil has two windings of wire, one upon the other. The inner one, called the primary, or low-tension winding, is connected with a 6- or 12-volt battery. This primary winding is made up of relatively few turns of thick, insulated wire. The outer winding is of many turns of thin wire. It is called the secondary or high-tension winding, and it is connected with the sparking plugs.

The high-tension current is induced when the low-tension circuit is suddenly broken. To break the circuit regularly at the times when a spark is needed in the cylinders there is a pair of tungsten metal points, which open and close by the action of a fibre bumper riding on a cam.

The sudden break causes a rapid decrease in the induced magnetic field of the inner or primary coil. This decrease will in itself induce a voltage in the inner coil and tend to keep the current flowing in it and thereby cause sparking across the contact breaker points. To obviate this a condenser is connected across the contact points and the induced current flows into the condenser instead of sparking across the points.

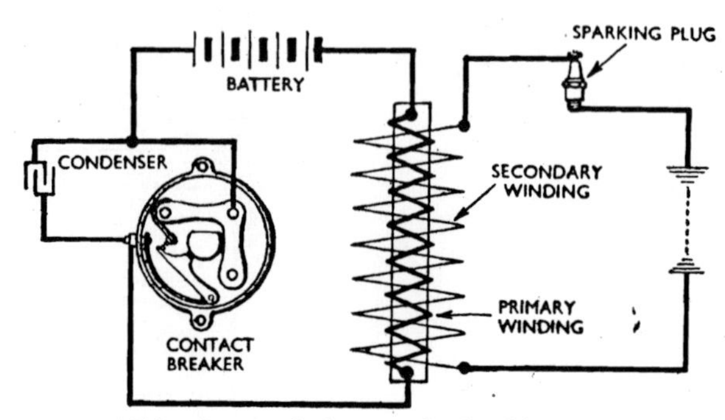

Wiring diagram of a battery and coil ignition system.

Moreover, the condenser voltage eventually becomes greater than the voltage induced in the coil and this sends a current in opposition to the dwindling primary current and causes the field to collapse more rapidly across the secondary winding, and therefore raises the induced voltage.

Collapse of the primary magnetic field due to the opening of the contact breaker, and assisted by the condenser, generates a high voltage in the secondary coil. This voltage is high enough for the current to jump the small gap at the sparking plug points.

For an engine having more than one cylinder, a distributor is needed. The distributor is a revolving arm which makes brushing contact with insulated metal segments. Each of these segments is connected to the sparking plug of one of the cylinders.

DYNAMO GENERATORS

We have seen that when a coil of wire rotates in a magnetic field, a current is generated within the coil as it cuts the magnetic lines of force. In an alternating-current generator the ends of the wire forming the coil on the rotor are connected to a pair of slip rings, which are made of brass or bronze, and are shrunk on to the shaft over mica insulation. The rings are insulated from each other by mica or fibre washers and the external circuit is connected to the winding by brushes made of hard carbon. These brushes are brought to bear lightly on the slip rings.

A simple generator of this design will produce an alternating current. For half a revolution of the coil, or armature, the current will build up to flow in one direction. Then the current will fall off to nothing and, during the other half of the revolution, it will build up in the opposite direction.

A current of this kind will operate tractor lights satisfactorily, but current which is to be used to charge an accumulator must be uni-directional, and therefore a simple alternating-current generator cannot be used without modification. In the modified machine an alternating current is produced in the armature conductor, but the current supplied to the external circuit is uni-directional owing to the action of a device known as the commutator. Let us consider the loop which has already been assumed to rotate in a uniform magnetic field, and in which an alternating current is being generated. Instead of being brought to slip rings the two ends of the rotating loop are connected to a single ring which is split longitudinally at two points diametrically opposite. The direction of the current in the external circuit is reversed when the armature rotates through 180°, and the result is that the current in the external circuit is flowing only in one direction.

The split ring device causing the reversal of current in the external circuit is called a simple or two-part commutator, each of the conducting portions being termed a commutator segment. In practice, in order to obtain a current which is constant in quantity as well as continuous in direction, armatures with many coils, and therefore many commutator segments, are used.

When the generator on a tractor engine is used to charge the battery, an automatic cut-out switch has to be fitted in the battery-charging circuit so that when the engine is stationary or is turning over slowly the connection is broken and so the battery does not discharge itself into the windings of the dynamo.

There is so little difference between the construction of a dynamo and the construction of an electric motor that if a charged battery is connected to a dynamo it will tend to cause the armature of the dynamo to rotate. Whether it is actually able to rotate the armature and therefore the engine connected with it or whether it can only tend to make this movement, the dynamo will be providing a path for the current from the battery and will soon discharge it. Therefore, the charging circuit of the dynamo and accumulator has an automatic cut-out switch which closes the circuit only when the dynamo is being driven fast enough for it to be generating a current at sufficient pressure to overcome the pressure of the electricity stored in the battery. This automatic make-and-break is magnetic. Contact is made when sufficient electricity is passing through a coil to produce enough magnetism to operate a contactor. As the dynamo slows down the magnetism becomes weaker until at last the circuit is broken.

In the electrical system of a tractor the battery is really a buffer between the dynamo and the components which need electricity for their operation. Although its first purpose is to maintain a store of electricity for use when the dynamo is not working it also has a balancing function when the dynamo is being driven.

A discharged battery is hungry for electricity and will accept usefully all the current which the dynamo can generate. A fully charged battery does not need current from the dynamo and cannot make use of it. Excessive charging of a battery is undesirable. If it does nothing worse, it will drive away so much water as vapour from the electrolyte that

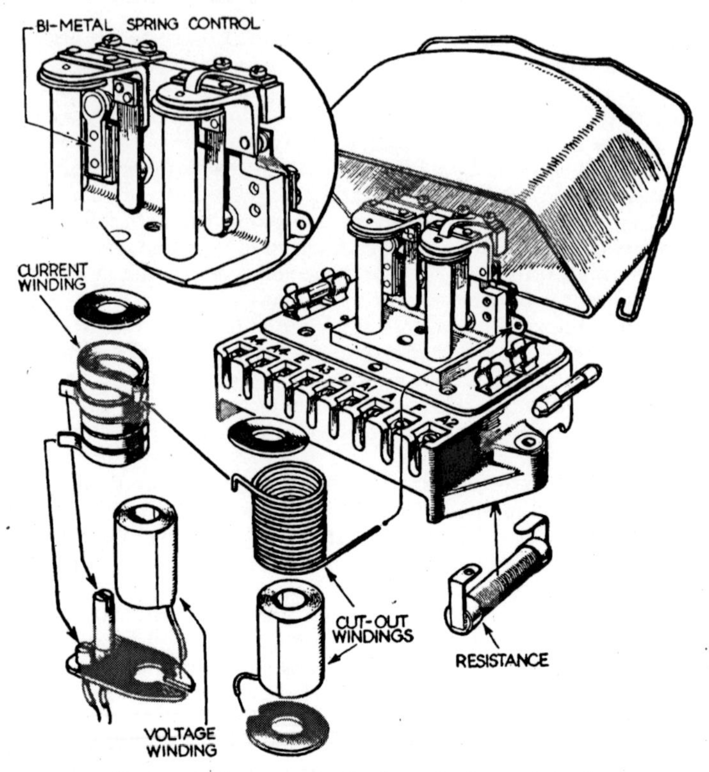

Components of a cut-out, junction box and voltage control unit.

frequent topping-up with distilled water is necessary to save the battery from deteriorating. Overcharging can very easily happen in a tractor doing several hours' steady work with no lamps or other apparatus to drain away the current. A voltage-control relay switch is often incorporated in the electrical outfit to prevent this overcharging. When the voltage of the battery itself is restored sufficiently to induce a back electromotive force high enough to operate the automatic switch, the output of current from the dynamo is reduced. This not only saves the battery, it also prevents the dynamo from getting hot unnecessarily.

The essential component of a voltage control unit is two electromagnets. Each has an iron core and each has a separate voltage coil, and each has a current winding. The voltage coils are in separate circuits, but the current coils are in series and form parts of one circuit although the one coil has thicker wire than the other. Each soft iron core has its armature and set of contact points, one being for the cut-out, the other for the regulator.

At the beginning the cut-out points are open and the regulator points closed. When the generator starts to produce current this flows from the positive brush, through the circuit to the baseplate which forms an electrical bus-bar, to which connections are made for the various arms of the circuit. Several paths are available for the current at the bus-bar, some flowing around the cut-out voltage coil and some round the regulator voltage coil, both currents flowing thence to earth. The remainder flows through the closed regulator points and so through the field coils of the generator. Current always flows along the path of least resistance and thus the current by-passes the resistance and flows straight through the closed regulator points. As the cut-out points are open at this stage, no current can flow from the bus-bar through the current coils.

The generator voltage increases with the speed of the armature until it is sufficiently high to charge the battery and as this increased voltage forces current around the cut-out voltage coil, the iron core of the cut-out acquires sufficient magnetism to attract the armature, thus closing the cut-out points which were previously open. With the cut-out points closed a further path is available for the generator current from the positive brush flowing to the bus-bar. It divides from this point, some flowing round the regulator voltage coil as before, some along the cut-out voltage coil, thus maintaining sufficient magnetizing force to hold the cut-out armature down against its spring, and some still through the closed regulator points. The bulk of the current, however, now flows from the bus-bar through the closed cut-out points and round the series coil, half of which is wound around the cut-out core and half around the regulator coil, to the battery.

As the generator output increases with a rise in engine speed, there is the possibility that the generator, at high engine speeds, will supply a current at too high a voltage to be continuously applied to the battery and system. When such a point is reached, the current flowing round the regulator voltage coil and the current coil magnetizes the iron core sufficiently to attract the armature. The contact points of the regulator make-and-break are arranged so that when the core attracts the armature it opens the regulator points and does not close them as in the case of the cut-out points. As the regulator points are now open, current which previously passed through the closed points straight to the field coils will now have to flow through the resistance and so back to the field coils of the generator. Since the current has to be forced through this resistance there is less of it flowing and, as this reduced current is being supplied to the field coils of the generator, the output is reduced suitably.

When the output has fallen to a predetermined level and, in the meantime, the magnetizing force of the regulator core has also decreased, a spring forces the regulator armature upwards again, closing the points. The current can again take its original path through the closed points. In practice the regulator points open and close very

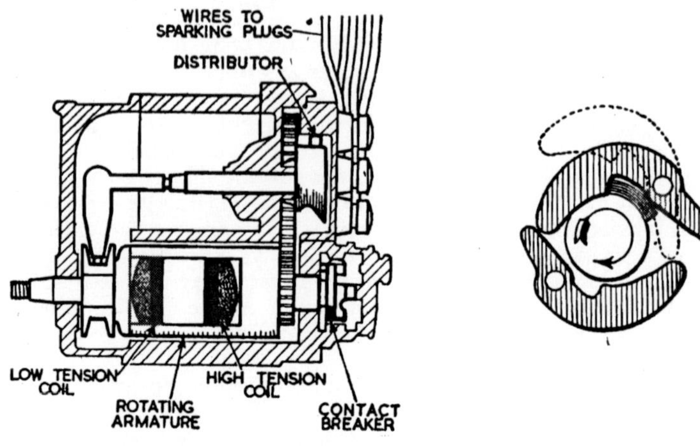

(Left) Section of a high-tension magneto, showing the position of the rotating armature. (Right) An impulse starter on a tractor magneto. The pawl disengages when the speed of rotation is high enough to throw the arm into the position shown by the dotted line.

quickly, constantly regulating the output current to the desired amount.

THE MAGNETO

In many tractors the current for ignition is generated not by battery and coil, but by magneto. Generally speaking, it can be said that the magneto combines two instruments, though this is not the whole of the story. It is a generator for making a low-tension current, and it is also an induction coil for transforming this low-tension current to a high-tension current capable of jumping the gap at the sparking plugs.

Its low-tension coil and high-tension coil are both wound on to the same armature. This armature, driven from the engine, rotates between the poles of a horseshoe magnet, and the rotation induces a current in the low-tension winding.

The magneto armature is usually not driven directly from the engine. The drive passes through an impulse-starting mechanism. This device causes the armature to jerk round in a series of quick movements when the engine is being turned slowly. It contains a coil spring. Part of each revolution of the magneto-driven shaft does not move the magneto armature at all; it merely winds up the spring. Then the spring suddenly releases and flicks the armature round so fast that a good spark occurs at the plugs even when the starting handle of the tractor is turned only very slowly. As soon as the engine starts and gathers speed, this impulse-starting mechanism goes out of action, and the drive to the magneto becomes solid.

The impulse mechanism makes it possible to start the engine by pulling up the handle without cranking it round and round. This lessens the risk of injury to the driver.

SPARK TIMING

The occurrence of the spark has to be timed so that the mixture of fuel and air shall be ignited just at the point in the cycle of the operations when the mixture is most highly compressed and when the explosion will have the greatest effect in forcing the piston downwards. If the spark (and therefore the explosion) occurs too late, some of its efficacy will be wasted. If, on the other hand, the explosion occurs before the piston has finished its upward stroke, there will be conflict between the movement of the piston due to its upward movement resulting from the previous power stroke and the explosion which is trying to push it back down the cylinder bore. Such conflict produces a knock.

When, however, the engine is running fast the spark must be made to occur while the piston still has quite an appreciable distance to travel to the top of its stroke. This is to allow time for the jumping of the spark and the effective burning of the compressed mixture of fuel and air. The faster the engine is running, the more must be the lead given to the spark.

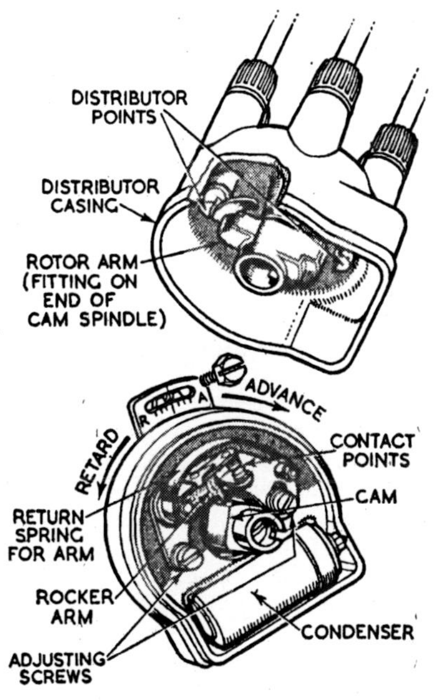

A distributor for a four-cylinder coil-ignition engine, with arrangement for advance and retard.

The generating of the spark is, as we have seen, caused by the breaking of the low-tension circuit by the separating of the ignition points. The timing of the opening of these points is determined by the position of the cam on the contact breaker of the magneto or coil. In order to ensure that the spark shall not be too far advanced while the engine is running slowly, and too far retarded when it has gained speed, an adjustment is provided so that the timing can be altered within a small range while the engine is actually running. In many tractors this adjustment is effected by a hand lever. In others the advance and retard is operated by a centripetal device which automatically advances the magneto timing as the engine gathers speed, working in very much the same way as a centrifugal engine-speed governor works.

So far it has not been found possible to design a magneto which has both the impulse-starting mechanism and a centripetal advance-and-retard mechanism, and the makers have to choose between one or the other. For this reason the automatic advance-and-retard device is used chiefly on coil-ignition tractors.

FLYWHEEL EFFECT

Of the four strokes of the piston in a cycle of a four-stroke engine, and the two strokes in a two-stroke engine, only one is a power stroke. For the other strokes, the engine has to drive the piston, instead of the piston driving the engine. Therefore, a flywheel has to be fitted to the crank-shaft to store momentum to be used for driving the piston on its idle strokes.

Most tractors, however, have more than one cylinder, and the firing is timed so that the power impulses are as equally spaced as possible, consistent with the balancing of the moving parts. In a four-cylinder four-stroke engine, always one or other of the pistons is performing a power stroke while the other three pistons are doing idle strokes. Therefore, in a multi-cylinder engine the crankshaft is driven more smoothly than in a single-cylinder engine, and the flywheel need not be so large.

To make the flow of power smoother and overcome vibration caused by the motion of the engine parts, an attempt is made to counteract vibration tendencies by balancing piston and crank throw. This means that in some multiple-cylinder engines a choice must be made between balanced firing and balanced piston and crank throw. In two-cylinder engines it would be necessary to have both pistons moving to and fro together if the engine fired every full revolution. The moving of both pistons together might cause vibration even though the firing were balanced. Smoothness of operation is therefore obtained by balancing the piston and crank throw so that one piston is moving inward while the other is moving outward, even though this arrangement means that uneven firing is used. A heavy flywheel can compensate for the uneven impulses.

In four-cylinder engines the firing is arranged evenly, so that a power stroke occurs every half revolution of flywheel travel. Both the firing order and piston and crank throw are balanced. Usually the front and back pistons move together in one direction, while the two centre pistons move together in the other direction.

STARTING THE ENGINE

The internal combustion engine is not self-starting in the same way as a steam engine. The sequence of operations has to be set going by some outside agent. Petrol and paraffin engines, and most of the smaller

Diesel engines, are merely turned by hand through a crank until the first explosion occurs in one or other of the cylinders and the engine begins to run on its own account. Even in tractors having a smaller size of engine, however, an increasing number of models are being fitted with electric starters.

The electric starter motor must exert a high torque in order to turn the engine. Therefore, the starting motor has to take a heavy current. The drive from the electric motor to the tractor engine has to be taken through a low reduction gear, and it is necessary to have some means of throwing the gear in and out of engagement. The most usual drive

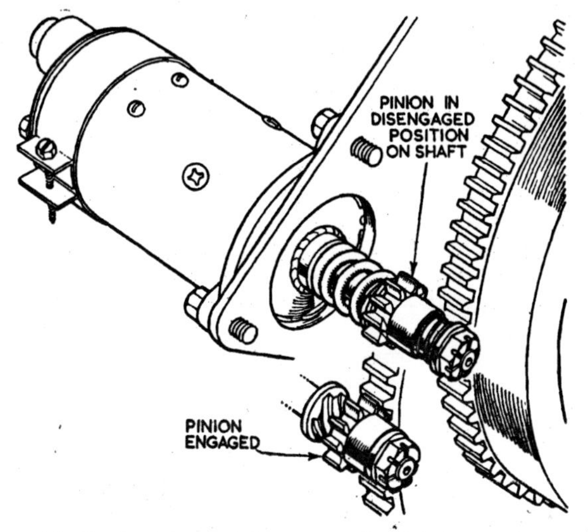

The fitting of the Bendix pinion carried on a coarse threaded sleeve on the armature shaft of the electric starting motor.

is of the Bendix type. A small pinion on the armature shaft of the starting motor meshes with teeth on the flywheel of the tractor engine.

The method of engaging and disengaging the pinion must be such as will not injure the starting motor when the engine begins to fire or if, through the ignition being too far advanced, the engine should kick back. In the Bendix drive the pinion is carried on a coarse-threaded sleeve on the armature shaft. Behind the pinion is a coil spring. As soon as the starter switch is closed and the armature begins to rotate the pinion moves forward along the armature shaft and engages the teeth on the engine flywheel. As soon as the engine fires and turns the pinion faster than the armature shaft is turning it, the pinion is carried back out of engagement by the threads. The spring has to absorb the shock as the pinion comes into mesh with the flywheel or when the engine kicks back.

The gear teeth on the flywheel are usually not cut on the flywheel itself but on a band which is either shrunk on to the periphery of the flywheel or is made in sections and fixed in position with countersunk screwpins. Thus the gear ring can be renewed if the teeth become worn or chipped.

The heavy current needed by the motor makes a direct switch difficult. For one thing, the switch could not be placed at any distance from the motor unless the wires leading to it were very thick indeed, and for another thing a very large hand switch would be needed. Therefore a relay switch is employed. The small starting button closes a low-amperage circuit which energizes an electromagnetic switch situated near the motor. This solenoid switch moves a plunger which puts the main starter current into circuit.

Some large Diesel tractors which have no electrical equipment have a small petrol engine which can be clutched in and out of engagement with the large Diesel engine. With the clutch disengaged the petrol engine is started and run for a few minutes until it is warm and then it is thrown into engagement with the Diesel engine, which it rotates until the Diesel takes up the work.

There are, however, devices for helping the starting of large Diesels by hand. On some tractors, arrangements are provided for heating the combustion chamber by conduction from the heat of a blow-lamp outside the cylinder. Other tractors use a heating cartridge for helping the first explosion. The cartridge consists of a wad of slow-burning paper. This wad of paper is held in a plug which can be screwed into and out of the cylinder. The paper is set burning by a match and then the plug is screwed into the cylinder. One tractor using the heating cartridge has also an automatic compression release device. This allows the engine to be turned through about six revolutions with very much less than full compression of the gas in the cylinder. At the end of the six revolutions the compression release device is automatically put out of action and the engine ought to fire.

Some tractor engines, however, have a carburettor and electric ignition in addition to the heavy-oil injection apparatus. These engines are started as petrol engines and are only turned over to Diesel operation when the engine has become warm. This would not be any advantage if the compression ratio remained the same, but this system of starting is made useful by the incorporation of a compound combustion chamber.

The main chamber has a compression ratio suitable for fuel injection operation. Alongside the main combustion chamber is a smaller one which can, at will, be thrown into communication with the main one. When the valve between these two chambers is open the engine has a low compression ratio, and is therefore easy to start on petrol. The small chamber has a sparking plug and a petrol carburettor. When the engine has been started on petrol and has been running for two minutes or so the valve connecting the main and auxiliary chambers is closed by a hand lever. This lever also brings the Diesel fuel pump into operation and the engine becomes a compression-ignition unit. At the same time the electrical ignition and petrol carburation systems are thrown out of commission.

The semi-Diesel or hot bulb engine has a compression ratio of only six to one, and is easier to turn over. The air is not heated sufficiently by the compression stroke alone to cause spontaneous ignition of the injected fuel and the necessary extra heat is provided by an uncooled part of the cylinder head in which heat from earlier firing strokes is stored. Before the engine can be started from cold, this hot spot in the cylinder head has to be heated up by means of a blow-lamp.

To facilitate the cold starting of full Diesel engines at low crankshaft speeds, some engines have a device for injecting di-ethyl ether into the intake air. This produces a chemical reaction in the cylinder evolving sufficient heat to cause ignition; the device first provides a rich ether-air mixture enabling the engine to fire for a few strokes, after which a weaker mixture is supplied until running on normal fuel supervenes. The ignition-promoting fluid is contained in an expendable plastic capsule which is pierced automatically when the holder is inserted.

Bi-fuel Carburettors

Vaporizing oil engines are difficult to start unless the chamber of the carburettor contains only petrol. If the vaporizing oil tap should have been turned on inadvertently, or if any vaporizing oil should remain in the chamber from when the tractor was last at work, it is difficult or impossible to start the engine from cold. To obviate this tendency to trouble, some tractors are fitted with a double-purpose carburettor which is, in effect, two separate instruments. There is one float chamber and pipe line system for petrol and a quite independent float chamber and pipe line system for vaporizing oil. Thus, the fuels cannot become mixed, and the large tank for vaporizing oil is never in any way connected with the small tank for petrol for starting.

FUEL FILTERS

All fuel pipe line systems have filters to hold back dirt which would block the jets of a spark ignition engine or choke and damage the components of a Diesel pump and injector. Water in fuel also can be held back by suitable filtering and settling devices.

Sediment bowls are usually made of glass so that inspection will show when enough dirt has collected to make cleaning worth while. The outlet from the bowl is higher than the inlet, so that most of the dirt, and most of any liquids which are heavier than the fuel, do not pass over into the pipe to the engine. Strainers of fine-mesh wire gauze are fitted at some of the pipe unions. With some sediment bowls a filter made up of annular discs is incorporated. The discs, shaped like washers, are packed close together, but the surface of each bears indentations which prevent complete contact of one disc with another. The fuel goes first into the bowl and then finds its way between the discs to the outlet pipe. The dirt falls to the bottom of the bowl.

SEALING AGAINST DUST AND GAS

Grit and dust from the soil cause much wear, and wherever possible the tractor is protected from their entry. Faces to be joined are carefully machined, and felt washers are fitted at wheel bearings where grit might work in, or oil work out. This making of oil-tight, dust-tight, and even gas-tight seals between moving parts is one of the problems of engineering.

The pistons would not, of themselves, fit the cylinders of the engine exactly enough to maintain the high pressure of gas on the compression stroke, and the low pressure on the suction stroke. Even if the piston could be made an exact sliding fit in the cylinder when the engine was cold, it would be nearly impossible to arrange the shapes and masses of the metals of the piston and cylinder so that expansion by heat would not make the piston either too tight or too loose. Therefore spring piston rings are used. The piston itself has grooves around it, and in these grooves are rings. These rings, when they are unconfined and are at rest, are slightly larger in diameter than the bore of the cylinder. Consequently when they are confined by the cylinder walls they press outwards and make a gas-tight seal.

Each cylinder has two, three or four piston rings, but not all the rings are primarily compression sealing rings; one or more of the rings is a scraper ring to wipe off surplus oil from the cylinder wall. Their functions, however, are not quite distinct, and really they are supplementary and help each other, but the rings are known as compression rings to seal against loss of compression, and scraper rings to regulate the degree of lubrication known to be required.

A piston ring must, in its closed state, conform exactly with the shape of the cylinder in which it operates and exert sufficient pressure on it to maintain the contact. Since the usual shape of a cylinder bore is a circle, the free shape of the ring has to be such that it will close to a nominal circle when it is constrained within the cylinder.

The cylinder wall pressure that a ring exerts varies roughly with the cube of the thickness; thus a small increase in ring thickness results in a large increase in ring pressure. Radial thickness also influences the stress to which a ring is subjected, either in opening over the piston or in closing in the cylinder. Radial thickness has much influence upon the tendency of a ring to flutter or vibrate near the free ends. Flutter results in blow-by, excessive wear on rings and cylinders, and possibly ring breakage. A high radial thickness minimizes flutter, but brings increased stress in opening or closing.

The size of the gap of the ring is important. With a large free gap the stress in fitting a ring over its piston is low, but the stress in the working position is high; with a low free gap the reverse is true. Therefore a compromise dimension for the gap is used. The pressure exerted on the cylinder wall depends on the radial thickness of the ring, the modulus of elasticity of the materials, the contact area of the ring and the number of rings used. The unit pressure controls the thickness of the oil film, and the total pressure affects frictional power losses.

Narrow rings bed in more quickly to the cylinder than wider rings, and they minimize the tendency to ring flutter, though unless they are of special material or specially treated they tend to wear excessively. A large number of narrower rings provide a better seal than a smaller number of wider rings.

A ring must be free to float in its groove and exert its normal pressure against the cylinder under all running conditions. Thus its expansion at working temperatures, together with dimensional changes of the piston and cylinder in which it is working, have to be taken into account when the size is chosen. A large proportion of the heat dissipated from the piston travels to the cylinder through the rings and the rings are thus hotter than the cylinder wall and therefore liable to great expansion.

The cylinders of the engine themselves cannot be actually sealed against dust, because every suction stroke draws in, with the fuel, a charge of air from the atmosphere. Every gallon of fuel burnt in the tractor needs 10,000 gallons of air for its combustion. That amount of the air in the cloud that travels with a tractor working in a dusty field can contain enough abrasive matter to cause much wear in the engine.

Some of the dirt that enters the engine mixes with the solid carbon products of combustion to help form the black deposit which accumulates in combustion chambers and on piston tops. This part of the dirt does not do any permanent damage to the engine. Another part of the dirt may be blown straight out through the exhaust valve. Some small part of the dirt and grit from the air, however, finds its way into the lubricating oil where, if it is left to flow around with the oil into the bearings, it can do much harm, both by direct abrasion and by blocking oil ways.

Air-cleaners

Tractors are fitted with air-cleaners to hold back as much as possible of the dust from the air which is being drawn into the cylinders of the engine. The development of these air-cleaners, suitable for the very difficult conditions of agriculture, has been a long process. The cleaner must remove the dust effectively, but it must not unduly restrict the intake of air into the carburettor.

Many types of cleaner have been tried. The first was the dry type, either just a sieve or filter, or a series of vanes and baffles which caused the heavier dust particles to be thrown out of the air stream. Another type had a water bath through which the ingoing air was bubbled. Most of the present models are oil air-cleaners. In one the inlet air impinges on a fibrous element soaked in oil; in others, the air bubbles under or over the surface of oil held in a cup, atomizes some of the oil and carries it on through a metal mesh which holds back most of the oil and any of the dirt sticking to the globules of oil.

The water air-cleaner had the disadvantage that the water was liable to freeze in cold weather and burst the container. It had, however, one great advantage: it caused the air going into the carburettor to be saturated with water and this had a good effect upon the running of the engine. The water seemed to act as an anti-knock additive.

Some oil air-cleaners allow a slight oil-mist to be carried over in the air going into the engine, and this may have some good effect as an upper-cylinder lubricant.

Primary Cleaner

The simple dry type of cleaner is not used any more as the main defence against the dust coming into the engine with air. It is, however, used in many tractors as a pre-cleaner to remove large particles of dust and pieces of chaff and similar rubbish, so as to lighten the duty of the air-cleaner proper.

The pre-cleaner is sometimes situated at the top of the oil air-cleaner; in other makes it is at the entrance to an air inlet pipe which is extended vertically for 3 ft or more, so that the entrance is above the thickest part of the dust cloud raised by the churning of the wheels of the tractor and the working of the implement.

Pipes connecting the air-cleaner and the carburettor air-intake must be free from leak. If they are not, some of the dust-laden air can by-pass the cleaner. Other places where dust-laden air can enter the engine, without going through the cleaner, are worn carburettor valves. For example, the shafts that hold throttle valves and choke valves usually pass through bearings on the body of the carburettor, and if the shafts do not fit snugly into the bearings, air and dirt can enter the engine. Dusty air can get in also through defective gaskets on the carburettor flange and on the intake manifold.

LUBRICATION SYSTEMS

Every engine has to have a reliable system for ensuring that oil shall flow to all the bearings so that there shall be no metal-to-metal rubbing contacts.

In some engines oil is actually pumped to the main bearings of the crankshaft, and to the big-end bearings of the connecting rods, and is sprayed on to the rest of the moving parts of the engine. In other engines, the big-end bearings dip into the oil in the crankcase, and splash the oil upon themselves, and they also stir up a mist of oil that lubricates the cylinder walls and reaches to the gudgeon pin in the little-end bearing. The circulation of the oil is achieved by the rotation

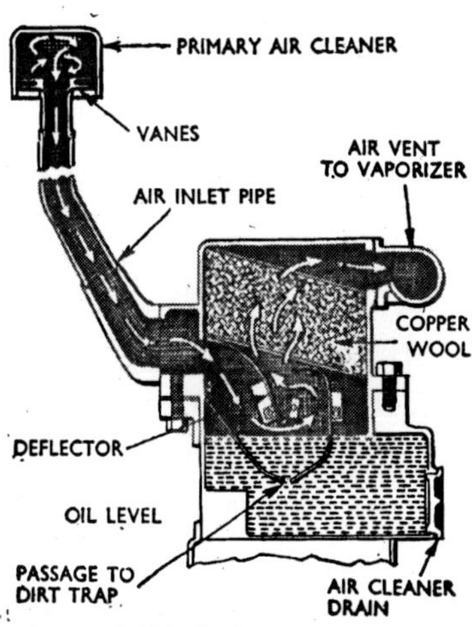

A type of oil-bath air-cleaner in which the globules of oil, holding the dirt, are retained in a copper-wool filter pad from which they can drip back down into a dirt trap.

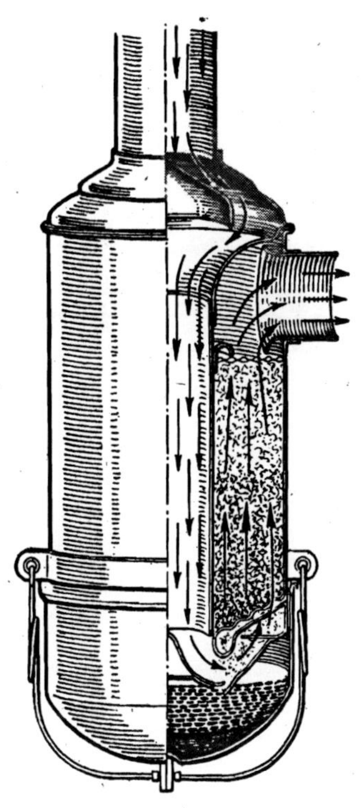

(Above, right) A type of oil-bath air-cleaner in which the air bubbles under and over the surface of oil held in the cup, atomizes some of the oil and carries it on through baffles which hold back most of the oil and the dirt sticking to the globules of oil.

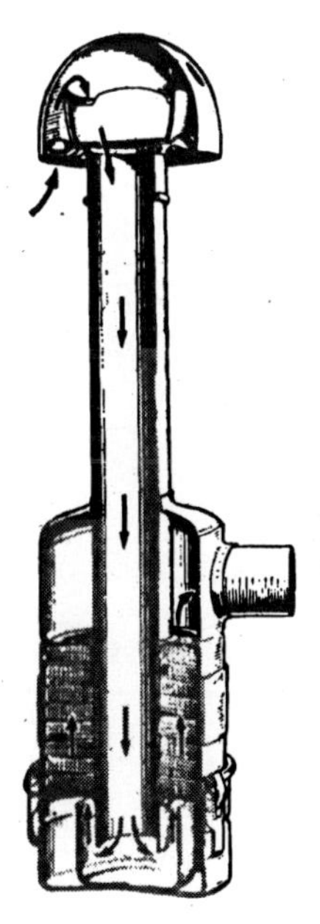

Parts of an air-cleaner, with oil bath, and a swirl-type primary air-cleaner.

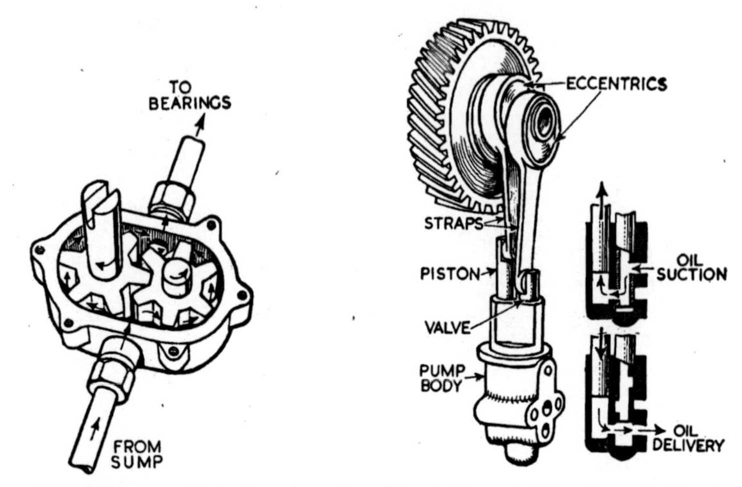

(Left) A gear-type lubrication pump. One pinion is driven, and the other meshes into it. Oil trapped between the teeth of the pinions is expelled at high pressure along the pipe to the bearings. (Right) A piston-type lubrication pump with mechanically operated valve. The piston and valve are worked by eccentrics.

of the fly-wheel, which causes the oil to be thrown off, always to one side, by centrifugal action. Channels and piping are usually provided to gather some of the oil that is splashed round the inside of the engine and to conduct it by gravity to the main bearings and to some of the accessory parts of the engine such as timing wheels.

The pressure circulation system uses a pump in the crankcase to circulate oil through pipes and drilled passages in the crankshaft to the main bearings and connecting rod bearings and then through drilled connecting rods to the gudgeon pins. Oil escaping from the bearings is thrown on to the cylinder walls and furnishes lubrication for this part of the engine. One kind of pump has a plunger or piston, and a mechanically operated valve; the piston and valve are generally operated by a connecting rod mounted on an eccentric bearing. Gear pumps also are used; usually one pinion of the gear pump is driven, and the other meshes into it. Oil trapped between the teeth of the pinions is expelled at high pressure along the pipe to the bearings.

There is also a system of lubrication which employs both these methods. The cylinder walls and gudgeon pins are lubricated by splash or throw-off from the connecting-rod bearings, while the rest of the engine is lubricated by pressure circulation.

Whether the engine lubrication is by splash or by force-feed pump, the circulation brings cool oil to the bearings and also makes possible the interposing of a filter screen to collect any grit or hard pieces of carbon that might damage the bearings.

Oil in Petrol Lubrication

In the air-cooled two-stroke engines fitted to some small walking tractors, lubrication of the bearings and the piston is achieved very simply by mixing a little lubricating oil with the petrol which is used as fuel. This method is possible since the mixture of fuel and air goes first into the crankcase.

When the piston descends and compresses the mixture, most of the oil mixed with the petrol is thrown out of suspension. A mist of oil settles on the bearings, and, in some makes of engine, forms a pool at the bottom of the sump from which ordinary splash lubrication can operate.

The oil and petrol in the tank must be well mixed to avoid a concentration of oil entering the carburettor. If the outlet pipe is higher than the level of the bottom of the tank, the carburettor is not likely to receive an undue concentration of oil.

PRINCIPLES OF ENGINE LUBRICATION

Internal combustion engines make great demands upon lubricating oil.

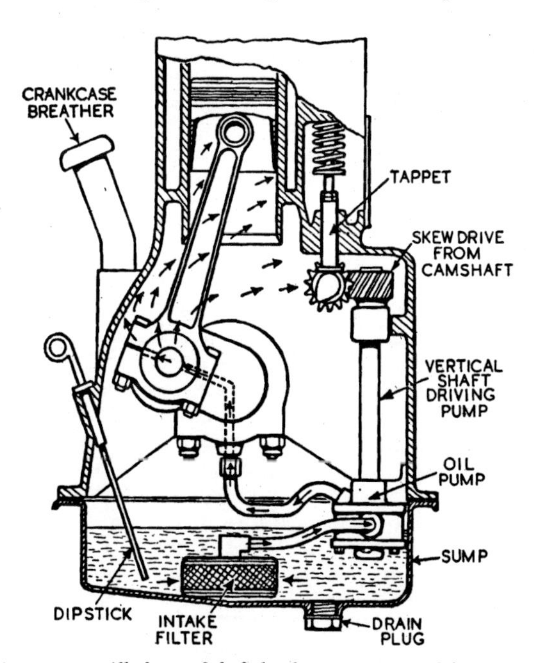

A lubrication system. All the crankshaft bearings are pressure-fed, but in this engine the camshaft, tappets and little-end bearings are fed by oil mist which also covers the cylinder walls.

What is required of a lubricant is that it shall form a separating film between the metal surfaces which are rubbing together. The difficult problem is to maintain this separating film in all conditions of load, speed and temperature.

For example, a film of oil may be separating the moving surfaces of a bearing perfectly satisfactorily while the spindle in the bearing is running freely under no load, but may be squeezed away from the parts of the bearing in closest contact when the spindle is under load. Again, high speed may cause the film of oil to be thrown away from the surfaces, and also at high temperatures the oil may become so thin that the film it forms is quite inadequate to protect the bearings.

In view of these possibilities it might be thought that the thicker the oil the better it is, for it is more likely to form an adequate separating film. This is not quite true, however, because if the oil is too viscous, power will be lost by friction set up in the oil film itself. Moreover, a thick oil will not circulate around the moving parts quickly enough to carry away the heat generated in the engine.

For most of the time, in its circulation through the engine, the temperature of the lubricating oil may not be much higher than the water jacket temperature, or the temperature of the cylinder in an air-cooled engine, but when the oil is in a film between the cylinder wall and the piston it may be exposed to the combustion flame. This is particularly likely to happen on worn engines.

In all bearings there is a working clearance to allow room for the film of lubricating oil. It is often said, and truly said, that lubricating oil does not wear out. In the temperatures and other conditions of our tractor engines, the oil itself is nearly indestructible. Nevertheless, the oil does eventually become unsuitable for further service. This is because it becomes contaminated. The combustion of the fuel brings water and soot, which become absorbed by the lubricating oil if they come into contact with it, as indeed they must. Oxidation of the lubricating oil also frees some carbonaceous matter, and other foreign matter comes in the dirt drawn into the engine through the air intake and through the opening in the crankcase called the breather. It is to be noted, however, that the crankcase breather, though it may possibly admit a little dust, is essential to keeping down the formation of oil-water sludge. Without some opening in the crankcase the gas that escapes past the piston rings cannot get away into the atmosphere and it becomes absorbed by the oil. Most of this gas is water vapour which condenses and mixes with the oil in a sludge emulsion. Some soot also comes from the gas. The contact between the products of combustion and the lubricating oil in an engine with a good piston ring seal is small, but in an engine with worn cylinders enough of the burnt gases find their way into the crankcase to contaminate the oil in it seriously, and an inefficient crankcase breather will prevent the gases from getting away.

Operating conditions which do not permit an engine to reach its optimum working temperature may cause undue wear of the cylinders. For every pound of fuel consumed a pound of water in the form of steam is produced. Any steam not expelled from the combustion chamber will condense on the cold cylinder walls and by corrosion accelerate the rate of wear. An oil that clings to the surface of the cylinders can offset some of the effect of unburnt fuel when combustion is not complete.

Oil in an internal combustion engine is continually coming into contact with the hot metal surfaces of the piston and cylinders, and at these high temperatures it can become oxidized. The breathing of the crankcase allows fresh oxygen to come into contact with the hot oil, some of which is in the form of mist. Such conditions will cause oil to deteriorate and acidity may develop which can corrode certain bearing metals, for example copper-lead bearings. Oils that lack stability will often polymerize to form resinous compounds and to produce heavy carbon deposits in the combustion chamber, thereby retarding transfer of heat to the cooling system. The severe formation of such deposits will result in detonation or pre-ignition. In some engines working under very high loads a very hard deposit from the oil settles in the piston ring grooves, causing the rings to stick and spoiling the compression of the engine. Some impurities in the fuel, particularly sulphur, cause the deposit to be very hard.

The formation of some of these undesirable products in lubricating oil can be delayed by mixing various chemicals with the oil to reduce oxidation. Most tractor engine brands of oil have these additives, and are called inhibited oils.

Oil Viscosity

Engines get hot, and this makes important the property of change of viscosity of oil with temperature. We saw that if the film of oil is too thin it will get pressed out from between the bearing surfaces when the bearing comes under load, and we have seen that if it is too thick, or viscous, the oil will produce friction. This latter point is very important indeed in an engine; an engine having thick oil in its sump is very difficult to rotate by hand. Therefore it is clear that we need an engine oil which does not lose its body quickly with rise of temperature.

This resistance to change of viscosity with temperature is the mark of a high-quality engine oil. The following table shows the results of tests on a first-quality oil and a second-quality oil. The viscosity has been measured at three temperatures, 70°F, 140°F, and 200°F, by recording the length of time in seconds required for a certain quantity of the oil to drip through an aperture of a certain diameter.

Viscosity at	*1st Quality*	*2nd Quality*
70°F	3400 sec	3600 sec
140°F	350 ,,	255 ,,
200°F	106 ,,	75 ,,

From this table it can be seen that the second-quality oil is thicker than the first-quality oil at the temperature at which the engine would be started on a summer's day and therefore the engine would be slightly more difficult to turn when it had the second-quality oil in the sump than when it had the first-quality, but as soon as the engine had become hot enough for the oil to be at 200°F the second-quality oil is very much thinner indeed than the first-quality oil.

The rate at which the viscosity changes with temperature is expressed as the viscosity index. This index is a figure on a scale derived from the relation of the viscosity of the oil measured at two temperatures, usually 140°F and 210°F. Originally the scale ran from zero to 100, the higher the figure the lower being the rate of change of viscosity with temperature, but it is now possible to produce lubricant which comes out to have a viscosity index higher than 100.

Grades of Oil

Most manufacturers agree that any good make or brand of lubricant is suitable for their machines, but they rightly insist that in the matter of grade (that is viscosity, nature of additives, and so on) the farmer will do well to follow closely the recommendation of the men who made the tractor. Indeed the various bearings are designed with some specific lubricant in view, and if a very different oil or grease is used then the machine cannot run well, and wear will take place more quickly than it need.

In engine lubrication it is usual to use a heavier, or thicker, oil in summer than in winter, since both starting and operating temperatures are likely to be higher.

The maker's handbook generally has a chart consisting of a diagrammatic representation of the lubrication points of the tractor, with a legend giving the frequency of replenishment needed at each point and the grade of oil or grease required. Sometimes the actual named grade of lubricant is given. In others some more general description is given such as the S.A.E. number. The letters S.A.E. stand for Society of Automotive Engineers, and are given to a method of denoting viscosity by numbers. For example, the recommendation may say that the engine should have S.A.E. 20 oil in winter and S.A.E. 30 oil in summer. The implement dealer or oil supplier will suggest a lubricant that satisfies the recommendation.

In general, the harder an engine works, the more necessary it is to use a lubricating oil which has additives to combat the various kinds of contamination. For example, some heavy-duty engines are particularly prone to resinous deposits from the condensation of oxy-acids produced by the partial oxidation of lubricating oils. These resinous deposits help to form the undesirable lacquer or varnish which occurs in piston ring grooves. Detergent additives neutralize the oxy-acids before they condense and polymerize to varnish. The detergent also absorbs small particles of any deposit that may be forming, and prevents those small particles from joining to make larger conglomerations, which might be abrasive or might build up into blockages in oil-ways. The small particles can pass harmlessly through the lubrication system. They are small enough to pass through a filter element, so the oil soon becomes dark in colour.

Oils which contain this detergent in considerable quantity are called Heavy Duty (H.D.) oils. In Britain they are sometimes called also Premium oils, but the American Petroleum Institute makes a distinction between Premium oil and Heavy Duty oil, and classifies oils as Regular, Premium and H.D. The Regular oil is a straight mineral oil generally suitable for use in engines working only moderately hard, and Premium oil is an oil having oxidation stability and corrosion prevention properties. Heavy Duty oil has the dispersant character given by a detergent as well as the properties of oxidation stability and corrosion prevention.

Fuels refined from some sources of supply contain rather much sulphur. Nearly all Regular oils, many Premium oils and even some Heavy Duty oils deteriorate rapidly when they are in contact with the products of combustion from fuels containing a large quantity of sulphur. Gums, varnishes and sludge are formed. In some conditions of operation, the products of combustion contain enough sulphur to form corrosive acids which break down the film of oil on the cylinder walls and piston rings. To remove the sulphur from the fuel during the refining would be a very expensive operation and would raise the price of the fuel so that it became too dear for use in an agricultural tractor. It has been found, however, that some of the bad effects of the sulphur can be mitigated if suitable lubricating oil is used in the engine. When the fuel being used is known to have a sulphur content of more than 0·5 per cent a Premium or H.D. oil is desirable.

Since oil is nearly indestructible at ordinary engine temperatures, it becomes as good as new if the contaminating matter can be cleared out of it completely. Several firms reclaim waste oil by re-refining it. The re-refined oil is in one respect better than new, because some of the varnishes are totally deposited during the few hours the oil is used, and they will not be deposited again when the re-refined oil is put back into use.

The re-refining process removes additives, and these have to be replaced if a Premium or H.D. type of oil is desired.

Oil Cleaners

Most tractor engines have gauze strainers placed in the oil flow, and many have also an oil filter with an element which can be cleaned or replaced.

One type of oil filter has a tight pack of absorbent cloth through which the oil passes, and another type has a felt or flannel cloth element arranged over a wire frame to offer a large area of filter surface to the oil. In these filters the element is enclosed in a chamber so that oil flows from the outside to the inside, and impurities taken from the oil collect on the outside surfaces. A spring-loaded by-pass valve can cause the oil to flow around the filter element if it becomes clogged. This valve is fitted between the oil inlet and outlet lines. When the restriction of the filter is so great that the oil inlet pressure becomes larger than the spring pressure of the by-pass valve, oil will pass around the filter and not be filtered.

Another kind of filter uses an extremely fine metal strainer or edge-type metal filter. This consists of a flat wire wrapped edgewise round a cylindrical cage. There is a spacing of 3/1000 in between the edges. At first the element will remove only the larger particles, but when it becomes covered with a filter bed it begins to hold back finer particles. As the bed continues to build up, the filter becomes more efficient in removing fine matter. When the thickness of the material accumulated on the element restricts the flow of oil through it, the element has to be removed and cleaned. A by-pass valve is fitted in the base of the filter and this opens and allows the oil to continue circulating to the engine parts when the filter bed has become too impervious.

Paper filter elements have several times the filtering area of most types of metal element. When the paper element has been used for a specified length of time, it is removed and replaced by a new one.

In another type of filter the by-pass is an essential part of the process of cleaning the oil instead of being just an emergency device. About one-tenth of the oil circulating through the engine is taken through a by-pass circuit of its own, and returned to the sump. Although only one-tenth of the oil is being treated at a time, that one-tenth can be cleaned thoroughly since only a small quantity is passing through the filter. Most of the filters used in by-pass systems have a replaceable cartridge element.

Impurities gathered by a filter from the oil of a new engine are swarf, sand and scale which could do very serious damage to the engine in its first few hours of working. From a run-in engine they consist of metal dust, metal scale from the oil pipes, carbon and varnish-like substances that form the sludge that wears the bearing surfaces, and gums up the oil-ways and valves.

ENGINE COOLING

The combustion of the fuel mixture in the cylinder produces great heat and raises the temperature of the engine. Although the engine runs at its best when it is fairly hot, the temperature must not be high enough to destroy the lubricating quality of the oil. Therefore, the cylinders have to be helped to keep cool.

On the single-cylinder engines usually employed on two-wheeled walking tractors the cylinder is air-cooled. Deep, thin, radial fins are cast on the outside of the cylinder to radiate the heat, and a current of air is blown on to the fins from a fan driven from the engine.

A four-cylinder Ford Diesel engine with overhead valves and pneumatic governor. The drive to the cams can be seen and a piston and gudgeon pin are shown in section. (*See Chapter* 1).

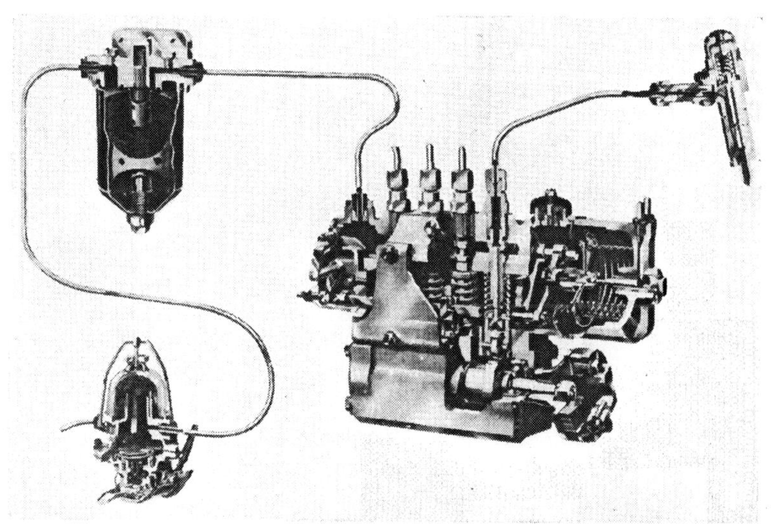

Fuel injection system of a Ford Diesel engine. Lower left, a lift pump to raise the fuel from the tank; upper left, a filter, through which the fuel passes to the injection pump which has a pneumatic governor at the right-hand end. From the injection pump, pipes lead to each of the four injectors. Only one of the injectors is shown, in the top right-hand corner. (*See Chapter* 1).

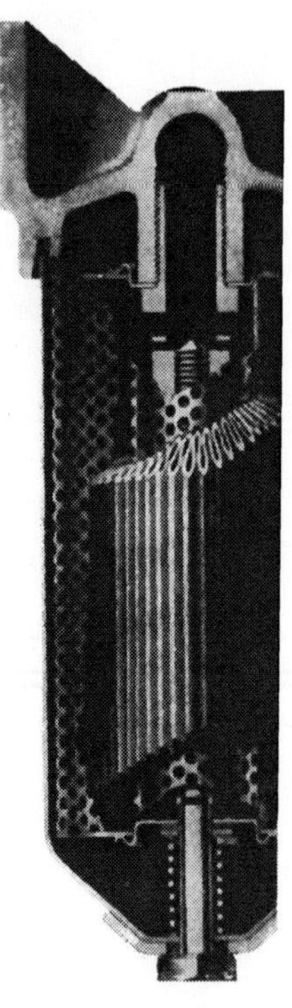

(Above) Mechanical lift pump operated by the engine, for raising fuel to carburettor or injection pump. The glass bowl has been removed and the detachable mushroom-shaped filter can be seen. Under the pump there is a hand lever so that fuel can be pumped when the engine is at rest. (*See Chapter* 1.)

(Right) Purolator Micronic oil filter having plastic impregnated paper folded in such a way that a large surface is in use. (*See Chapter* 1.)

(Below) A starting capsule, containing ether, being inserted into the C.A.V. cold-starting device for Diesel engines. (*See Chapter* 1.)

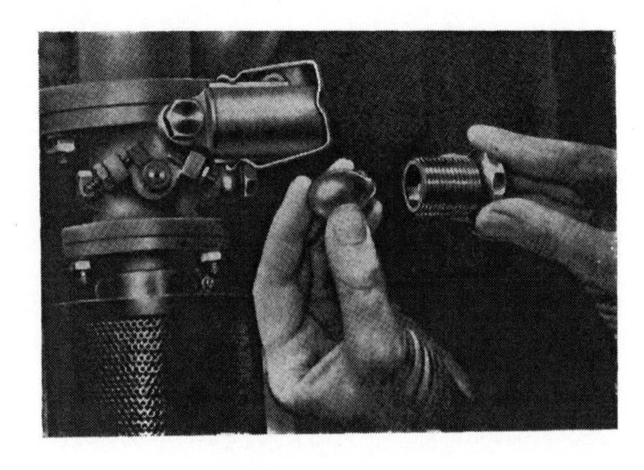

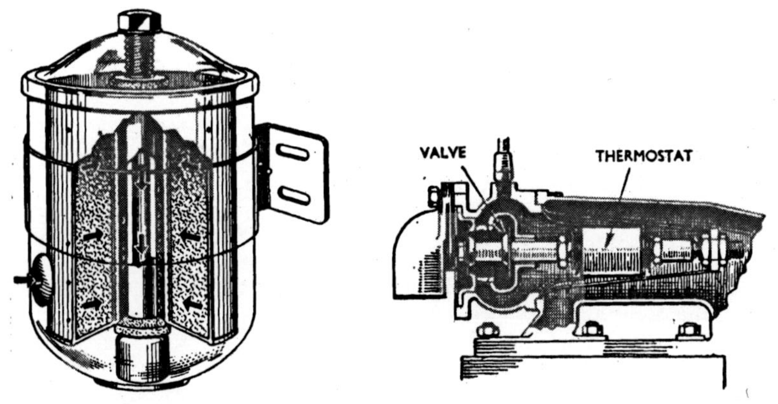

(Left) A Fram cleaner used in the lubricating oil circulation system of an engine. The arrows show the path of the oil through the filter. (Right) A thermostat which operates a valve on a horizontal spindle in the water cooling system of an engine. The valve opens and closes as the thermostat capsule expands or contracts according to changes in the temperature of the water.

Multi-cylinder engines are nearly always water-cooled. A water-jacket surrounds the cylinders. When the water in the jacket becomes hot it rises and passes up a pipe to the top of the radiator. To reach the bottom of the radiator it has to pass through narrow, finned, tubes, or through the cell divisions of a honeycomb, which present a large surface to the air. From the bottom of the radiator the cooled water passes to the bottom of the water jacket and circulates again and again. To hasten the cooling, air is drawn through the radiator by a fan. In some makes of engine a pump is used to speed the circulation of the water.

Pressure Cooling Systems

Some tractors have a gas-tight radiator system. The filling cap is a close fit and there is no overflow pipe. It is worth noting here that when the engine is hot pressure is generated in such a radiator system, and care must be exercised in removing the cap lest hot water spurt over the operator.

Some tractors have also a thermostatically-controlled valve which can cause part of the water to be by-passed without going through the cooling passages of the radiator. This device is arranged so that the water does not circulate completely until the engine is hot. While it is getting hot only a small part of the water is involved. This thermostat is operated by a gas-filled capsule in very much the same way as the control on an egg incubator. The device has one rather serious drawback; in very cold weather it is possible for water, untreated with anti-

D

freeze compounds, to freeze in the part of the radiator temporarily blanked off, even when the engine is running.

ENGINE SPEED GOVERNOR

A tractor engine has no direct throttle control like the accelerator pedal of a motor car. Instead, the opening of the carburettor throttle is controlled by a governor which ensures a constant engine speed even when the load upon the engine varies.

Some governors work by centrifugal force. When the load on the engine is reduced (for example, when the plough is lifted out of work at the end of a bout) the engine begins to go faster, but the increased centrifugal force on the spring-loaded governor weights causes them to move outwards from the spindle upon which they rotate. Their movement outwards actuates a lever which closes the carburettor throttle and reduces the engine speed. On the other hand, if the load on the engine is increased (for example, by the plough encountering a specially stiff patch of soil) the engine tends to slow down, and the governor weights close in nearer to their spindle. This causes the lever to open the throttle and so provide more fuel to the engine to enable it to maintain its speed even now it is more heavily loaded.

The setting of the governor is controllable, so that the engine speed can be regulated from a 'tick-over' to its full speed. The lever for changing the setting is placed within each reach of the driver.

Pneumatic Governors

When a centrifugal governor is used on a Diesel engine the linkage is connected with the regulator scroll of the injector pump, but many Diesel engines have a pneumatic governor instead of a mechanical one. The intake manifold vacuum is employed to control the fuel delivery from the injection pump. The pneumatic governor consists of a throttle control unit in the intake manifold, and a diaphragm governor unit mounted on the injection pump in such a way as to control the movement of the rack. These two units are connected by two pipes.

When the engine is idling, the depression or vacuum in the induction manifold is sufficiently high to cause the spring-loaded diaphragm to be moved against the spring tension. This causes the control rod of the injection pump to move to minimum fuel delivery position. When the engine speed increases the manifold depression gradually drops. This causes the diaphragm to be moved by the spring and the rate of fuel delivery is increased.

An adjustable damping valve is usually fitted to the governor unit to prevent irregular running at idling speeds.

CHAPTER TWO

TRANSMISSION, STEERING AND BRAKES

ALTHOUGH the power of the engine is often used through a belt or power take-off shaft, the chief purpose of a tractor is, as its name implies, to draw a load. To convert engine power to traction a train of gearing to the wheels that run on the soil is needed and, even then, useful traction depends on the grip between wheels and soil.

The internal combustion engine, unlike the steam engine, cannot produce full power at low speed. It is not flexible. Therefore, a change-speed gear box has to be used in the transmission of the tractor so that the engine speed can be kept fairly near its best speed even though it is desired to change the forward speed of the tractor.

From the crankshaft the power generated by the engine is transmitted through a clutch to the change-speed gear box, and thence through right-angle gearing, usually differential, to the driving wheels, or, in a track-laying machine, to the sprockets that drive the track.

In some tractors the final drive to the rear wheels is by roller chains running in an oil bath. Chains or secondary gears can reduce the gear ratio. Therefore the differential axle gears can be made to run at high speed, and so they can be made small and light.

THE CLUTCH

The engine clutch consists of two discs, or two sets of discs, held into contact with each other by powerful springs. One of these two discs is driven by the engine: the other drives the gears of the gear box. Pressing the clutch pedal of the tractor forces these two discs apart, against the pressure of the springs, so that the disc connected to the engine can rotate without driving the disc that is connected to the gear box. When the clutch is held out of engagement the gear shift lever can be used to select one of the several speed ratios even when the engine is running. Then, when the clutch is let into engagement again the drive to the rear wheels or the tracks is complete, and the tractor moves.

This taking up of the drive through the clutch causes much friction between the discs while they are passing from the free position to the full contact position, and the clutch plates have to be lined with a material which has a high friction value but does not burn or deteriorate when it gets hot. Some clutch plates run in oil.

DIFFERENTIAL GEARING

The differential gear allows one of the driving wheels, or one of the tracks, to go faster than the other when the tractor is turning a corner, without the drive to either side being interrupted.

Although it allows one of the back wheels to travel faster than the other when a corner is being turned, the differential gear allows the drive of the engine to be operating on both wheels instead of only one, as would be the case if some simple pawl and ratchet device were used, such as in the land wheels of a mower. In some tractors the differential is in the rear axle; in others it is in an intermediate axle. The intermediate axle system allows a reduction in gear ratio to be achieved between the differential and the driving land wheels. Sometimes the final drive is through a train of reduction gears, in other designs chains are used. The final reduction gear at this stage allows the differential gears to be run at a higher speed and therefore to be made of lighter construction. Another possibility taken advantage of by some makers is that a final reduction gear allows the tractor to be built with greater ground clearance, since the intermediate axle with its differential can be put higher than the centre of the land wheels.

Differential gearing has one disadvantage. If one of the rear wheels should lose its grip on the soil, and spin round helplessly, the other rear wheel, which may be on perfectly firm ground where it could get a grip, will remain still. Therefore, on some tractors a differential lock is provided. When this lock is engaged, the differential action is put out of use, and the axle becomes solid so that the rear wheels are positively connected with one another, and if either of them is on firm ground the tractor can get a grip.

Hydraulic transmission of power from engine to driving wheels can take the place of the clutch, change-speed gearbox and differential gearing, and it can allow the engine to adjust itself automatically to the required load.

GEAR LUBRICATION

Lubricating oils for the gearbox and back axle are usually thicker than engine oils because the oil is not pumped to the bearings and gears or thrown to them by high-speed movement.

The oil for transmission gears must not, however, be too thick. If it is very thick its drag can absorb much power. This will cause heating of the oil, and to some extent the heat will correct the viscosity of the oil by making it more fluid, but this is not a desirable way to achieve the right viscosity. With the correct grade of oil, a well-designed transmission system need never get really hot.

Some designs of transmission gears need an oil with a very high film strength to withstand the very great pressure exerted by one gear face upon another. These special oils, called extreme pressure oils, contain an additive to produce a film tenacious enough to prevent the teeth making metal-to-metal contact and becoming scored. Extreme pressure lubricants should not be mixed with other oils. The other oils may contain constituents which will cause the additives in the extreme pressure oil to come out of solution.

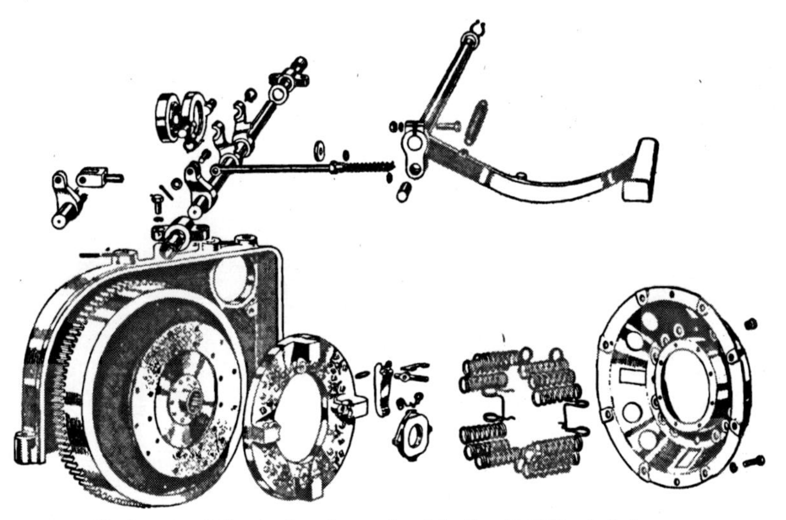

Components of the clutch and control pedal of a Nuffield wheeled tractor.

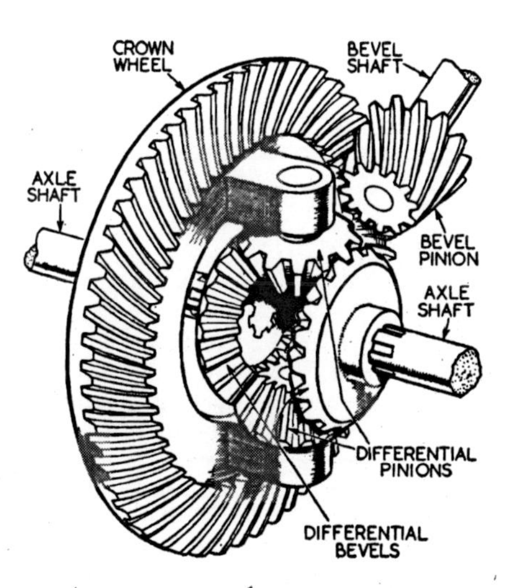

Principal components of a differential gear.

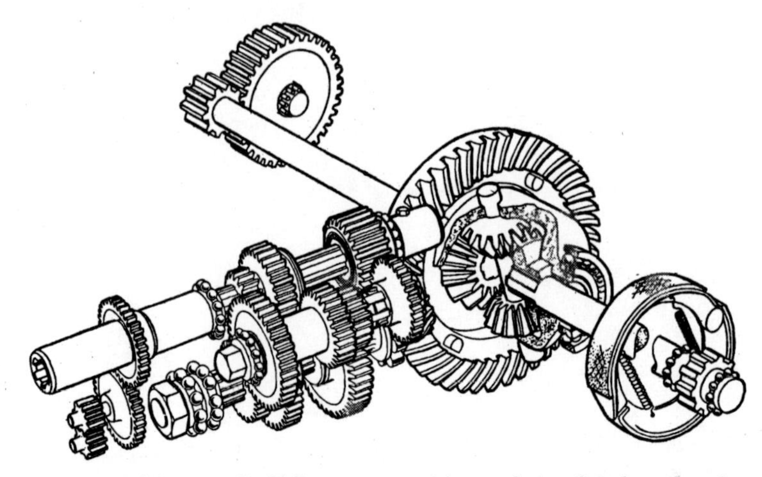

Change-speed gears on a David Brown tractor, giving a selection of six forward speeds. The differential gearing, with brake on differential shaft, is also shown.

Some gearboxes are designed to take quite a light grade of oil, little heavier than engine oil, and thick oil in these gearboxes would cause overheating. In fact, for all components of the tractor, the grade of oil recommended by the tractor designer should be used. The various bearings are designed with some specific lubricant in view, and if a very different oil or grease is used then the machine cannot run well, and wear will take place more quickly than it need. For example, if oil is used in a steering-arm bearing in which the designer intended to have grease, the oil may soon run out; and if grease is used in a steering gearbox made for lubrication by oil, the gears may merely cut a channel for themselves through the grease, and very soon be running in a dry space surrounded by solid grease.

TRACK-LAYING TRACTORS

The amount of power available for the cultivation of the soil is often considerably smaller than the horse-power of the engine. This difference is made up of losses in the transmission gears, the rolling resistance of the driving wheels in the soft surface of the land, and the loss of power due to slip between the driving wheels or track and the soil. The amount of slip varies with the type and condition of the soil, and with several factors in the tractor itself.

The grip of a wheeled tractor depends upon the weight of the tractor and the design of the driving wheels. If they are iron wheels their grip varies with the shape and size and number of the spade lugs attached to the rim of the wheel. If they are pneumatic-tyred wheels their grip varies with the size of the tyre, with the weight upon it, and, to a small extent, with the pressure of air in the tube of the tyre.

There are some soils upon which at some seasons of the year a wheeled tractor cannot obtain enough grip to transmit the power of a large engine, and this led designers, even in the early days, to try to contrive some form of track-laying device. The track, an endless belt made up of steel links shod with strakes, runs over two sprockets. One of these sprockets, the rear one, is driven from the engine. The other is an idler, and its position can be moved to adjust the tension of the track.

Types of Track-layer

In the early track-laying tractors, the pins and bushes forming the joins of the links of the tracks were made of materials which had to be lubricated. The bearing surfaces, however, could not be kept free from soil in which the track has to work, and the grit from the soil mixed with the oil to make a grinding paste which very soon wore out the pins and their bearings. Nowadays the track joint is formed by a carburized and hardened steel pin, which fits closely in a carburized and hardened steel bush. This bearing is run dry, and it lasts remarkably well. The idler sprocket, and the three or four track rollers arranged along the bottom run of the track to carry the weight of the tractor, run on dust-sealed bearings lubricated by oil gun.

Tractors having this kind of track are called crawlers or full track-laying machines. There is, however, a half-track machine which is in the form of a conversion of a wheeled tractor. The driving wheels are removed, and sprockets are fitted on the axle ends. These sprockets drive the tracks. Behind each sprocket is an idler pulley. The front wheels of the tractor are retained, for steering, so the machine becomes a half-track tractor.

The track used on this conversion is different from the track used on conventional track-laying tractors. The links are not free to hinge in both directions; in one direction their movement is restricted so that they lock into one another and form a continuous girder. This girder is not, however, in a straight line; it has a curvature on a radius of about 10 ft so that the track in contact with the ground is imitating a wheel of 20 ft diameter.

When the track links are in contact with the ground they are interlocked, and so there is no movement of the pin joints while the weight of the tractor is on the track, nor indeed do the pins have to take any driving load while the weight of the tractor is on them. This is because the lower run of the track is in compression, not in tension, since the driving sprocket is in front of the idler.

The half-track attachment allows the tractor to be operated in conditions where wheels would not grip at all, and in nearly every condition it permits a higher drawbar pull.

In another kind of half-track attachment, each track consists of six pin-jointed sections. In these sections is a double rail, on which the driving sprocket runs. Between the rails are rollers, which take the drive of the sprocket teeth. The track sections are connected by roller chains which pick up each section after the tractor has rolled over it. This particular kind of half-track has been used successfully on tractors employed to make ditches and lay drains.

Types of Track Link

Articulated pin-jointed tracks usually have spade lugs or grousers on every ground plate, so that contact with the soil surface is made through these. In favourable conditions, about 90 per cent of the weight of the machine can be taken off at the drawbar in pulling the load, but there are conditions of weather and soil surface such that the tractor will dig itself in until the weight is supported on the body of the machine, and the tracks idle round without propelling the machine. The track is fitted with a spud plate to every link, and therefore several of them are in contact with the ground at any time. If the machine is correctly balanced when it is exerting a drawbar pull, all the plates in contact with the soil should carry about an equal proportion of the weight of the machine. Therefore the load imposed on each lug or grouser is not very great, and may not be sufficient to force it into hard soil. The performance of the track usually improves with the degree of penetration of the grousers, when it is operating over a soft surface. If the soil is so hard that the grousers cannot penetrate, the drawbar pull is small, but since the use of tracklayers ought to be confined to soft surfaces, this does not matter.

With a flexible pin-jointed track, the pressure on the ground cannot be quite equally distributed. Pressure is high immediately below the weight-carrying rollers, and low in places between them. This brings a cockling effect in the track which absorbs power and disturbs the soil in a way which lowers its shear strength.

In the driven girder track normally used as a half-track, penetration is to some extent assisted by the fact that the track has a curvature, and also there is no rocking or cockling of the grouser in the soil after penetration has taken place. The drawbar pull obtainable with a machine fitted with the girder type of track can be 130 per cent of the weight imposed on the track.

In some tracks the links are held together by rubber pads which interlock with the edges of the steel links. The rubber-jointed or elastic girder track is intermediate in type between the flexible track and the driven girder track. The elastic girder track is built up on a curvature, and requires the weight of the machine to flatten it in contact with the ground. Thus, the ground pressure is equally distributed under the weight-carrying rollers and elsewhere. The rubber-jointed type of track wears only very slowly, it is noiseless and it can run at high speed.

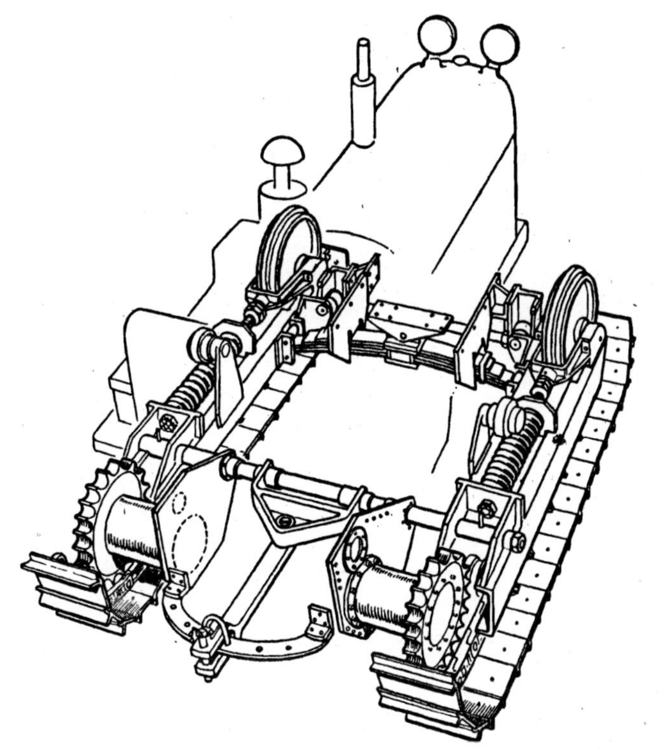

Layout of a David Brown track-laying tractor, showing the articulation of the track frames at the rear and the transverse springing of the frames at the front. At the rear are toothed driving sprockets, while plain guide wheels are at the front.

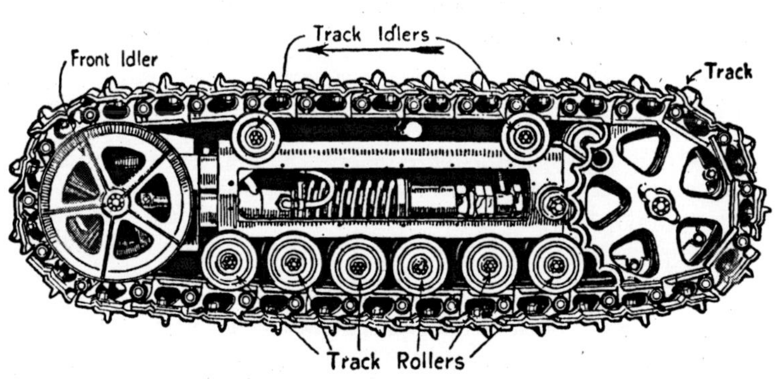

The track assembly of a crawler tractor.

Adhesion is not as good as with the driven girder track, though in some circumstances it is better than with the articulated pin-jointed track.

Track Wear and Adjustment

On all kinds of tracks, one idler sprocket is adjustable so that it acts as a tensioning device. Usually the idler at the end of the track opposite to the driving sprocket can be moved backwards or forwards by means of a screw thread on the track frame. A large coil spring keeps the idler sprocket as far away from the driving sprocket as the screw thread adjustment allows. This spring also acts as a cushion to absorb shocks.

Track frames on full track-laying vehicles are supported at the front through a suspension system which allows independent vertical movement of each track. This is necessary, otherwise the tracks could not both follow the contour of the land when the ground is very uneven. The vertical movement is damped by springs, either coil springs or leaf springs.

In soils which are not very abrasive, tracks have a long life, but they must be kept in good adjustment, to take up the effects of wear.

It is to be noticed that wear of the track has two effects. The first effect is just to make the track or chain too long, and this can be put right by tightening. The other effect is to change both the pitch of the chain and the pitch of the sprocket. If they change by the same amount and in the same direction no harm is done; if they change unequally or in opposite directions, there will come a time when the track will tend to ride up the sprocket teeth.

When tracks wear and become longer, each link comes to have a longer effective length. At the same time the teeth on the driving sprocket are wearing, and it is interesting to enquire whether the combined result of the wear in the track and of the wear in the sprocket decreases or increases the relative pitch. To make this enquiry we have to assume that the pitch circle diameter of the chain wrapped round the arc of the sprocket wheel equals the pitch circle diameter of the sprocket wheel itself, and also that the wear (that is 'extension') of the chain, divided by the number of its links, equals the wear in each tooth of the sprocket wheel. This may not be quite true because (*a*) the rollers which touch the gear teeth can rotate and therefore can distribute the wear around their circumference, (*b*) the track will wear in the link joints as well as in the drive rollers, and the results of these two kinds of wear must be added together when the track is in tension and must be subtracted from each other when the track is in compression, and (*c*) the hardnesses of the metals in contact are not likely to be equal. The assumptions are, however, nearly enough correct.

Now, wear in the chain will have effect in a lateral direction and will directly increase the pitch of the chain links. In that part of the chain on the arc of the sprocket the change in dimension will be circumferential. But when we come to consider the effect of wear on the

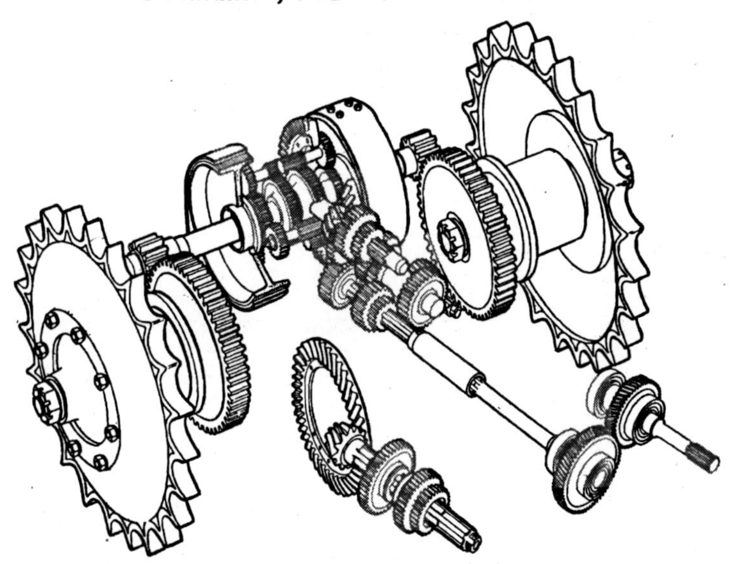

The transmission system of a David Brown track-laying tractor, showing the first reduction gear from the engine, the change-speed gear box, the spur gear epicyclic differential gear, the brake on the high-speed differential axle and the final reduction gears to the sprockets which take the tracks.

sprocket teeth we have to note that, unlike the case of the articulated chain, the wear will remain in the place where it was formed. The direction of the wear will be somewhere between radial and circumferential. Let us resolve it into (*a*) a radial component and (*b*) a circumferential or tangential component. Now the circumferential component will not affect the pitch of the teeth; it will merely move the summit of each tooth a little way round the circle. But the radial wear will decrease the pitch of the teeth by reducing the pitch circle diameter of the sprocket. Since the number of teeth remains the same, a reduction in diameter, and therefore in circumference, will reduce the pitch of the sprocket teeth. Let us suppose that a half of the actual distance of wear is effective in a radial direction. Then let us say that each link, and each tooth, has worn by a distance x. Then, as we have seen above, the effective distance caused in the chain is x for each link, but the effective increase in sprocket pitch caused by the same amount of wear is only $\frac{x}{2} \times \frac{1}{2\pi} = \frac{x}{4\pi}$. So that, for every increment of one unit on the pitch of the sprocket teeth there is an increment of 4π on the pitch of the chain. Therefore if the sprocket and chain start by having 1 : 1 pitch, any wear will produce overpitch.

For most tractors the track adjustment is correct when the track has a sag of 1½ to 2 in at a point half way between idlers and sprocket. This is measured by placing a flat bar along the top of the tracks or by placing the pinch bar on top of the track carrier roller and being able to lift the track 1½ to 2 in above the roller.

To adjust the tracks, remove the covers at the back of the idlers or in front of the sprocket. Loosen the clamp nuts and adjust the track to the correct tension, drive the tractor back and forward to equalize adjustment, then recheck tension.

If the track becomes longer and longer due to the increasing wear in the pins and bushes the adjusting bolt has to be turned more and more to take up the slack. This cannot, of course, go on for ever and in most instruction books there will be found an indication of how much adjustment can be made before the track must be overhauled so that the slack in the hinges can be taken up by new pins or bushes or by turning the bushes into another position.

Although the pins must not be lubricated it is most important that the bearings of the track roller frame should be lubricated regularly with semi-fluid lubricant at every 20 hours of use. The track carrier rollers and the outer bearings of the roller frame must also have the lubricant replenished every 20 hours. Semi-fluid lubricant is correct in all these places and also for the large bearings which carry the idler wheels.

Rubber-jointed Tracks

We have seen that joints having pins and bushes made of carburized steel last remarkably well, but some wear of these hinges is inevitable when the tractor is used in abrasive soils. With much wear of the joints, elongation of the track becomes serious. Moreover, in order to reduce the friction of the pin joints it is necessary to run the tracks with the minimum of tension, and this condition permits cockling between the weight-carrying roller. Therefore, a compromise must be made. It is possible to put so much tension on the tracks that the friction makes the tractor immovable.

Because freely jointed tracks can go into reverse curvature, it is difficult to provide even pressure on the ground over the whole area with which the tracks come into contact. The pressure transferred to the ground at the point immediately beneath the rollers which carry the weight of the machine is greater than it is elsewhere along the track. This is true also of rubber belt tracks.

What is wanted in a full-track tractor is that the track shall remain a flat rigid platform over which the weight-carrying rollers can run with low rolling resistance. In rubber-jointed tracks, the ground plates are held together, but at the same time separated from each other, by rubber blocks. The hinge movement between adjacent ground plates takes place by a distortion of the rubber. There is considerable girder effect in rubber-jointed tracks, and this helps towards uniform and continuous contact with the ground. The track, when not supporting a load, has considerable curvature, and resists being put into reverse curvature.

Rubber-jointed tracks provide a slightly cushioned drive, and this is probably advantageous to the transmission system.

WHEEL AND TRACK GRIP

Very often the performance of the tractor is limited by the soil grip rather than by the power of the engine. Tractors more often stall from wheel slip than from engine overload. Even when the wheels or tracks are gripping, the rolling resistance caused by the strakes of spade lugs as they enter or leave the soil can lose more power than is lost in the whole of the transmission gearing from engine to back axle.

Soils are rarely in the right condition to allow wheels or tracks to grip by friction alone. Sometimes they are too dry and powdery. More often they are too moist. Therefore cleats or spade lugs on steel wheels and tracks, or deep tread bars on rubber wheels, are needed to help the grip by imprinting a rack along the surface of the soil. When the lug has penetrated into the soil, the rotation of the wheel or track presses the lug against the wall of the hole into which it has pushed itself. Grip then depends upon the resistance of the soil to shear, and this varies very much indeed with the kind and condition of soil.

Occasionally muddy soil packs around the spade lugs and reduces their effective length. Open or skeleton-type driving wheels have been used to do away with this difficulty. These wheels have no broad rim; they consist merely of a disc with lugs fitted at the circumference and so they give less lodgment to mud.

The grip of wheels and the rolling resistance they offer in their passage over the soil vary with the size and shape and number of the lugs or bars, the overall diameter of the wheels, and the weight on them.

On most full track-laying tractors the diameter of the driving sprocket is smaller than would be the diameter of the driving wheel on a wheeled tractor of similar power. The track assemblies themselves are heavy and compact, and the whole construction can be arranged so as to keep the centre of gravity of the machine low. This low centre of gravity gives a feeling of safety to the driver when he is turning the outfit on a steep headland, or is doing sideling work.

The half-track machine has a larger driving sprocket, but the idler is small and near the ground, and the machine has quite good stability. However, it should be noted that the large diameter of the driving sprockets is not inherent in the half-track principle. It is used on the present representatives chiefly because the tractors were designed primarily as wheeled tractors, and if the diameter of the driving sprocket were made very different from that of the driving-wheel it replaces, the forward speed of the tractor in its various gears would become unsuitable.

The upkeep cost of track-layers is higher than that of wheeled tractors largely because of wear in the pin-joined tracks and in the sprocket wheels that drive the tracks. The amount of wear in these places depends very largely upon the kind of soil in which the track has to work; it is particularly heavy in some sandy soils.

Rolling Resistance

Probably one of the reasons why tracks have, in general, a lower rolling resistance than spade lugged wheels is that the cleats or grousers do not need to be so large on tracks and they are subject to less movement while they are actually embedded in the soil. A long spade lug on a small-diameter wheel must be swinging quite fast all the time it is in the ground, but the lugs on a flexible track are probably nearly stationary for most of their stay in the soil.

Methods of measuring rolling resistance with any accuracy are difficult. Slippage can be measured directly and easily. We can calculate how far a tractor would travel for a certain number of revolutions of its driving wheels or driving sprockets if there were perfect ground grip and no slip whatever; and we can measure how far it actually does travel. The difference between these two distances represents loss of forward speed due to slip. But measuring the loss of energy due to rolling resistance presents many problems. We can tow the tractor and measure its resistance to being towed, but we have no assurance that the resistance of the wheels or tracks when they are being pulled over the soil is the same as their resistance would be when they are doing the work of transmitting power. However, though we cannot directly measure rolling resistance in the driving tracks or wheels, we can deduce it if we can first measure the overall efficiency of the track or wheel.

In the case of wheels, one of the factors very much affecting their performance is the weight upon them. The weight upon the driving wheels when the tractor is at work may be very different from the weight upon them when the tractor is at rest. The pull of the implement can cause weight to be transferred from the front axle to the back axle. This transfer of weight can be used, by judicious adjustment of the height of the drawbar hitch, to help the wheel grip. Also, it can be used experimentally to measure the efficiency of the driving wheels, and indirectly, to estimate their rolling resistance.

The overall efficiency of a driving wheel or track is the ratio between the energy available from it as drawbar pull, and the energy that was put into it by the mechanism of the tractor. In the case of a wheel the input of energy can be calculated from measurements of the reaction tending to lift the front of the tractor off the ground when a load is being pulled at the drawbar. Once the overall efficiency of the wheel has been determined, the rolling resistance can be estimated by assuming that all the loss of energy is due to slip and rolling resistance.

Much work has been done on factors affecting the rolling resistance

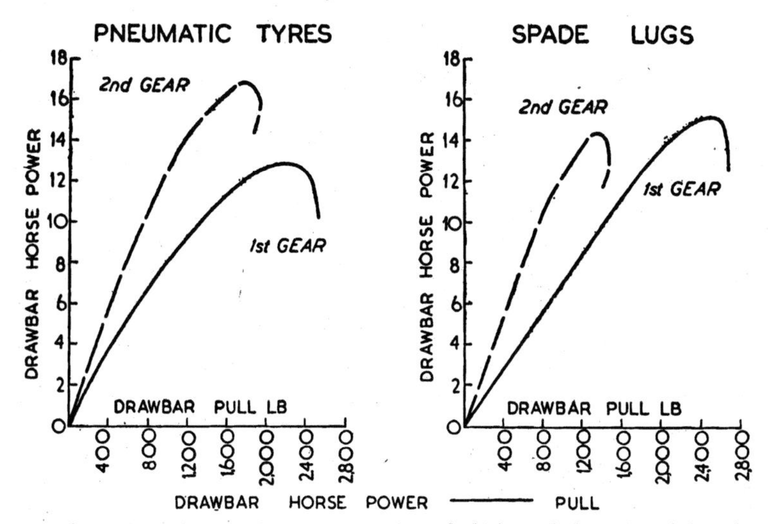

Difference in performance between pneumatic-tyred driving wheels and spade-lugged wheels on tractors. Soil and other conditions were kept the same. Pneumatic tyres are more efficient in second gear but spade-lugged wheels are more efficient in first gear.

of pneumatic tyres when the wheels are driven wheels, and are being drawn over the soil. This work has been done in connection with the development of tyres for carts and implements, and although its findings cannot be used directly for application to driving wheels, they can help in a consideration of the best kind of wheels to use on the front of wheeled tractors and half-track tractors.

It is likely that in some soil conditions the present front wheels can offer quite high resistance to the forward movement of the outfit. Indeed, it may be that one of the reasons four-wheel drive machines are good in some soft soil conditions is that the front wheels can pull themselves out of the depression they have made for themselves, instead of having to be pushed out by work transmitted through the rear wheels.

For most experiments on traction it has been found better to use dynamometer loading cars rather than actual agricultural implements. The size of the load is then always under control, and changes in its value can be made in small, accurately gauged steps. But this way of loading causes the tractor to be working in conditions rather different from those it would encounter if it were ploughing, cultivating or harvesting. For one thing, if a large number of runs has to be made, and great stretches of land are not available, the outfit has to travel repeatedly over the same strip, and the soil may acquire a different character after each trip. Moreover, ploughing, which after all is the biggest job of most tractors, means that a small, wheeled tractor is put to work in a very special condition, one difficult to imitate when a loading car is used.

The land-side driving wheel is travelling on the ordinary field surface, but the furrow-side driving wheel is travelling on a platform of newly cut soil. This newly-cut face usually provides a much better surface for the grip of wheels than does the ordinary field surface. The tilt of the tractor also takes weight off the land-side wheel, and transfers it to the furrow-side wheel. A medium-sized tractor may have a weight of 14 cwt at each back wheel when the tractor is on an even keel. If, however, the tractor is tilted so that one side is 8 in lower than the other, as it may be when ploughing, the weights at the two back wheels can change to about 11 and 17 cwt respectively.

That particular condition, however, is confined to small and medium-sized tractors. High-powered tractors usually plough sufficiently wide to allow the tractor to ride wholly on the land, without necessity to offset the plough to cause side draught. Indeed, in the case of crawlers, whether track-layers or half-tracks, it is most undesirable that the tractor should have one track in the furrow; the tilt causes overmuch side thrust on the tracks.

In the case of wheels much research work has been done. It is now known that a large overall diameter for driving-wheel pneumatic tyres is desirable for work in every kind and condition of soil, and that in nearly all conditions of soil a narrow-section tyre grips just as well as a wide-section tyre.

STEERING GEAR

Many different methods of operating front wheel steering are employed for tractors. A usual method is worm-and-wheel. The worm is on the steering wheel shaft, and the worm-wheel is on the steering arm. The steering arm moves the link rod and imparts movement to the two stub axles which carry the bearing for the front wheels. The steering arm moves through only about a quarter of a circle, and therefore only one quarter-sector of the wheel is ever engaged. Thus the teeth on a quarter of the sector of the wheel become worn while three-quarters get no wear at all.

In some tractors only a sector of wheel is provided; in others a whole wheel is fitted, and an unworn sector can then be brought into use merely by turning the worm wheel to a new place on the spindle of the steering arm. In makes which have only one sector provided, other means have to be used for taking up the effects of wear. Usually the mesh of the teeth can be adjusted by moving the position of an eccentric housing which holds the bearings of the worm wheel.

Another system of steering, called cam-and-lever steering, has a quadrant studded with four or more tapered pins which engage in a cam which is in the form of the thread of a coarse screw on the steering wheel shaft. It can be adjusted to take up wear by removing shims to

A six-speed Ford gearbox in which there are three ratios to be selected by the main gear lever (top, left) and a primary high-low selection lever (top, centre). At the right-hand side of the picture is the engine flywheel with the clutch and with the ring gear for the starter pinion. In the bottom of the gearbox is the drive for the power take-off shaft. (See Chapter 2.)

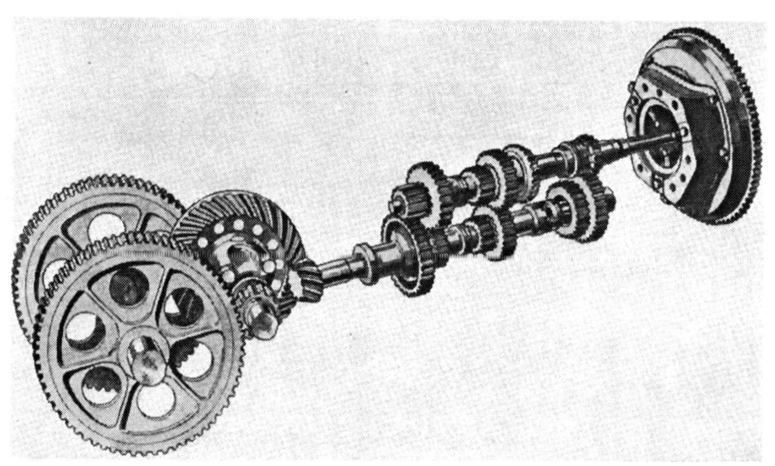

Transmission system of a Fordson wheeled tractor, showing the engine flywheel and clutch, the change-speed gears, the differential gearing and the final reduction gears. (See Chapter 2).

An experimental tractor, designed by the National Institute of Agricultural Engineering, with hydraulic transmission in place of gearbox and differential rear axle. No clutch and change-speed gears are required, and one control lever is all that is necessary to start, stop and reverse and also to change speed. The engine automatically adjusts itself to the load. (*See Chapter* 2.)

The pump, driven from Fordson Major power unit, which provides the oil pressure for the N.I.A.E. hydraulic transmission. (*See Chapter* 2.)

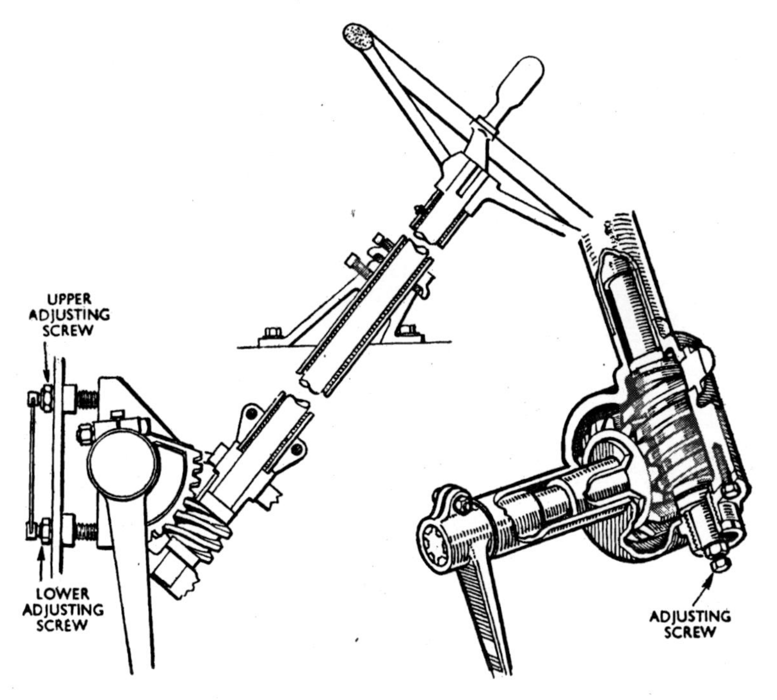

Two forms of steering gear, showing method of adjustment. Left, worm-and-sector; right, worm and full circle wheel.

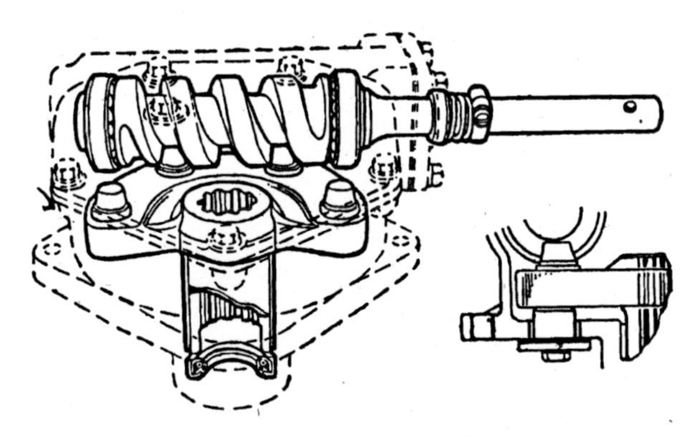

Ross cam-and-lever steering. The lever is in the form of a quadrant with four tapered studs. To compensate for wear, shims can be removed to alter the position of the thrust pad which keeps cam and studs in contact.

E

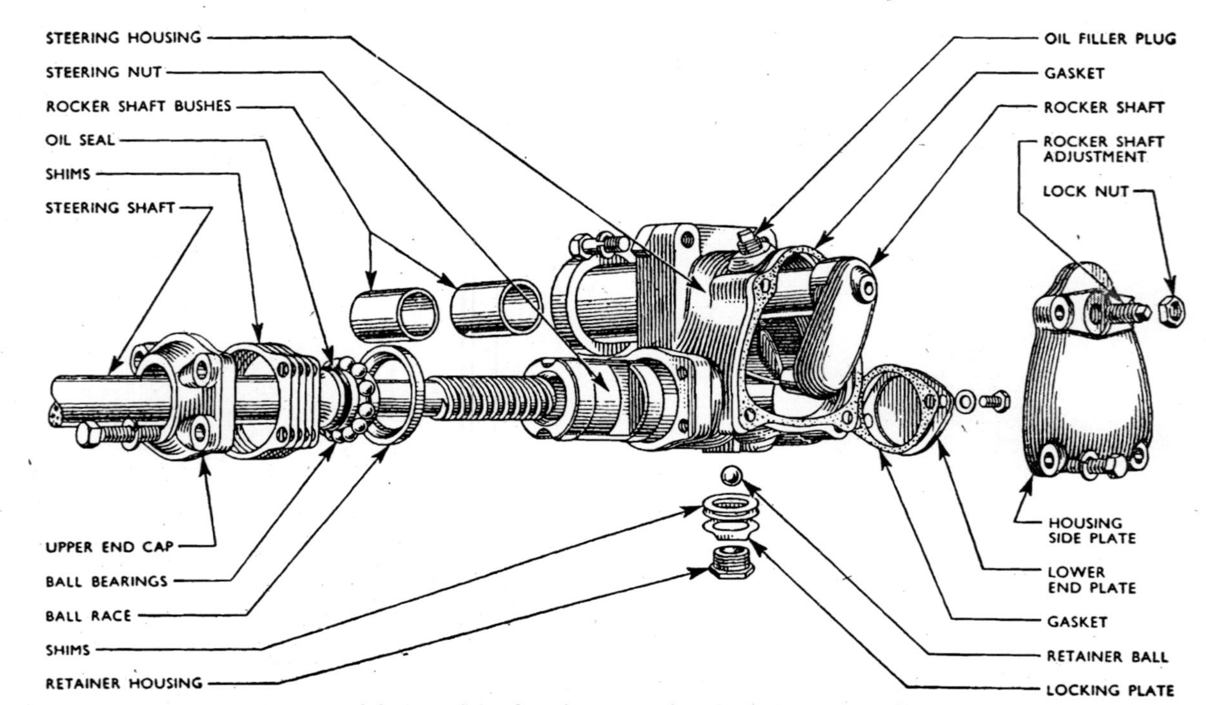

The parts of the nut-and-thread steering gear used on the Fordson Major tractor.

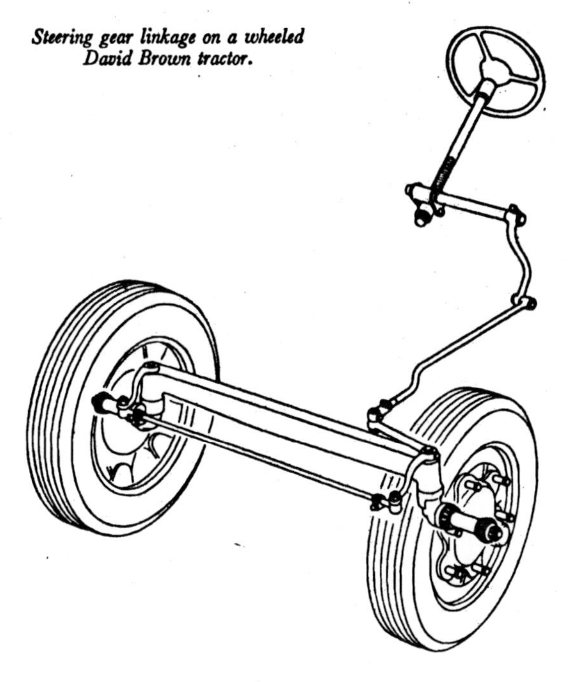
Steering gear linkage on a wheeled David Brown tractor.

alter the position of a thrust pad which keeps cam and studs in contact. In a third popular type of steering, the threaded extension of the steering wheel shaft is caused to move a nut along the thread. A rocker shaft, actuated by the movement of this nut, is linked with the stub axles.

Steering by Brake

A track-laying tractor has to be turned by slowing down one or other of the tracks. In some track-layers this is done by having separate driving clutches, one to each track. In tractors of this type no differential gearing is needed. In other makes, the slowing of the track is accomplished by a brake, and, through the differential gearing, the other track is driven at a correspondingly higher speed.

This differential-brake steering can be used, too, to help the steering of a four-wheeled tractor, if each rear wheel has an independently operated brake. Sometimes, when it is pulling a heavy load, a tractor will not answer to full lock steering of the front wheels, because the front wheels skid over the soil. The turn is greatly helped by braking whichever of the rear wheels is on the inside of the curve.

Steering by interfering with the equality of the driving effort to the wheels is particularly well used in some makes of two-wheeled walking tractors. These small tractors, which have directly attached implements, are steered by handles in the same way as is a horse plough or a horse hoe, but, to help the operator to make the sharp turns necessary at headlands, one of the two wheels can be braked or de-clutched. This device, when the operator has by practice become familiar with its application, can take much of the hard work out of handling a heavy two-wheeled tractor.

In large track-laying tractors the steering is carefully contrived to give accurate and sensitive control without excessive wear on brakes, clutches or gearing, and to provide as full traction as possible when the tractor is making a turn.

In models that steer by separate driving clutches, brakes are usually incorporated as well as clutches. The first part of the travel of the steering lever releases the clutch; further travel applies a brake to the track. Two steering levers are arranged one on either side of the driver. When he pulls the left-hand one the drive to the left-hand track is disconnected, and since the right-hand one continues to be driven the tractor will turn towards the left. If the tractor is pulling a heavy load the turn will be made sharply, but if the tractor is running unloaded, the turn will be too gradual and the driver will have to pull the lever harder and so apply the brake on the track. This ability to brake as well as de-clutch is particularly important when the tractor is travelling down a hill steep enough for the engine to be retarding the movement of the tractor. Releasing the clutch on the left-hand track can make the tractor turn to the right instead of to the left if the right-hand track continues to be held back by the engine. Pulling the lever back farther so as to apply the brake corrects this.

Differential steering, by braking one track, changes the effective gear ratio between the engine and the track. With simple differential gearing, applying the steering brake stops one track completely and causes the other track to go twice as fast. To correct this, some heavy track-layers are fitted with a compensated gear, called a controlled differential, and this prevents excessive loss of pull when the tractor is making a turn, because it keeps the braked track in motion, running at half the speed of the other track and contributing to the pull.

BRAKES

Internal-expanding brakes are in general use in tractors. A pull on the brake rod turns an actuating cam into a position such that it forces the brake shoes apart and presses their fabric-covered surface against the brake drum. The tendency of the brake shoes to turn with the drum is resisted by the torque plate to which the shoes are anchored.

On some wheeled tractors the brakes work directly on the rear wheels. On others, they are at some point in the transmission where the shafts are running at a higher speed than the driving wheels. For example, in tractors having a final reduction gear or chain after the differential gearing, the brakes are often on the differential shaft.

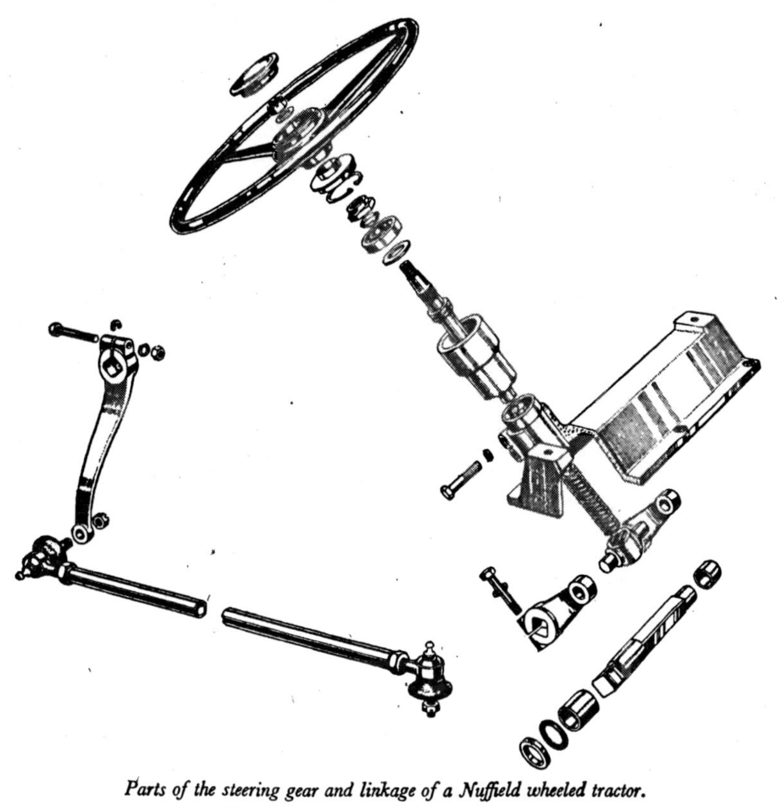

Parts of the steering gear and linkage of a Nuffield wheeled tractor.

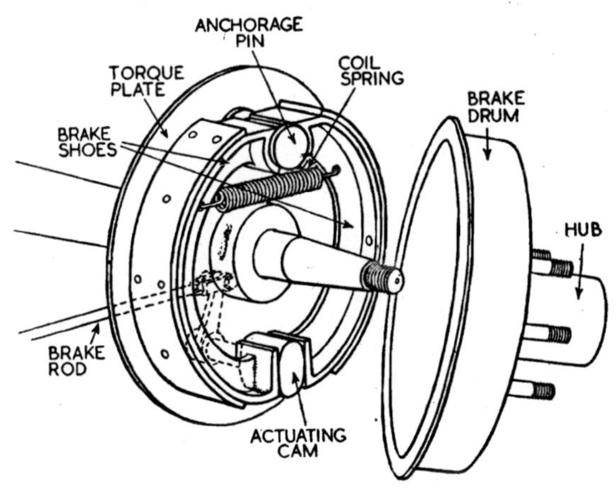

One kind of brake in use on tractors. A pull on the brake rod turns the actuating cam into a position such that it forces the brake shoes apart and presses their fabric-covered surface against the brake drum.

Because the drums are revolving at a higher speed, the brakes can be made smaller than if they were on the driving wheels, and yet can still have as much retarding power. They work less smoothly, however, because their stopping action has to be transmitted through gears or chains. Moreover, if the final drive should fail, the brakes become inoperative, and this might be serious, especially when the tractor is used on the roads.

FOUR-WHEEL DRIVE

Some tractors have four-wheel drive. That is to say, the front wheels as well as the rear wheels receive power from the engine. This arrangement calls for a differential drive to wheels which can be moved on a stub axle for steering. This is not easy, and the presence of the front differential casing can restrict tilt of tractor when the tractor is ploughing with one wheel in the furrow, because it has been found more difficult to devise a high-clearance power-driven front axle than a high-clearance power-driven rear axle.

Much development and research on four-wheel drive tractors has been carried out. Tractors of equal weight, track width, height of drawbar, height of centre of gravity and with the same tyre dimensions have been subjected to comparative tests. Tractors with front-wheel drive, tractors with rear-wheel drive with and without differential, four-wheel drive tractors with all wheels of equal size and with weight concentrated on front axle, and four-wheel drive tractors with conventional wheel diameters and weight distribution, all have been subject to trial. Measurements of drawbar pull have been made with these tractors on level ground at 10 and 35 per cent slip, as well as on ploughing and on the flat with variations between the slip coefficient of the individual driving wheels; the steering, manoeuvrability and behaviour on gradients were also compared. It was found that a higher drawbar pull could be obtained from four-wheel drive tractors under nearly all conditions and that working across the slope was easier. With the same lock, the outer turning radius of four-wheel drive tractors on dry, firm ground and without a load, is greater than that of conventional tractors, but it is smaller on rough ground and with a load. If, however, it is possible to switch off the drive to the front axle, this type of tractor is at no disadvantage on road transport.

A four-wheel drive tractor in which the weight is concentrated on the rear axle cannot show to advantage. Experiments have been made in which an implement was hitched alternatively to the rear drawbar and to a hitch between the two axles in front of the estimated centre of gravity. Several farm operations were carried out on hilly land of heavy clay and the best results were in all cases obtained with the optional front-wheel drive engaged and with the implement hitched at the forward hitch point. By this means the weight is about equally distributed on all four wheels and slip as well as soil compaction are kept at a minimum. The same conditions can be achieved with a four-wheel drive tractor which has the weight concentrated on the front axle and the implement hitched at the usual drawbar at the rear.

Some four-wheel drive timber-hauling tractors are built with all four wheels steerable. The effect of steering the rear wheels as well as the front is to reduce the turning radius for a given wheel base without unduly increasing the maximum angle of articulation of the wheels. The drive to the wheels is through propeller shafts and universal joints situated immediately above the pivot point of the axles and followed by final spur reduction beyond the universal joints. The rear steering can be locked for road use where a tight steering radius is not required, and locks are also provided for the differentials.

In some road transport tractors steering is confined to the articulated front wheels. The drive is taken from a centrally situated gearbox through differentials to propeller shafts and through universal joints direct to the four wheels. Another timber-hauling and industrial tractor has only the front wheels articulated for steering and the drive to these is by cross-wise propeller shafts and universal drive. This machine has an internal gear final drive which has the advantage of compactness and the possibility of a high reduction without excessively wide centres. This tractor employs three differentials which are of a special design incorporating worms and wheels in place of bevels or spurs. Two German examples have the steering confined to the front wheels, and both have suspension systems incorporating a centre pivoted front axle which ensures an even load on each wheel no matter how uneven the ground may be. Another German machine employs centre pivoted axles both front and rear controlled by coiled springs and hydraulic shock absorbers.

An American design has four rigidly mounted wheels, as close together in a longitudinal direction as can be conveniently achieved while ensuring proper clearance between the tyres. All four wheels are positively driven by spur gearing and the steering is accomplished by making use of the clutch and brake mechanism as used on track-layers. With the laterally rigid four-wheel machine, since there is no angular deflection of the front wheels, there is no curve to which the wheels can form a common tangent, and skidding of the tyre tread is necessary when the tractor turns a corner.

COMPARISONS OF WHEELED AND CRAWLER TRACTORS

From tests made in Britain we can deduce a comparison of the performance of four-wheel drive, two-wheel drive and tracks, though the surfaces over which the tractors travelled were not field conditions.

The British Standards Report No. BS/N.I.A.E./54/11 describes a test of the County four-wheel drive tractor. This machine has a Fordson Diesel engine unit, is fitted with 13–24 pneumatic tyres on its front wheels as well as at the rear, and is steered by clutch and brake instead of by front axle. In the drawbar tests, a maximum operating horsepower of 34·3 was attained at 4·29 m.p.h. with a pull of 3000 lb, and 30·6 at 1·63 m.p.h. with a pull of 7050 lb. The best fuel consumption was 16·4 drawbar horsepower hours for a gallon of fuel.

The four-wheel drive tractor tested happens to have the same size and kind of engine as that fitted in a two-wheel drive tractor and a crawler tractor for which nearly similar tests have been made. Figures are therefore available which help in an assessment of the usefulness of four-wheel drive mechanisms in tractors, though since the British Standards drawbar test of pneumatic-tyred tractors was carried out only on a tarmacadam surface, the figures give no certain guide to the performance to be expected from the four-wheel drive tractor when it is working in adverse agricultural conditions.

A Fordson Diesel two-wheel drive tractor with 14–30 rear tyres was the subject of an N.I.A.E. test reported as Test No. 96; and the County Crawler tractor with Fordson Diesel engine was tested and reported upon as Test No. 88.

The forward speeds in the various gears at the rated engine speed of 1600 r.p.m. are a little different in the three tractors, but nevertheless an interesting comparison of power outputs can be made. The operating maximum drawbar horsepower recorded for the two-wheel drive tractor was 31·4, for the crawler tractor 33·3, and for the four-wheel drive tractor 34·3. The two-wheel drive tractor on tarmacadam pulled 4700 lb at 1·94 m.p.h.; the crawler on dry grassland pulled 9500 lb at 1·24 m.p.h., and the four-wheel drive tractor pulled 7050 lb at 1·63 m.p.h. in the B.S.I. test. The maximum sustained pull recorded for the two-wheel drive tractor was 5350 lb, for the crawler tractor 10500 lb, and for the four-wheel drive tractor 7750 lb.

The best fuel consumption recorded for the two-wheel drive tractor was 14·8 drawbar horsepower hours per gallon at a pull of 3600 lb, and this was equalled by the crawler tractor although this was at 6700 lb pull. The four-wheel drive tractor gave 16·4 drawbar horsepower hours per gallon; the range of pulls over which the consumption was within 10 per cent of this figure being 2850–6100 lb.

In the four-wheel drive tractor the tyres were 75 per cent water-ballasted. The weight of the tractor and operator was 9240 lb, and the weight on the rear wheels was 4510 lb. The two-wheel drive tractor had its rear tyres 95 per cent water-ballasted and it was also fitted with wheel weights. The total weight of the tractor and operator was 7420 lb and the weight on the rear wheels was 5320 lb. The weight of the crawler tractor with operator was 8395 lb.

In the conditions of the tests described in these three reports, four-wheel drive seemed to give a better performance than two-wheel drive, though the high pull of 10500 lb recorded for the crawler on dry grassland was not approached by the four-wheel drive on the B.S.I. test surface, in spite of the heavier weight of the four-wheel drive tractor.

CHAPTER THREE

FUELS FOR TRACTORS

THE fuels at present in general use for tractors are petrol, Diesel oil, paraffin and liquid petroleum gas. The Diesel oil category includes grades of gas oil which will fire in compression ignition engines, and the paraffin category includes products ranging from lamp oil to the distillate used in America, and the vaporizing oil used in Britain.

Important qualities of any liquid fuel are its initial boiling point and its end point. The initial boiling point of fuel is the temperature at which it first starts to boil and change into a vapour. The end point is the temperature to which it must be heated in order to vaporize every portion of it. Pure water boils at 212° F and will completely boil away at this temperature. Petroleum fuels, however, differ from water in that only a small portion will boil off at any given temperature. It is to be noticed that in order to boil off and vaporize any petroleum fuel, the fuel must be heated to a temperature considerable higher than the temperature at which it first starts to boil and vaporize.

The temperature of vaporization of any fuel is important because a fuel will not mix with air and burn completely in an engine unless it is thoroughly vaporized. For example, referring to the following table which gives end points as well as initial boiling points, petrol starts to vaporize at about 100°F but must be heated to about 400°F to vaporize completely. Hence, we see that petrol will partly vaporize at ordinary air temperatures and an engine can be started on it without difficulty.

Fuel	*Initial Boiling Point °F*	*End Point °F*
Petrol . . .	95 to 105	350 to 435
Vaporizing oil . .	200 to 350	440 to 520
Diesel fuel . . .	380 to 460	575 to 725

However, until the engine is thoroughly warmed up, even petrol is not completely burned and used up as it should be. The table shows that vaporizing oil must be heated to 200° F or higher before vaporization starts, and complete vaporization requires a temperature of 500° F or higher. This explains why a carburettor-type engine cannot be started on vaporizing oil but must be started on petrol and warmed up first. Furthermore, a higher engine temperature must be maintained to vaporize the fuel completely and convert it into power. This additional heat for distillate-burning tractors is obtained by maintaining a higher cooling water temperature and by hot-spot or heated manifolds. The table explains also the cause of oil dilution in vaporizing oil engines. If a high enough temperature is not maintained to vaporize or burn the vaporizing oil, the unvaporized portion works its way down into the crankcase and dilutes the oil. In a Diesel engine, the temperature of combustion is very high, and there is less dilution than might be expected from the high boiling point of the fuel.

Oxygen is necessary in the combustion or burning process. With engines, this oxygen is obtained from the air; that is, air is drawn through the carburettor and mixes with the vaporized fuel in a certain proportion to form a mixture which burns rapidly and expands. The expansion can be used to generate power. Oxygen is the important element supplied by the air. Air consists of a mixture of about 22 per cent oxygen and 78 per cent nitrogen, but the nitrogen is inert and goes through the engine unchanged.

All petroleum fuels contain carbon and hydrogen and are called hydrocarbons. For our investigation here, the hydrocarbon can be represented as CH_4 and the combustion process can be illustrated by the equation: $CH_4 + 2O_2 + \text{ignition} = CO_2 + 2H_2O$. The oxygen derived from the air, O_2, is mixed with the vaporized fuel and ignited by a spark. A very rapid burning process occurs which results in a chemical change in the materials. The by-products of this combustion or explosion are carbon dioxide (CO_2) and water (H_2O). In other words, if the carburettor of a tractor engine is correctly adjusted, the exhaust materials should consist only of carbon dioxide and water. This would be a perfect fuel mixture, and if we know the weights of the various elements, carbon, hydrogen and oxygen, the exact proportions of these materials can be computed. Such a computation will show that for every pound of fuel burned by an engine, about 15 lb of air must pass through the carburettor. If the exhaust residue consists of only carbon dioxide and steam, no visible material comes from the exhaust when the engine and exhaust pipes are hot. However, water in the form of steam is frequently visible on a cold day when an engine has just been started.

Incorrect mixtures of fuel and air cannot give perfect combustion. A rich mixture, containing too much fuel and too little air, results in sluggish engine operation, lack of power and wasted fuel. Poisonous carbon monoxide is produced when rich fuel mixtures are employed. A weak fuel mixture, with too little fuel and an excess of air, brings uneven running and overheating of the engine.

FUEL ENERGY VALUES

All petroleum fuels have heat energy available to be converted into power in an engine. If the chemical composition of the fuel is known, the number of heat units available in a given quantity of the fuel can be calculated exactly. A comparison of this figure with the actual

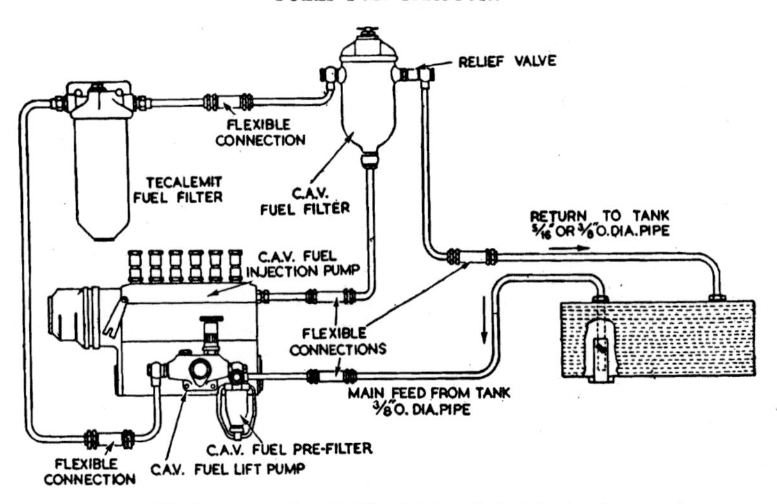

The fuel system of the Perkins P6 six-cylinder Diesel engine.

recorded energy provided by the engine gives data for calculating the thermal efficiency of engines. The energy available in a pound of petrol or paraffin or Diesel oil is about the same, but it must be remembered that petrol is much lighter than the other two.

The burning of the carbon proceeds in two stages, most of the heat being set free at the second stage. The burning of the hydrogen is achieved in one stage. In terms of the amount of heat released, the two stages of burning the carbon can be set out as:

(1) Carbon + oxygen = carbon monoxide + 4500 B.Th.U.
(2) Carbon monoxide + oxygen = carbon dioxide + 10,000 B.Th.U.

The burning of the hydrogen yields far more heat, and the single stage of its combustion can be set out as:

Hydrogen + oxygen = water + 62,500 B.Th.U.

In some circumstances a form of pre-ignition, known as detonation, can occur when the last part of the charge is burning in a spark ignition engine. It produces a high-pitched noise and it causes a falling-off in power and overheating.

In correct combustion, the flame front advances steadily across the cylinder head until it has covered the whole space of the combustion chamber. In a knocking combustion, the whole of the unburnt charge remains ahead of the flame front and explodes just before the flame reaches it. The explosion sets up vibration in the gases and the material of the cylinder, and causes the high-pitched noise. In extreme cases, the explosion can cause fracture of the head of the piston or can spoil connecting-rod bearings. The vibration set up in the gas scrubs the hot gases against the walls, and there is greater heat transfer to the cylinder head and the piston. This can cause piston ring sticking by carbon and gum. Detonating combustion tends to wipe the lubricating oil film off the cylinder walls.

Detonation is promoted by use of fuel of low anti-knock value. It can be caused also by overheating of the fuel charge, or by too high an induction manifold pressure.

Considerable volumes of carbon dioxide and water are produced by the operation of the engine. Each gallon of petroleum fuel burned produces about a gallon of water in the exhaust gases.

PETROL

All petroleum fuels come from treatment of crude stock, and in thinking of the various fractions of crude petroleum used as engine fuels let us consider first the motor spirit fraction. Motor spirit, or petrol or gasoline, is the fraction with a boiling range between 95° F and 435° F. It contains molecules of five to 15 carbon atoms, and is a very volatile and satisfactory engine fuel.

A desirable property in engine fuels is resistance to knocking. The better the fuel is in this respect the higher the compression ratio that can be used in the engine: compression ratio being the ratio between volume of the cylinder when the piston is at the bottom of its stroke and the cylinder volume when the piston is at the top. High compression ratio gives an efficient engine, but unless the fuel has high anti-knock properties a spark ignition engine will not run smoothly because of pre-ignition. The resistance of the fuel to knocking depends on its chemical composition. One particular fuel, called iso-octane, has a high resistance to knocking. The word octane is used in the unit for measuring anti-knock properties; iso-octane, a derivative of petroleum, is assigned by definition an octane rating of 100. At the other end of the scale, normal heptane, which first came from the Jeffry pine tree, but is now synthesised, has a very low anti-knock value and is assigned an octane number of 0.

Laboratory tests to determine the octane number of petrol are conducted in a single-cylinder engine. The fuel to be tested is used in a set of standard conditions prescribed by the petroleum industry.

The severity of knock is measured by means of a bouncing pin. This is installed through the head of the engine into the upper portion of the combustion chamber. Different rates of pressure or knock in the combustion chamber move the pin upwards. This upward movement is measured by electric impulses flowing through a knock-meter, calibrated in arbitrary numbers from 0 to 100. Thus, varying intensities of movement caused by a certain fuel can be indicated.

When the knock intensity of a fuel being tested is known, various mixtures of iso-octane and normal heptane are blended until a similar knock intensity is produced on the knock-meter. The percentage of iso-octane in that matching mixture is called the octane number of the petrol. If, for example, a sample of motor spirit under test gives the same engine performance as a blend of 70 per cent iso-octane and 30 per cent normal heptane, then the octane number of the sample of motor spirit is said to be 70. By means of these octane number determinations, all grades of motor spirit can be matched against the standard reference fuel and can, therefore, be compared for engine performance.

The octane number of petrol can be improved by the addition of certain chemical compounds, such as tetra-ethyl-lead. Highly efficient

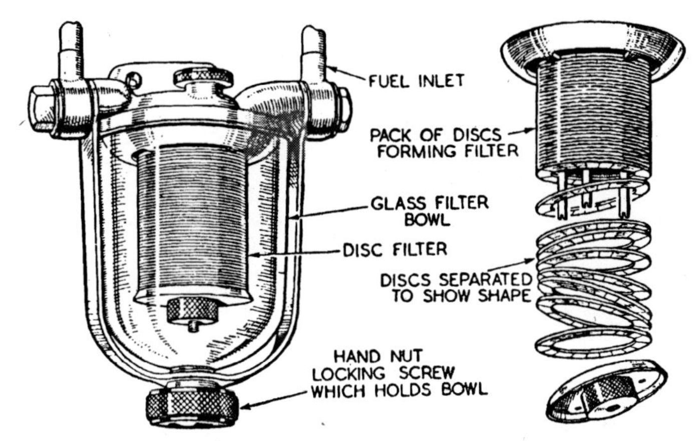

A fuel filter in which circular discs, packed together, are used instead of wire gauze. The fuel goes into the bowl first and then finds its way between the discs to the outlet pipe. The dirt falls to the bottom of the bowl.

engines can be built if the designers can be sure that the machine will always be used with fuel with high anti-knock properties.

Another property always provided in motor spirit is high volatility. This makes for easy starting of the engine. Too-high volatility, however, may be troublesome in making the fuel pipes prone to vapour locks. High volatility can also, on very cold damp days, cause freezing of the water in the air being drawn into a carburettor, forming a coating of ice which will upset the performance of the carburettor. The refiner has to provide sufficient light fractions to give the required good starting qualities yet not an excess of light fractions such as would cause vapour locks or carburettor freezing while the engine is running.

The anti-knock characteristics of petrol do not affect the starting of an engine. A low-octane petrol may even provide just as quick starting.

What it cannot do is to provide full power in a high-compression engine.

An engine can use plain non-treated petrol without injury if the maximum power is developed without pinking or detonation; and economy of running cannot be improved by using treated petrol. If pinking or detonation occurs, the engine cannot operate economically, and petrol of higher octane rating should then be used, say 75 octane rating. Valve deterioration has resulted from the use of leaded petrol in engines not requiring the lead for control of detonation.

The octane rating of petrol may also be raised by blending with alcohol, as we shall see later in this chapter. The following table gives the approximate increase in octane rating of petrol due to blending with alcohol:

Fuel	*Octane Number*
Base fuel (regular petrol)	56
Base fuel plus 10 per cent alcohol	65
Base fuel plus 20 per cent alcohol	80

Benzol, also mentioned later in this chapter, has an octane rating of 90, and is sometimes blended with petrol to raise its anti-knock value.

VAPORIZING OIL

The next fuels to consider are distillate and vaporizing oil. In the distillate of petroleum the kerosene fraction within the boiling range of 150° F to 300° F, after refinery treatment and blending, provides two main commercial products, burning oil for heating and lighting, and a vaporizing oil for use as an engine fuel. The different uses of these two products demand fuels having different compositions. To prevent the formation of smoke, burning oil contains very little aromatic hydrocarbon. For an engine fuel, however, it is essential to have present a quantity of aromatic hydrocarbons to give the fuel good anti-knock properties. Lamp oil must have a low flash point so that it shall not be dangerous when used in lamps and stoves, and its smoke point, a test that is carried out in a standard lamp, must show that a good flame height can be attained without smoking. In some countries vaporizing oil and distillate are not available, and lamp oil is used as a tractor fuel. A tractor which is to run on lamp oil should have a low-compression engine with a specially hot vaporizer plate.

For vaporizing oil the important properties to be considered, apart from the distillation range and the flash point, are the volatility and the octane number. Vaporizing oil has to be converted to vapour before it can be used satisfactorily as a fuel in a piston engine. The volatility of the fuel is therefore important, since the more volatile the fuel the more quickly will it be possible to change over from petrol after starting a petrol paraffin tractor from cold, and the less likely it is to cause dilution of the lubricating oil. The anti-knock property of vaporizing oil is measured by the octane number in the same way as with motor spirit. An octane rating of 50 is usual.

DIESEL OIL

Diesel oil has a boiling range between 380° F and 725° F. The best fuel viscosity for Diesel engines is between 32 and 40 seconds Redwood and 9° F and 100° F. With fuel of a lower viscosity than 32 seconds there is a tendency for dribbling to take place at nozzles, while a fuel viscosity above 40 seconds Redwood 1 at 100° F has bad atomization and is imperfectly burned in the engine cylinder. On a cold day the starting of a Diesel engine may be greatly affected if the oil has too high a pour point, or setting temperature.

Octane numbers are not quite applicable to Diesel fuels. Instead, a cetane number is used. This gives a measure of the combustion characteristics of the fuel. The hydrocarbon cetane when used in a Diesel engine shows little delay between the moment of fuel injection and the start of combustion. Tests for cetane rating are made in an engine by comparing a sample of a particular Diesel fuel with blends of cetane and methyl naphthalene, which is a product of poor ignition quality. The cetane number of a Diesel fuel indicates the percentage of cetane in a blend of cetane and methyl naphthalene that has the same delay time as the fuel under test when run in the same engine under the same load and temperature conditions.

Freedom from sulphur in Diesel oil is important because of its harmful effect on the engine, as we have noted in discussing lubricating oils, and also on the pipes of the fuel line and the pumps and injectors.

Diesel oil is not sufficiently volatile to work in a carburettor engine, and it can only be used in a compression ignition engine.

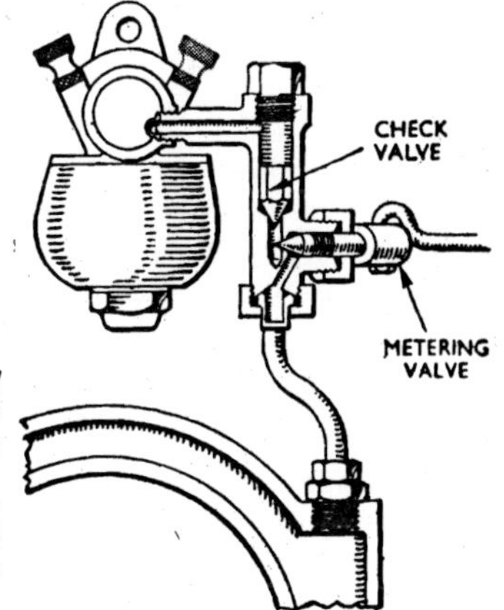

A device for introducing water into the inlet manifold through a metering jet and check valve, operated by inlet pressure. This attachment, used on tractors which have to run on lamp oil or other low-grade fuels, allows water to enter when the engine is running under heavy load, but prevents it entering under light load conditions. Added water keeps down flame propagation so that the power stroke becomes a steadier push.

LIQUID PETROLEUM GAS

Liquid petroleum gas (L.P.) is much used as a tractor fuel in the USA. This fuel consists of propane, or a mixture of butane and propane. It is compressed into a steel cylinder. A similar product is sold in Britain for domestic heating and lighting in rural districts. The fuel has a calorific value of about 21,500 B.Th.U/lb.

Iowa State College, Ames, Iowa, has done considerable work in comparing the performance of L.P. gas with other tractor fuels. It has been found that it is clean-burning, leaving little carbon deposit, that the compression ratio can be raised to 8 to 1 without detonation and that at Iowa prices it is probably cheaper to use than distillate fuels.

L.P. gas can be used by vapour-withdrawal carburation or by liquid-withdrawal carburation. With vapour-withdrawal carburation the fuel is drawn off the upper part of the cylinder in vapour form and passed to a carburettor to be mixed with air. This is satisfactory in ordinary weather, but in the winter at Ohio it gave trouble owing to the fact that the evaporation cooled the contents of the cylinder so much that the liquid would not continue to vaporize freely.

With liquid-withdrawal carburation the liquid fuel is drawn from the bottom of the cylinder, passed through a pressure regulator, and piped to a heat exchanger to be vaporized before passing to a carburettor to be mixed with air.

OTHER TRACTOR FUELS

Possible alternative fuels to petroleum for farm use are producer gas, methane gas generated by the anaerobic decomposition of farmyard manure, alcohol and benzol.

Portable gas producers have been made for providing fuel for tractors. They are much the same as those designed for lorries and buses and they use charcoal, anthracite or coke. The gas is made by drawing steam through glowing fuel. With anthracite, 10 lb of fuel yields the same amount of power as about one gallon of petrol. In the models that have been made for tractors the fitment, without fuel and water, weighs some 4 cwt and is placed mostly to one side of the tractor. In practice it has not been noticed that this extra weight has any detrimental effect. Indeed, in some cases it has helped the grip of the driving wheels.

It is essential that the gas should be washed and cooled before it is admitted to the cylinders of the engine. In the early experiments particles of coke found their way into the engine and caused damage. If. however, the gas is properly cleaned, an engine run upon producer gas can have a longer life than one run on vaporizing oil. There can be no dilution of the lubricating oil by unvaporized liquid fuel, and this ought to mean longer life for the bearings and the cylinder wall.

An engine designed for vaporizing oil will run on producer gas without any alteration being made to the engine; and it will run

Hydraulic motor built into a rear driving wheel of the N.I.A.E. hydraulic transmission tractor. The motor has five radially spaced cylinders rotating around a fixed eccentric member. (See Chapter 2.)

A Fordson tractor fitted with girder-type half tracks instead of rear wheels. (See Chapter 2.)

The driving seat and controls of a Platypus track-laying tractor. The cranked levers are for steering, the control lever for the main change-speed gearbox is in front, and the lever for the auxiliary high-low box is behind. (*See Chapter* 2.)

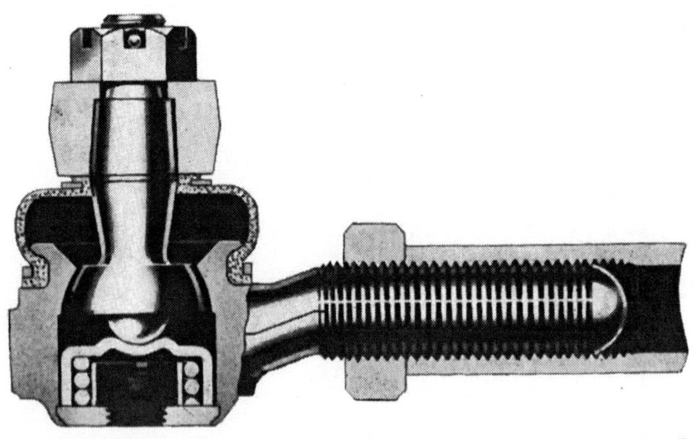

Thompson steering joint incorporating automatic adjustment by spring to take up the effects of wear. (*See Chapter* 2.)

An N.I.A.E. dynamometer loading car used in testing drawbar pulls and fuel consumptions of tractors on a tarmacadam surface. (*See Chapter* 4.)

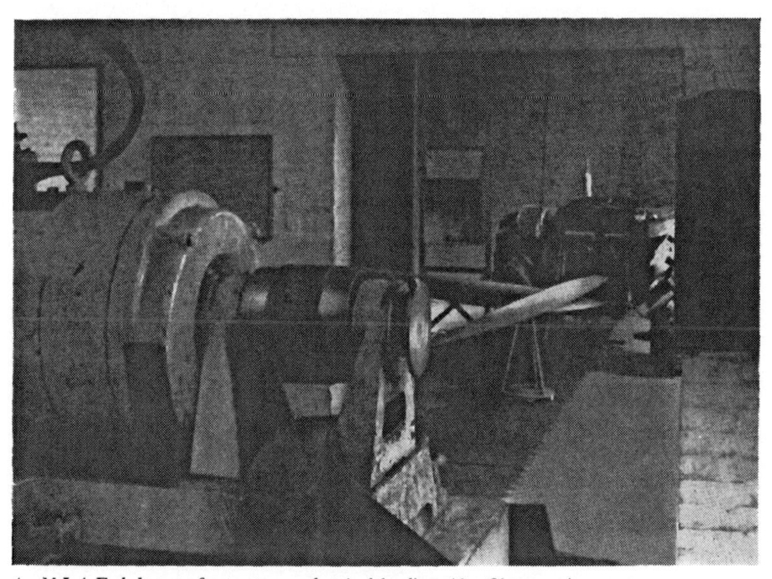

An N.I.A.E. belt test of a tractor on electrical loading. (*See Chapter* 4.)

PLATE 8

An Italian four-wheel-drive tractor, the Slanzi, which has a 17 h.p. Diesel engine, with three-speed gearbox. (See Chapter 4.)

A Fiat 24 h.p. Diesel-engined crawler tractor. (See Chapter 4.)

efficiently for the amount of fuel consumed, but its absolute output of power will be smaller by about 40 per cent than if it were running on vaporizing oil. The output of power on gas can be increased by fitting a different cylinder head to raise the compression ratio of the engine.

More skill is needed to work a producer-gas tractor than one running on liquid fuel. From time to time the firegrate must be cleared of clinker and the filters for the gas must be regularly kept free. The heat dissipated by a producer-gas generator is fairly high, and on a hot day in summer the tractor seat can become an uncomfortable place.

Methane gas is one of the products of decomposition of farmyard manure and the other is carbon dioxide. As much as 60 per cent can be methane if the decomposition is carefully controlled. Methane gas has a calorific value of about 700 B.Th.U/cu.ft. It can be compressed into 1/500 of its volume at normal temperature and pressure.

The plant for making methane on the farm is filled from a mixing tank. The waste materials, whether weeds, potato haulms, chaffed straw, or solid and liquid manure, are thrown into this tank. A pump mixes it into a slurry (no liquid is used apart from the liquid manure or other material, and the water absorbed by or naturally contained in the mix). Then the pump moves the sludge into the fermentation silo.

To start the bacterial fermentation it is necessary to maintain a temperature of between 86° and 90° F. The sludge is heated by hot water, by indirect coil. The duration of the digestive process depends on the temperature at which the mass is kept, but no worth-while increase in speed of production is obtained above 95° F, and very high temperatures are actually unfavourable to the development of methane bacteria. The digestion of about half of the organic matter takes four days at a temperature of 86° F.

The digested sludge is said not to lose any of its fertilizing property. Indeed, claims have been made that the losses are lower than occur in the orthodox methods of maturing farmyard manure. The operation of the plant needs some skill.

For stationary engines the gas can be used as generated; for tractors it must be compressed into a liquid to be stored in cylinders. Compressing pumps are expensive and the only cylinders so far used successfully to withstand the high pressures are made of a costly nickel steel.

There is a large experimental gas-producing plant on a 250-acre farm at Allerhop, in Germany, and a small plant has been in operation in Britain (in Surrey) since 1918, though this provides gas only for domestic use in the farmhouse. Large plants which use public sewerage as their material are in successful use by some local authorities in Britain, notably at Croydon and Worcester.

Alcohol can be produced from farm products such as potatoes, but so far this has not been much done on a large scale in Britain. As an engine fuel it has a high resistance to compression, thus allowing the use of higher compression ratios in internal-combustion engines. It gives a smoother power flow at wide throttle openings, due to anti-

F

detonation characteristics, and gives the engine a slightly higher thermal efficiency, mainly because of the lower flame temperature. Its calorific value is about 12,000 B.Th.U/lb. An alcohol-burning engine will run relatively cool, and little carbon is formed. Larger quantities of heat are, however, required to vaporize it in the induction system than is necessary for petrol. Alcohol is a homogeneous fuel and possesses no low boiling-point fractions. It is therefore best used in admixture with benzol or petrol, so as to eliminate starting difficulties. One way in which alcohol can be used as an auxiliary fuel is as alcohol-water injection. In this process, the anti-knock qualities of alcohol and the effect of the high vaporization temperatures of alcohol and water combine to lower the intake mixture temperature and raise the effective octane number of the fuel.

The energy obtainable from a given weight of alcohol is a little less than that from the same weight of petrol, and a larger jet is therefore required when alcohol is used as the main fuel. The carburettor float must be adjusted to correct the fuel level for the higher specific gravity of alcohol. Alcohol is slow-burning, and the ignition must be well forward.

Alcohol has no effect on the metal of pistons, rings, cylinders, valves or bearings, but it does dissolve certain substances. It cleans out fuel tanks and pipes, and the matter removed will at first clog filters and carburettors in an engine which has not previously used this fuel.

The use of water alone to lower the temperature of the inlet mixture improves the running of paraffin engines. The water is introduced into the inlet manifold through a metering jet and check valve, operated by inlet pressure, which allows water to enter when the engine is running under heavy load, but prevents it entering under light load conditions. Added water improves the running when an engine has to be run on very low-grade fuel, such as lamp oil. It keeps down pre-ignition, and slows down flame propagation so that the power stroke becomes a steadier push, as in a compression ignition engine.

Benzol is a spirit derived from the distillation of coal tar. As a fuel it is superior to petrol or paraffin, but higher compression ratios are required. A commonly used mixture is petrol and benzol.

CHAPTER FOUR

CHOOSING A TRACTOR

IT is rare for a completely fresh start to be made in equipping a farm with implements and machines. Nearly always the new piece of machinery being bought has to be fitted into the existing economy of the farm, and has to be capable of being worked in with the machinery already on the farm. Therefore it is not often that a farmer has complete freedom of choice in buying a new machine. What he buys today is influenced by what he bought last year and will influence what he buys next year. Nevertheless, some purchases do provide decisive signposts to further purchases. The most decisive of all purchases, particularly for a small farm, is a tractor, since the size and kind of tractor will determine the size of implements which can be economically handled and the type of implement which can be fitted. For example, the kind of tractor determines whether we shall have a two-furrow or three-furrow plough, and whether it shall be trailed or attached directly to the tractor.

The kind of tractor can influence also even the barn operations. If the farmer has a tractor with a belt pulley he may think it worth while to use quite a large grist mill and get the grinding of the corn into cattle food done quickly each week rather than have a smaller mill with its own engine or electric motor.

The influence of the tractor on the choice of implement can be seen clearly if we take extreme examples. The man who has bought a powerful track-laying tractor will not be interested in a light two-furrow plough for it, and the farmer who has decided that a walking tractor will do his work will waste no time in considering whether he can do his own deep ploughing. The difficulty comes in deciding between much finer shades of possible methods of mechanization.

Extreme examples of range of tractor power do, however, emphasize one important point, and that is that in many respects the small farm is much more difficult to mechanize efficiently than the large farm. The small farm cannot have implements scaled down to meet its size. The large farm does not automatically need a heavy track-layer, and the small-holder a miniature tractor. The labour of the owner of the small-holding is as valuable to him as the labour of workmen is to the manager of the large farm; the small farmer puts himself at a great disadvantage if he is taking three times as long to do an acre of ploughing as the man with a large farm and a large tractor.

The greatest obstacle to simplifying the mechanization of a small farm is the fact that, for every crop grown, exactly the same operations have to be performed as would be performed on a large farm. Therefore, the small farmer cannot reduce the range of the machinery he instals, compared with the large farmer, without leaving much work to be done by hand. This he does not want to do because since most small farms are family-operated, with a very little hired labour, saving of hand work is of very direct personal interest. To the larger farmer saving in labour means saving expenditure and increasing profits; to the working farmer it means fewer hours of toil.

SMALL-HOLDINGS

In considering the smaller farms a distinction should be made between small general holdings and the specialized market garden holdings which, although they are small in acreage, are often very large as labour employers and as producers of food by intensive cropping. For these market gardens a whole series of mechanical aids has been devised to reduce the hand labour of digging, sowing and transplanting, and the high value of the crop makes the mechanical aids worth while on very small acreages.

A small-holding is usually an even more general farm than the large highly mechanized holding, which so often specializes chiefly in one branch, such as grain growing or dairy farming. Therefore the small farmer with his limited capital has to face even more complexity of machinery than does the big farmer.

The ultimate object of all farm mechanization is to save hand labour, and the greatest help of all is the tractor. The tractor enables one man to control a larger piece of work than if he were driving horses. It may mean that he can plough three furrows with a tractor in the same length of time that he would take to plough one furrow with horses, and that is not the whole of the saving, for the horses need time spent on them in feeding and bedding, whereas the tractor needs very little attention when it is not actually working. Also the horses themselves need food, and part of their time they will be working only to provide food to feed themselves; and the land used for growing horses' food could be growing a cash crop or be feeding other cattle.

It is clear, therefore, that the small general farmer has an even greater need to do away finally with his horses than the large farmer has, but to do so completely the tractor will have to be of a kind suitable to do many jobs besides heavy cultivations. It will have to mow, sweep hay, haul wagons, and plant and hoe row crops.

Not all these light jobs of carting and hoeing are done by horse. Often the tractor is used very inefficiently. Rolling, chain harrowing and hay tedding can probably all be done more cheaply by horse than by tractor, reckoning the job as a unit on its own. Yet these jobs do not justify the keeping of a horse all the year round, and a little extravagance in using the tractor for light work is offset by its efficiency when it is doing heavy cultivating.

The benefit that the tractor has in enabling one man to control a large piece of work tends to be lost when the farmer buys a small tractor, scaled down to the size of his farm.

There is another consideration, perhaps even more important than this one of the time of the worker. The man on the small farm may be quite willing to spend long hours working in the field, but however hardworking he is and however ambitious he is to run his farm at a low cost so that he can build a reserve of money, his plan of work will more easily be upset by bad weather than will the plans of his neighbour with a large tractor. This is the reason why one-tractor farmers so often buy a larger tractor next time than the tractor they are using at present; they feel how useful it would be to have a little extra power.

We have seen that ploughing and heavy cultivating are not the only work to be done by tractor, and even if a large tractor with a large capacity to get all the fields ploughed at exactly the right time could be afforded, a lighter tractor might still be needed to do row-crop work, for transport and for quite a number of the odd jobs which make up so much of the year's work. The farmer large enough to run a fleet of tractors can include big tractors in his fleet and small ones, but on a small farm every machine and implement must be versatile. For example, the large farm can have one kind of implement to pick up the hay crop and another kind to pick up green grass, but the small farmer ought to have a dual-purpose machine. Even then the machine will have so little use each year that, provided that it is well looked after, its life will be very much longer than the life of the same machine on the farm which has a large acreage to be dealt with.

Rate of Depreciation

This factor of slow rate of depreciation of machinery is in most ways an advantage to the small farmer; in some ways it can be a disadvantage. A life of seven years is what most economists put down for a tractor, and if we are to assume that the tractor actually will last only the seven years, whether it is working on a farm with an arable crop of 20 acres or on one where it has to look after 100 acres of arable, then the depreciation cost, which may form a very large part of the total running cost of the tractor, is going to fall five times as hard on the small farmer as on the larger farmer; but metal bearings do not wear out when no shaft is rotating within them, and the rubber on tyres, though it may perish with the passing of years, will deteriorate only very slowly when they are not at work. There is, in fact, no reason at all why a farmer who uses machines for a small acreage each year, and drives them carefully and looks after their maintenance regularly, should not be able to make his machines last two or three times as long as they would on a large farm.

It must be remembered, however, that the rate of interest on the capital cost is just as high for the small farmer as it is for the large farmer. Moreover, machines do become obsolete, or at any rate new ideas come along; the farmer would like to take advantage of these new ideas but feels he must not buy the machine incorporating them until his present one is worn out. Fifteen years, which might be the life of a farm implement used in this careful way, is a long time in the history of an industry developing as quickly as that of the manufacture of agricultural machinery. It would be very hard to have to go on using an obsolete implement.

Ploughing, or some other method of stirring or turning the soil, is likely to last for a long time to come, and it is the amount of heavy cultivation work which ought to influence most the decision about the size of tractor, particularly when there are only one or two tractors on the farm.

The first step is to decide how many acres of ploughing will have to be done each year. Then we must decide how many weeks of the year can reasonably be spent in doing that ploughing.

RATE OF PLOUGHING

The number of acres per hour a tractor can plough while it is actually in the bout, turning the furrows, can be estimated roughly by multiplying the speed of the tractor in miles per hour by the width of the ploughing in inches, and dividing by 100. Time must be allowed, also, for the turning of the outfit at the headlands, the marking out of the lands and the travelling to and from the field. Substituting 120 for 100 in the above calculation will allow pretty fairly for this; so that the miles per hour multiplied by the width of ploughing in inches and divided by 120 is a reasonable forecast of a tractor and plough's capability.

The width of ploughing possible will, of course, depend partly on the kind of soil, but mostly it will be governed by the power of the tractor. A medium-sized wheeled tractor will pull a three-furrow 10 in plough in most soils at 3 m.p.h. A two-wheeled walking tractor will plough a single furrow; a baby four-wheeled tractor will pull one or two furrows; and a heavy track-layer will pull a six-furrow plough.

DEMONSTRATIONS

A prospective buyer ought not to be at all reluctant to ask an agent to demonstrate tractors on his farm. It is only by seeing the tractor actually at work on the very soil on which it will be used, and doing one of the jobs for which it is being bought, that a confident decision can be made. The agents will be glad of this opportunity. They do not want dissatisfied customers. Moreover, the expenses of such demonstrations are reckoned to be part of their running costs, and form part of their recognized selling costs.

PERFORMANCE FIGURES

Before asking for demonstrations, however, it is possible to do much sorting out of the makes and kinds of tractor available, using advertisements and catalogues to provide the data for the classification. The catalogues will give such details as sizes of engines, and number of gear ratios, and type of tyre equipment, and they may also contain actual figures from test reports, for independently controlled tractor tests have been carried out regularly for many years.

In looking at manufacturers' catalogues and leaflets, confusion can arise from the fact that in this quickly developing subject, different names are sometimes used by different makers for the same types of machine and for the same accessories. The British Standards Institution has endeavoured to provide a universal basis for nomenclature, and the use of this could obviate a cause of uncertainty in comparing specifications, and of confusion and delay in purchase of spare parts. A glossary of terms relating to agricultural machinery and implements has been prepared as BS 2468 : 1954, and the following definitions, which will be useful to tractor owners, accord with those laid down in the Standard.

Term	*Definition*
Agricultural tractor	A self-propelled power unit with wheels or tracks.
Drawbar	A member fitted to a tractor for the attachment of hauled implements. The drawbar may be swinging, fixed or adjustable.
Belt pulley	A wheel to transmit power to another machine(s) by means of a belt.
Power take-off	An external splined shaft to transmit power to other machine(s).
Power lift	A mechanism driven by the power unit to control an attached implement.
Hydraulic lift	A mechanism driven by the power unit to control an attached implement by hydraulic means.
Variable speed governor	A mechanism to maintain the speed of the power unit at any selected value over a given range.
Power take-off shaft	A shaft for the transmission of power from the power take-off to a machine.
Toolbar	An attachment to carry cultivating or other equipment.
Wheeled tractor	A tractor with either three or four wheels.
Skid ring	A circumferential flange on the rim of the front wheel(s).
Wheel weight	An attachable weight to improve wheel adhesion.
Steel wheel	A steel wheel fitted with spade lugs or cleats.

Term	*Definition*
Spade lug	A wedge-shaped metal projection fitted to the rim of a steel wheel.
Cleat	A flat or angle strip fitted to the rim of a steel wheel.
Cleated wheel	A steel wheel fitted with cleats.
Grass stud	A round or pyramid-headed bolt fitted to a steel driving wheel in place of a lug or cleat.
Extension rim	A narrow rim with a single row of lugs or cleats fitted to the side of a steel driving wheel.
Roadband	A rim fitted to a steel wheel to prevent contact of the wheel lugs or cleats with the ground.
Skeleton wheel	A steel wheel to give a minimum ground contact area.
Pneumatic-tyred wheel	A wheel fitted with a pneumatic tyre.
Strake	A form of metal lug or cleat attached to a pneumatic-tyred wheel.
Track-laying tractor	A tractor which lays and picks up its own track(s).
Track	A series of jointed bearing plates or a flexible band forming an endless weight-carrying rail to transmit the drive to the ground.
Grouser	A transverse strake, incorporated in or attachable to a track, to assist adhesion between the track and the ground.
Half-track tractor	A tractor with front wheel(s) but with driving wheels replaced by track-laying equipment
Walking tractor	A tractor, usually with either one or two wheels, controlled by a walking operator.

Estimating Power Output

In most tractor specifications two figures for horsepower are given. One, the higher one, is a belt horsepower available for driving machines such as threshers, mills and saw benches. The lower one is the drawbar horsepower. As we have seen, the difference between the two figures is made up of loss of power in the transmission gears of the tractor, in rolling resistance between the wheels and the soil, and in the slip of the wheels on the soil.

If no performance figures are available we may be able to get some estimate of the tractor's capabilities by working out the power output from figures of the dimensions and speed of the engine, though to do this we must know the mean effective pressure, and this factor depends upon the design of engine. Calculations are the same for all internal combustion engines whether spark-ignition or compression-ignition.

The following formulae are for four-stroke engines:

Let N = number of cylinders.
S = stroke of piston in inches.
D = bore of cylinder in inches.

Then

$$\text{Indicated horsepower} = \frac{\text{m.e.p.} \times S \times d^2 \times 0{\cdot}7854 \times \text{r.p.m.} \times N}{12 \times 33{,}000 \times 2}$$

$$\text{Brake horsepower} = \frac{\text{b.m.e.p.} \times \text{r.p.m.} \times N \times S \times d^2}{1008 \times 10^3}$$

$$\text{or b.m.e.p.} = \frac{\text{b.h.p.} \times 1008 \times 10^3}{\text{r.p.m.} \times N \times S \times d^2}$$

$$\text{Litres of cylinder capacity per b.h.p.} = \frac{12970 \times N \times S \times d^2}{\text{b.h.p.} \times 1008 \times 10^3}$$

$$\text{B.h.p. per litre of cylinder capacity} = \frac{\text{b.h.p.} \times 1008 \times 10^3}{12970 \times N \times S \times d^2}$$

$$\text{Litres of fuel mixture per minute} = \frac{N \times S \times d^2 \times \text{r.p.m.} \times 6489}{1008 \times 10^3}$$

The following formulae are for two-stroke engines:

Let N = number of cylinders.
S = stroke of piston in inches.
D = bore of cylinder in inches.

Then indicated horsepower (i.h.p.)

$$= \frac{\text{m.e.p.} \times S \times d^2 \times 0{\cdot}785 \times \text{r.p.m.} \times N}{12 \times 33{,}000}$$

$$\text{B.h.p.} = \frac{\text{b.m.e.p.} \times \text{r.p.m.} \times N \times S \times d^2}{504 \times 10^3}$$

$$\text{or b.m.e.p.} = \frac{\text{b.h.p.} \times 504 \times 10^3}{\text{r.p.m.} \times N \times S \times d^2}$$

$$\text{Litres of cylinder capacity per b.h.p.} = \frac{12{,}970 \times N \times S \times d^2}{\text{b.h.p.} \times 504 \times 10^3}$$

$$\text{or b.h.p. per litre of cylinder capacity} = \frac{\text{b.h.p.} \times 504 \times 10^3}{12{,}970 \times N \times S \times d^2}$$

$$\text{Litres of fuel mixture per minute} = \frac{N \times S \times d^2 \times \text{r.p.m.} \times 6{,}489}{504 \times 10^3}$$

The following formulae apply to both four-stroke and two-stroke engines:

$$\text{B.h.p.} = \frac{\text{b.m.e.p.} \times \text{i.h.p.}}{\text{m.e.p.}}$$

$$\text{B.h.p.} = 0{\cdot}0001904 \times \text{r.p.m.} \times \text{torque (pounds feet)}$$

$$\text{or Torque (pounds feet)} = \frac{\text{b.h.p.} \times 5{,}253}{\text{r.p.m.}}$$

Estimating Drawbar Pull

The pull at the drawbar, or the pull available for use in a directly attached implement, depends upon, among other things, the speed at which the tractor is moving. Other things being equal, if a tractor could pull a two-furrow plough at 2 m.p.h., it could pull a single-furrow plough at 4 m.p.h. The power exerted by the tractor is the product of the drawbar pull and the speed, and it can be expressed as horsepower by using the formula:

$$\text{Drawbar h.p.} = \frac{\text{Drawbar pull in lb} \times \text{speed in m.p.h.}}{375}$$

The importance of using the tractor fully loaded is stressed later in these pages. The above formula shows that the loading can be increased to use up the available horsepower either by improving the drawbar pull, by doing a greater width of work at each bout, or by increasing the speed of the tractor.

Although small changes in forward speed can be made by changing the engine governor setting, the main stepping up or stepping down must be done by changing into a higher or lower gear. The gearboxes of most makes of tractor provide for three, four, five or six forward speeds, and one or two reverse speeds.

Tractor Tests

For most tractors, however, test figures will be available, since tractors have been tested regularly ever since 1920 when the Legislature of the State of Nebraska in the U.S.A. made a law providing that all tractor manufacturers who wished to offer tractors for sale within Nebraska must first submit a stock model to the state authorities to be tested by a board of engineers acting under the supervision of the State University. The scheme was mapped out with great skill and foresight, and the tests soon came to be recognized far outside the State of Nebraska.

Nebraska Tests

The Nebraska tests have two parts, the testing of the brake horsepower of the engine and the testing of the drawbar horsepower. The drawbar pull readings are made with the tractor on a test track towing a loading car of which the braking effect can be exactly regulated.

In 1930, world agricultural tractor trials were held in Britain under the auspices of the Royal Agricultural Society in conjunction with the Institute of Agricultural Engineering and the University of Oxford. The testing used in these trials followed the Nebraska methods, but it included also some actual field test of ploughing and cultivating. These agricultural tests were, however, in the nature of general assessments of performance and reliability rather than of actual measurements. All ploughing tests were carried out on hay stubble at a depth of about 4½ in; the same land was later used for the cultivating tests.

In later R.A.S.E. tests this field work was given more significance and field drawbar measurements came to be used in calculating results. The drawbar tests were all carried out on ordinary agricultural land, all drawbar measurements were made while the tractor was pulling a plough or cultivator, and tests were made on more than one kind of soil.

N.I.A.E. Tests

After the Second World War, the National Institute of Agricultural Engineering prepared a new test scheme which was put into operation in 1946. The N.I.A.E. scheme retained the R.A.S.E. provision that all drawbar tests were to be carried out on ordinary agricultural land and not on a testing track or other artificially treated surface, but it contained the provision that drawbar measurements need not necessarily be made while the tractor was pulling a plough or cultivator. The reasons for this concession were that it allowed latitude in the season at which the tests could be carried out.

This test was modified in 1948 to provide that pneumatic-tyred tractors were tested on tarmac as well as on the types of soil conditions originally specified. This was done because some manufacturers wished to be able to state the peak performance of their tractors in their dealings with overseas buyers. More extended tests and trials, including ploughing tests, also were provided to cover versatility and reliability.

British Standard Tests

The basic British Standard Test of 1951 retreats a long way from the agricultural emphasis of the R.A.S.E. trials. The rapid growth of the British tractor industry made it apparent that there was a need for a standardized test giving the performance of a tractor under controlled and repeatable conditions, particularly in connection with the tractor export trade. The committee drew up the standard with this object specially in view, and the range and scope of the tests laid down is limited to those which the committee thought necessary to that end. Provision is made for stationary tests of the performance of the tractor on belt work and for field tests of drawbar performance when pulling a load on a good level surface.

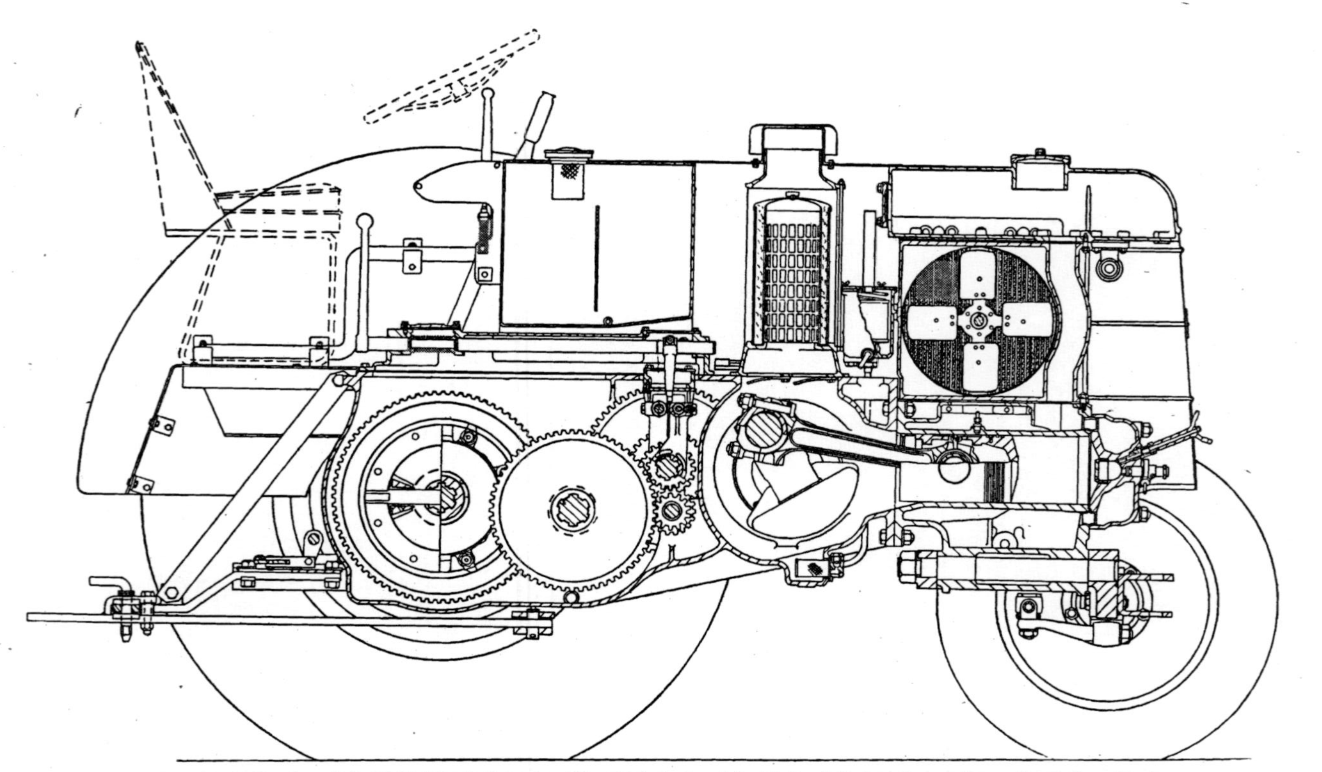

A sectional elevation of the Field Marshall tractor. The single horizontal cylinder, bolted to the crankcase, forms the support for the front axle and the radiator which is sideways on to the axis of the tractor. The crankcase and the main gearbox are parts of the same casting. The engine is a two-stroke Diesel, using crankcase compression.

The test itself is divided in the way which has come to be almost classical, into a section of belt tests and a section of drawbar tests. There is an additional third section which includes general remarks on the final examination of the tractor when dismantled by the makers after the tests, a note of repairs and adjustments done to the tractor during the test, and the specification of the tractor itself and of the fuel it was using.

The dynamometer belt test gives the horsepower, the fuel consumption and a record of the temperature of the air, the fuel and the cooling water. Results are given for a two-hour test at the absolute maximum power of the engine, running on a rich mixture, then a one-hour test at the operating maximum power on a normal mixture of fuel and air. Next there is a one-hour test with the engine running at 85 per cent of the absolute maximum power, the mixture now being set normal; and finally there are runs at various loads on normal mixtures.

Drawbar tests on pneumatic-tyred tractors are usually carried out on a level dry tarmacadam or similar surface, and drawbar tests of steel-wheeled and track-laying tractors on dry, level, mown or grazed grass-land or heavy clay. The tractor is coupled through a drawbar dynamometer to some convenient form of drawbar load, and measurements are made of drawbar pull, speed, slip, drawbar horsepower maximum sustained pull, and fuel consumption.

For these tests the tractor may be equipped by the manufacturers with any wheel or track equipment which is commercially available for the tractor, including additional ballast. In the case of pneumatic-tyred tractors the total weight of any tyre must not exceed the permitted maximum laid down by the tyre manufacturers.

Results for the drawbar horsepower, drawbar pull, forward speed, slip of driving wheels or tracks and fuel consumed are given at the absolute maximum drawbar powers with a rich mixture and at the operating maximum drawbar power with a normal mixture. A ten-hour test is carried out at 75 per cent of the operating maximum drawbar power.

The basic B.S.I. Test is a fair and complete trial of the tractor as a power producer, but it does not directly tell the purchaser how much of that power will be usefully available for work on any particular soil. However, we can turn drawbar figures into ploughing figures that will provide a very helpful guide even if the test does not include figures of a ploughing trial which can be transposed to the width and depth of ploughing required by the purchaser on his own land.

Estimating Plough Resistance

First of all, we make an estimate of the ploughing resistance of the soil on the farm. This is, of course, a rather doubtful figure because the resistance depends not only on the nature of the soil, but on the weather and on previous cultivations and cropping and manuring; but it can be said that the approximate figures in the following table represent the likely resistance to movement of a well-set plough in average conditions.

Nature of Soil	*Ploughing Resistance (lb/sq.in. cross-section of furrow slice)*
Very light land	5
Light to medium land . .	8
Medium to heavy land . .	12
Very heavy land . . .	16

Next, if the test figures do not include ploughing figures, we must estimate the power that will be lost in the additional wheel slip and rolling resistance when the tractor is working on undulating soft soil instead of on level tarmac or firm grassland. To allow for this we can take away one-tenth from both the pulls and the speeds recorded in the test report. This will be writing down the drawbar horsepower by nearly one-fifth.

Suppose the report we are studying shows that the tractor being considered registered a pull of 2000 lb in its ten-hour test in second gear at 3 m.p.h., at 75 per cent of operating maximum drawbar power. Then we can say that we can be sure of $2000 - 1 \times \frac{2000}{10} = 1800$ lb pull in good condititions on ordinary agricultural land at $\frac{3-3}{10}$ $= 2{\cdot}7$ m.p.h.

Suppose the farm where the tractor is to be used is on medium heavy land, of 12 lb/sq.in. ploughing resistance. Then the 1800 lb pull will manage a total cross-section of ploughing of $\frac{1800}{12} = 150$ sq.in. at 2·7 miles per hour.

If we are going to use a three-furrow plough with furrows 10 in wide, we shall be able to plough to a depth of $\frac{150}{3 \times 10} = 5$ in.

If this is not deep enough, or if the 30 in bout of work is not sufficiently wide to get the ploughing done quickly enough at 2·7 m.p.h., then we shall pass by this particular tractor and look for one with a higher pull at ploughing speed.

We can express the formula for this calculation as:

$$\text{Depth of ploughing} = \frac{9/10\ \text{(pull from ten-hour test report)}}{\text{ploughing resistance per sq.in. of furrow} \times \text{width of bout in inches}}$$

To find how quickly the work will be done we can use the following formula which gives the answer in acres per hour:

$$\frac{9/10 \text{ (speed from ten-hour test report)} \times \text{width of bout in inches}}{120}$$

This formula makes an allowance for time spent turning at the headland and in making openings and finishes.

In the case of the three-furrow plough imagined to be travelling in the bout at 2·7 m.p.h., the acreage accomplished per hour would be

$$\frac{2{\cdot}7 \times (3 \times 10)}{120} = 0{\cdot}68$$

Tests as a Guide to Reliability

In most tractor tests, the tractor must develop its absolute maximum power for two hours, and then run one hour at 85 per cent of this maximum power. Load is then increased step by step, until the speed drops below the speed at which maximum torque is obtained. At each step in load variation, measurements of power and fuel consumption are made. The belt test lasts about eight hours, and about five hours of this consists of running under maximum power and overload conditions. This is a hard test, because in ordinary field work tractors are not run at maximum power for more than a few minutes at a time. For most of their life they are developing less than half of their possible power. Therefore, this bout of concentrated work will show whether the cooling and lubricating systems are adequate.

To find whether the tractor on test is efficient in fuel consumption, we can reckon that the figure for belt horsepower hours per gallon for a spark ignition engine ought to be above 11 and for a compression ignition engine it ought to be above 16. The figures for consumption at low loads should be particularly examined, because the tractor is to spend much of its life on work such as transport which may use only a quarter of the maximum power.

The torque figures should be looked at carefully. It is desirable that the maximum torque, or twisting moment, should occur when the engine is running considerably slower than its rated speed, so that the engine shall pull well at low speed. Most tractors are rated to run at about 1,500 r.p.m., but the maximum torque ought to occur at about 1,000 r.p.m.

In drawbar tests, there is usually a ten-hour test at 75 per cent of operating maximum power in the usual working gear, and this can be taken as the equivalent of an extremely hard day's work for a tractor on heavy continuous cultivations. A more complete indication of the reliability of the tractor is available when the tests include extended farm trials.

In most countries that have tractor tests, there is a test to find when the tractor stalls when hauling a load up a hill, and also a test to discover whether the brakes work well enough for the driver to bring the tractor safely back down the slope again after the tractor has failed to climb the hill.

Usually there are special regulations to ensure that the tractor sent for test is a standard product. The tractor is partially dismantled at the end of the test and this gives an opportunity to see whether it was standard. This 'stock model' feature of tractor testing means that a buyer can expect his own tractor to give the same performance as that indicated in the test figures. This has been borne out by a survey made of 35 tractors, of an average of 20 months old, in the counties of Bedfordshire and Hertfordshire. The average power of these tractors was only 10 per cent lower than the test figures for the particular tractor. The 90 per cent of the test power was developed with only 4 per cent greater fuel consumption than the optimum figure given in the test report relating to the tractors. This survey not only justified the methods of testing; it also showed that the power output of tractors deteriorates only very slowly as they become worn.

Correcting for Altitude

Figures for test results are usually corrected to standard sea-level atmospheric conditions. The purchaser of a tractor to be used on high land must remember that when tractors are operated at considerable altitudes, the performance is affected by increase in height above sea-level. The volumetric efficiency of the engine and the power output depend upon air density because the capacity of the engine is fixed, and so the weight of air to fill the volume of the cylinders can vary according to the density of the air. The air density and atmospheric pressure fall at high altitudes because there is less weight of air above. The purpose of the superchargers on aero engines is to increase the air pressure so that volumetric efficiency does not decline so rapidly with increase in altitude.

Figures of spark-ignition engine performance at sea-level can be converted to apply at any given altitude, by multiplying the figures by the atmospheric pressure at that altitude and dividing by the atmospheric pressure at sea-level, and the variation of atmospheric pressure with altitude may be taken as being 0·89 in of mercury for every 1,000 ft. Sea-level figures assume an atmospheric pressure of 29·92 in of mercury. An engine which gave 40 h.p. at sea-level and was then called upon to operate at a height of 1,000 ft above sea-level would give

$$\frac{40 \times (29{\cdot}2 - 0{\cdot}89)}{29{\cdot}92} = 38{\cdot}8 \text{ h.p.}$$

Temperature also affects air density. The lower the atmospheric temperature, the higher the density and therefore the greater the weight of the charge of air going into the engine. Normal air temperature for tractor tests is reckoned to be 60° F and the conversion factor to correct the h.p. to any other temperature is

$$\sqrt{\frac{520}{460 + \text{given temperature}}}$$

A Marshall 70 h.p. six-cylinder Diesel-engined tractor. (*See Chapter* 4.)

Single-cylinder two-stroke Diesel-engined Marshall track-laying tractor. (*See Chapter* 4.)

An air-cooled four-cylinder Diesel-engined tractor, the Allgaier 44 h.p. model. (See Chapter 4.)

Colwood single-wheel rotary cultivator. (See Chapter 4.)

Colwood single-wheel tractor. (*See Chapter* 4.)

A two-wheeled tractor, the Howard Rotavator Yeoman fitted with a 4 h.p. engine. It is shown here as a rotary cultivator, but tined implements can be fitted in place of the rotors. (*See Chapter* 4.)

Bower wheel attachment for pneumatic tyres. Strakes extended. (See Chapter 7.)

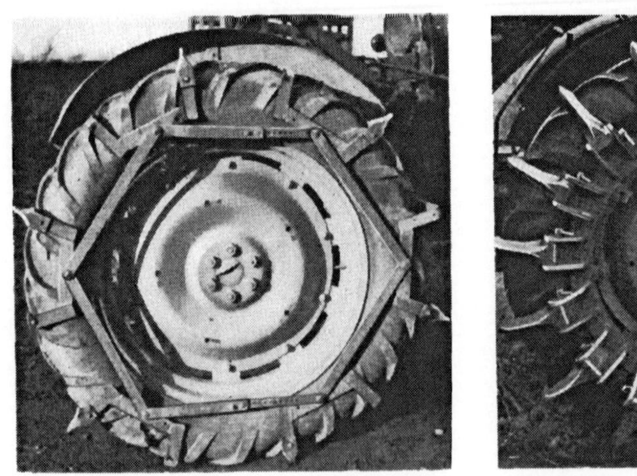

(Left) Donaldson Skidmaster girdle with spade lugs. (See Chapter 7.) (Right) Opperman strakes in working position. (See Chapter 7.)

An engine which gave 40 h.p. at 60° F would, at 30° F, give

$$40 \times \sqrt{\frac{520}{460 + 30}} = 40 \times \sqrt{\frac{520}{490}} = \text{about } 41 \text{ h.p.}$$

No similar formulae have been agreed upon for compression ignition engines, but it can be taken that the fall in power output with increase in altitude is likely to be lower than with spark ignition engines because the amount of fuel injected will be little altered.

Although the composition of air remains constant at higher altitude, its decreasing density means that the power output of the engine decreases with altitude whatever the oxygen requirement of a given fuel. (The oxygen requirements of petrol and vaporizing oil are about the same.) At any one throttle opening and jet setting a richer mixture will be obtained with increase in altitude. If the sea-level performance figures were obtained with normal mixture, the effects of the richer mixture at higher altitudes do to some extent offset the loss of power due to decreased atmospheric pressure. However, this is the wrong way to increase the power output, and the carburettor should be readjusted for the high altitude so as to bring the fuel mixture proportions back to normal.

It is useful to know that in most tractor tests figures for the slip of wheel or tracks are given as a percentage slip calculated from the formula (A—b) × 100 ÷ A, where A is the arithmetic mean of the distance travelled per revolution when running light and when towed, and b is the distance travelled per revolution under load.

Estimating Fuel Consumption

When the likely acreage per hour has been calculated, a rough estimate of the fuel which would be consumed per acre can be made. To do this we can make the assumption that the amount of fuel used per hour in our ploughing will be the same as the amount of fuel used in the ten-hour drawbar test in the standard trial. In the test result will be found the consumption in drawbar horsepower hours per imperial gallon of fuel. This figure divided into the drawbar horsepower registered during the test will give the consumption of fuel per hour. The consumption during our ploughing is not likely to be less than this even though the drawbar load is less, because the rolling resistance and slip which reduce the available power are a direct loss of fuel.

Let us suppose that in the case we have been considering the figure for drawbar horsepower hours per imperial gallon is 14. The recorded drawbar horsepower would be about 16, and therefore the consumption is $\frac{16}{14} = 1{\cdot}14$ gallons per hour. At the ploughing rate of 0·60 acres per hour this would mean a consumption of $\frac{1{\cdot}14}{0{\cdot}68} = 1{\cdot}7$ gallons per acre.

G

THE WALKING TRACTOR

In considering what size of tractor to buy we must not lose sight of the fact that the greatest benefit of a tractor is that it enables one man to control a large piece of work. This benefit tends to be lost when the small farmer buys a small tractor, scaled down to the size of his farm. This causes many smallholders to pass over the single-furrow class of tractor and buy one of the two- or three-furrow class, presently to be described. Nevertheless, it is true that on a very small holding a two-wheeled walking tractor, or a small three- or four-wheeled tractor of 6 h.p. or 8 h.p., can quite well manage all the cultivation work as long as its engine is powerful enough and its wheel grip certain enough to pull a single-furrow plough at all depths likely to be required. In thinking of furrow depth we must remember that a deep furrow slice must be cut wide as well as deep, since a narrow deep slice cannot be turned over neatly.

This means that at least a 4 h.p. engine in a heavy frame is needed on most types of soil to plough at a reasonable speed. This introduces a difficulty. The engine of 3 h.p. is quite sufficient for working hoes and seeders and small cultivating tines. Indeed a 6 h.p. two-wheeled tractor may be unnecessarily powerful and heavy for the job of hoeing since the draught of the tools for each row is low, and since a wide bout of work cannot very well be done accurately by a tractor having a narrow tread width. If a hoe at one end of a wide tool bar on a light two-wheeler happens to dig in deeper than its fellow hoes, the sudden uneven resistance can cause the whole tractor to slew out of the path it ought to be following.

Considerations of this sort have caused some small farmers to get their heavy ploughing done by contract and to keep the work of their own tractor to seeding and hoeing, and to performing the other lighter parts of cultivation and harvest tasks.

Nevertheless, many very small farms in Britain are at the present time doing all their work with a 6 h.p. two-wheeled walking tractor. On some larger farms, too, a two-wheeled tractor is the only power outfit belonging to the farm, while contract work is employed for many operations. Most of these farmers let only the ploughing be done by contract, but others employ contractors also for the harvesting and other operations which need expensive machinery.

One farm which was made the subject of an economics enquiry had been carrying on perfectly satisfactorily with all the work of its 40 acres of arable land, including the ploughing, done by a two-wheeled tractor, except that the drilling and the harvesting were handed over to an outside contractor.

Such a heavy programme as this means that the tractor is at work for very many hours in the year, and it is to be noted that the high number of hours of work usually brings a low overall cost per hour since the depreciation of the machine is not very much greater when it is working nearly every day than it would be if it had long idle periods.

If a small farmer can use a small two-wheeled tractor to free himself from the work needed in looking after horses in feeding, grooming and bedding them, he will have leisure that he never knew before. It is on small farms that the labour-saving possibilities of a tractor are particularly attractive. Most small farms are family-operated, with very little hired labour. Therefore, the saving of hand work is of very direct personal interest to the small farmer. To the large farmer, saving in labour means saving expenditure and increasing profits; to the working farmer it means not only bigger profits, but fewer hours of toil.

Walking tractors, with one or two wheels, and with handlebars for steering, will draw a directly attached single-furrow plough. The plough can be removed, and replaced by a tool bar which will carry cultivator points, spring-tined harrows, or hoeing blades. Seed-sowing

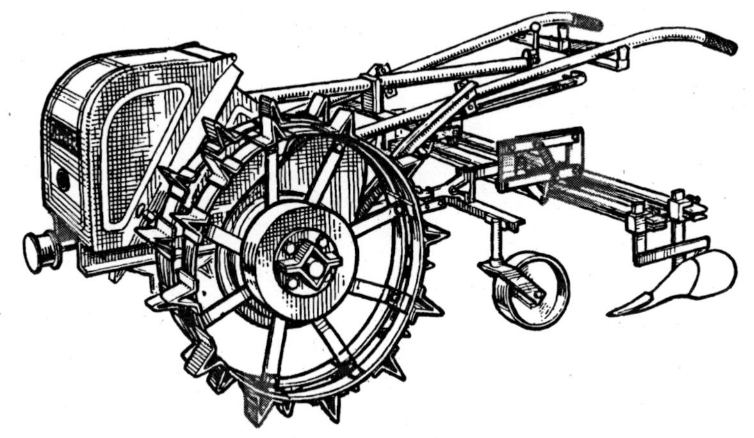

Two-wheeled walking tractor with change-speed gearbox. The change-gear lever can be seen near the handlebars.

boxes can be fitted to the tractor to make it into a self-propelled drill. Several two-wheeled tractors can have a power-driven rotary cultivator fitted. Some makers provide, as an extra, a riding carriage with a seat. This makes the tractor into a mechanical horse which will pull haymaking implements or a light trailer.

The smallest of these single-wheel and two-wheel machines are single-geared, though quite a range of speeds is available by varying the throttle opening. Some of the larger ones have a simple sliding two-speed gear which gives forward drive only, others have an automobile-type three-speed-and-reverse gearbox, and some have four speeds forward with two reverse.

The air-cooled petrol engines fitted to these walking tractors vary from 98 c.c. two-strokes to 550 c.c. four-strokes. Most of them are started by hand, many of them by a cord wound on a recessed pulley, but a few have a foot pedal starter.

SMALL FARM PROBLEMS

The greatest problem in the mechanization of a small farm is that every crop grown demands the same sequence of operations as it would if it were being grown on a large farm. Therefore the small farmer cannot reduce the range of the machinery he installs, compared with the large farmer, without leaving much work to be done by hand. When we remember that the small-holding usually contains an even greater variety of work than does the large highly-mechanized holding, which often specializes in some one branch, such as grain, we see that the small farmer with small capital has to face even more complexity of machinery than do some big farmers.

The small farmer, therefore, even more than the big farmer, should look for versatility when he is buying a tractor. For example, he should remember that a small tractor designed so that it can be fitted with a belt pulley becomes a portable engine which can go to the site of the work, hand its own power to drive a chaff cutter, or a mill or saw bench for cutting up logs. This is a very useful extension of the work of the small tractor, helping to do much more than replace a team of horses, and a belt pulley is a worth-while extra to types of tractors on which it can be fitted.

A difficulty is that however small and inexpensive the tractor is, some of the trailed implements to go with it are likely to be nearly as expensive in small sizes as they are in sizes suitable for a larger tractor. This difficulty, however, is largely overcome by the use of special mounted implements; for example, a mower cutter-bar can be bought as a directly-attached accessory for most small tractors, riding and walking, and is much cheaper than a trailer mower. The implements are quickly put on and off at ingeniously designed hitch brackets.

SMALL TRACTORS FOR SPECIAL JOBS

We have been thinking here of the case of a farm that is to have only one tractor. Before we leave this subject of small tractors we must mention their usefulness as auxiliary machines where large tractors also are employed, but where specialist jobs demand machines of small dimension. For example, a two-wheeled walking tractor can be used with ease in some circumstances which would be difficult for a large riding machine. It can be used right against fences for ploughing around headlands; and by steering it by one handle it can be directed under low branching fruit trees.

On some farms these specialist jobs, such as orchard work and small plot cultivation, are the chief work, and small riding tractors have been designed specially to do them. One of them, a miniature full track-laying tractor, having a 6 h.p. air-cooled engine and rubber-jointed tracks, is used by some growers for work in plantations of closely planted bushes and trees. It will pull a deep single-furrow plough, and is small enough to work inside a glasshouse.

SELF-PROPELLED CHASSIS

Some market gardeners may have more use for a self-propelled tool-bar than for a tractor. Self-propelled tool-bars range from walking-type 1 h.p. motorized hoes and motorized grass and bracken cutters to 8 h.p. machines which will drill, side-hoe, distribute fertilizers, and spray crops.

The riding machines developed from the need for a low-slung slow-moving self-propelled platform which would enable the operator to do hand work on row crops more easily than if he had to walk along the rows. In America some growers of tomatoes in the Southern States have for a number of years been using self-propelled platforms on which the workers lie down and pick the ripe tomatoes from the vines as they pass them, and put them in boxes on the platform. The British machines, however, are used mostly with mechanical hoes, and the nearness of the driver to the ground, and the slow forward movement, allow very accurate work on the plant rows.

In one of them the canvas seat, like a deck chair, can be suspended as low as 6 in from the ground. This particular machine consists of a light steel chassis carried on two front steerage wheels and a large-diameter driving wheel at the rear. This rear wheel is driven by chain, through a three-speed and reverse gearbox, by a 3 h.p. fan-cooled petrol engine. The forward speed can be adjusted from 2½ m.p.h. down to as slow as ½ m.p.h. There are two deck-chair seats, side by side, one for the driver and one for a second operator. The seats can be adjusted both horizontally and vertically. The outfit can be steered either by means of a foot-operated rudder-bar, or by an alternative hand lever.

The rudder-bar can be placed either centrally, for use with one operator, or on one side when there are two operators. It is sufficiently long to give delicate steering control if the operator keeps his feet one at each extreme end of the bar. This arrangement leaves him with both hands free for controlling the tool bar. The tool bar is made in halves, each 3 ft 6 in wide. Each section is pivoted separately, forward of the front wheels. The sections can be raised or lowered individually by hand levers.

The front wheels are adjustable for track width, in steps of 1 in from 4 ft 1 in to 5 ft 11 in, so that the machine can be used on rows spaced from 8 in to 36 in.

A more powerful machine, designed with much the same purpose in view, has an 8 h.p. water-cooled engine. In this machine two wheels are driven, through differential gearing, and these are at the rear. The steering is by a single small pneumatic-tyred wheel which can be deflected by a long-handled tiller.

The driver does not sit very low down on this machine, which has quite high ground clearance, but he is at the back where he gets a good view of the tools. The tools are mounted in the forward part of the frame, on a tool bar which can be easily lifted out, complete, and replaced by another bar carrying a different set of tools. Moreover, the individual tools themselves can quickly be moved along the tool bar to change their spacing. The tool bar is round in section, and at intervals of ½ in along its length it is grooved. Each tool arm, for example the stalk of each hoe, has a leaf spring pressing a key on the arm into the groove. When this spring is lifted the tool arm can be moved along the bar to a new groove. Each tool arm is independently sprung, but the lever for lifting the tools out of work raises all the arms together.

For seed drilling, six individual hopper seeders are mounted within the frame of the machine; while for spraying ground crops a tank and pump, driven from the engine, are mounted at the rear of the machine, and the spray nozzles are fitted to the front member of the frame. A drill for artificial fertilizer can be fixed to the frame to top dress crops at the same time as they are being hoed.

VERSATILITY

Having settled upon the suitability of the tractor for the main work we shall give it to do, we must go on to consider its usefulness for other work. For example, if we intend to use the machine for transport we shall look for a high top gear. If our fields are awkward in shape, or if we want to do row-crop work, we shall look for a small turning circle, and for high ground clearance.

But if the tractor is to be the only one on our farm, we must let the consideration of these details come only after we have settled the relationship between the amount of ploughing we have to get done each year and the length of time that any particular tractor will take to do that ploughing.

The importance of choosing a tractor big enough to finish all the ploughing in good time, even in a season of many wet days, is clear. It is equally important not to have a tractor that is too big. A tractor large enough to do the ploughing will also do all the heavy cultivating needed on the ploughed land. A tractor is wasteful of fuel if it is underloaded, and wasteful of capital if it is lying idle for too great a part of the year.

THE TWO-TRACTOR FARM

We have seen that where more than one tractor is to be used on the farm, the second one can be chosen with regard to its manoeuvrability in row-crop work, and to its use in high-speed hauling. A tricycle design of tractor, having a short turning circle, may be best here; and certainly it will be well to choose one having a high top-gear ratio and pneumatic-tyred wheels. A small machine, besides being useful for pulling a mower, and carting in harvest, will also help out at ploughing seasons, for it can do good work with a one- or two-furrow plough. It is handy for opening ridges and making finishes. This leaves the big tractor free to do only the straightforward ploughing within the bouts. An arrangement of this sort can bring great economy in time and fuel.

THE ONE-TRACTOR FARM

However, very many farmers have only one tractor, and therefore they have to choose one that can carry out as many operations as possible. That is why manufacturers make such a wide range of implements for their tractors, and why so many tractors are made only in the row-crop type, that is to say with high ground clearance so that they can straddle growing plants without damaging them. These tractors have adjustable track width, and a small turning radius, helped for wet soil conditions by independently operated rear brakes, one to either wheel.

CHOICE OF WHEEL EQUIPMENT

When the size of the tractor required has been decided upon, the next choice is in the wheel equipment. Indeed, we have to decide whether the tractor shall be a wheeled one or a track-layer.

Track-laying tractors are more expensive in first cost, and a little more expensive to maintain in order, than are wheeled tractors; but they will pull a heavy load on soils where wheeled tractors would fail. The choice between track-layer and wheeled tractor ought to be made to depend entirely on the contour and soil type of the farm where it is to be used, and not at all on the kind of work that the tractor will be called upon to do.

The low centre of gravity of most track-layers gives a feeling of safety to the driver when he is turning the outfit on a steep headland, or is doing sideling work on a hill. It is, however, on soft wet land that the real benefits of track-layers become apparent. The large area of contact between the track and the soil carries the tractor over patches where a wheeled tractor pulling the same load would dig itself in or would compress the soil so seriously as to interfere with the processes of cultivation or plant growth.

The extra upkeep cost in track-layers comes chiefly from wear in the pin-jointed tracks and in the sprocket wheels that drive the tracks. The amount of wear in these places is much influenced by the kind of soil in which the track has to work; it is particularly heavy in sand. In soils which are not unduly abrasive track-layer tractors can be used for several years without a major track overhaul and their use then becomes economic. It is fortunate that most of the heavy soils, on which a track-layer seems to be nearly essential, are non-abrasive clays.

Nevertheless, full track-layers are expensive machines and there is no point in using one in any circumstance in which a wheeled tractor could do the work efficiently. The track-laying tractor is a machine for the multi-tractor farm rather than for the one-tractor farm, and, even at the risk of its annual number of hours of work being small, it should not be used for jobs which ought to be done by a spade-lugged wheeled tractor.

It is, of course, true that heavily ballasted wheels can cause some reduction in the volume weight of the soil, and also a reduction in air and water permeability. A number of experiments on the volume weight changes caused under various conditions have provided some evidence that running tractor wheels in the furrow bottom at ploughing time can lead to formation of a pan which may extend 4 in below the furrow bottom. With wheeled tractors it is not often possible to let the tractor run with all its wheels on the unploughed land; but it is extremely doubtful whether the pan formed in the furrow can be very harmful to drainage and deleterious to plant growth.

The reduction of volume weight caused by running over soil with even an abnormally heavy tractor, under conditions favourable for compaction, seldom caused a reduction of as much as 10 per cent in volume weight. Under most of the conditions studied a single run-over caused insignificant changes in volume weight, though sometimes the soil appeared to be seriously puddled, and this indicated that changes in water and air permeability might be a more serious factor than compression. In adverse conditions, there were quite serious reductions of volume weight. In normal conditions, however, it would seem possible to make furrow bottoms at the same depth for ten or more seasons with little influence on the volume weight. It was found, with the tractors used, that the resistance of the soil to a penetrometer was increased to a depth of about 4 in beneath the furrow bottom only, even under severe conditions, so that if a sub-soiler were used to eradicate compaction of this type it would appear unnecessary for it to reach more than, say, 6 in below the ploughing depth.

No evidence of serious compaction behind spade lugs of tractors was found but the experiments were necessarily restricted to heavy soils. It was observed that spade lug and track-laying tractors puddled the soil less than rubber-tyred ones, and therefore probably had less effect on the permeability of the soil.

There is one circumstance in which the compaction of a very thin top layer of the soil, and the depression of the tread mark made by the pneumatic tyre, can make difficulty. That is when seed is being drilled on some types of soil, and the hardened cap of the soil prevents the seed coulter from penetrating easily to its correct depth. It is possible to use a tine, fixed behind the tractor wheel, to break up this soil cap; but this compaction of a seed bed certainly has to be borne in mind when heavy wheeled tractors are contemplated.

It is to be noticed that the degree of compaction and the effects of the compaction depend on the moisture content of the soil and upon the sequence of weather before and after the compaction took place. In some seasons plants may grow better in soil that has been compacted, while in other seasons the compaction will retard the growth. This is because in some seasons the soil may benefit from an increase in drainage properties, in others it may benefit from water retention. In all circumstances, compacted soil is bad for drilling.

A rough and ready method of finding the degree and location of compaction is to pour a suspension of precipitated chalk over the open slit or side of a soil testing borer after it has been withdrawn from the ground. The compacted zone, for example a plough pan, shows up white, but the zones with a good tilth and plenty of soil aggregates are unaffected. When soils are in good tilth, the non-capillary pore spaces allow the small particles of precipitated chalk to enter the soil freely. When the soil is compacted the soil pores will absorb the water but prevent the chalk particles from entering.

More accurate comparisons can be made with penetrometer readings and with determinations of volume-weight. A penetrometer has a probe which is forced into the soil in controlled conditions such that the energy needed to move the probe can be measured. Volume-weight or pore space determinations are made by weighings of soil blocks in as undisturbed a state as possible, or by determining the pore space by examining the soil under various air pressures and making a calculation on the assumption that the product of pressure and volume of a gas remains constant.

It is very rare indeed that the avoidance of soil compaction is sufficient justification for using a track-laying tractor instead of a wheeled tractor, and even where stability and grip are the deciding factors the merits of a dual-wheeled pneumatic-tyred tractor should not be overlooked. Tractors equipped in this way have been most successful in the work of reclaiming the slopes of Welsh mountains.

The choice between steel spade-lugged wheels and pneumatic-tyred wheels is nearly always decided in favour of pneumatics, but the choice is not easy where one tractor has to do all the work of the farm. Sometimes the right choice is to buy both kinds of wheel, and then use the iron wheels for ploughing and heavy cultivating, and change to pneumatic tyres when carting and harvesting have to be done.

If this course is taken it is important to choose a tractor in which it is easy and straightforward to change the rear wheels. The front wheels need not be changed, except for some row-crop work; those with pneumatic tyres can be left on for operations throughout the year.

TYPE OF ENGINE

Having decided upon the size and shape of tractor, the next point to look at is the type of fuel, and in practice this comes to a choice between Diesel oil, petrol and vaporizing oil.

The consumption of fuel for a given amount of work is about the same for petrol and vaporizing oil, but considerably smaller for Diesel oil, usually about three-quarters as much. The table overleaf shows the results of some tests on tractors of the same make and size.

One of them has a Diesel engine, another a petrol engine, and the third a vaporizing oil engine. The compression ratios are 16 to 1, 5·5 to 1 and 4·62 to 1. The engines all have four cylinders and all of them have the same bore and stroke. The engine in the vaporizing oil model was governed at 1,400 r.p.m., while the Diesel and petrol engines were governed at 1,600 r.p.m. Nevertheless, the conditions of work are nearly enough comparable for the tests to be informative.

COMPARISON OF DRAWBAR PERFORMANCE OF DIESEL, PETROL AND VAPORIZING OIL ENGINES

		Optimum fuel consumption		*Range of pulls (lb) over which fuel consumption is within 10 per cent of optimum*
Tractor	*Gear*	db.h.p.-hr/gal	lb per db.h.p.-hr	
Diesel	1	14·0	0·590	3,040–>5,000
	2	14·8	0·558	2,250–>5,000
	3	14·3	0·578	2,150–4,050
	4	14·3	0·578	1,500–2,800
Petrol	1	8·9	0·810	3,420–>5,100
	2	9·9	0·727	2,950–4,800
	3	10·0	0·718	2,520–3,750
	4	9·6	0·751	1,630–2,750
Vaporizing oil	1	9·3	0·886	3,600–>5,400
	2	10·3	0·802	3,000–4,950
	3	10·6	0·775	2,500–3,850
	4	10·6	0·780	1,750–2,750

It will be seen that the optimum fuel consumptions are: Diesel, 0·558 lb per drawbar h.p.-hr; petrol, 0·718 lb per drawbar h.p.-hr; and vaporizing oil, 0·775 lb per drawbar h.p.-hr. Therefore, if the three fuels were all the same price per lb, the Diesel tractor would have the lowest fuel cost. In many countries, however, differential excise taxes make petrol the dearest fuel, and Diesel and vaporizing oil about the same price as each other.

Diesel engines usually have better slow-speed pulling properties than spark ignition engines. Reliability of the two types is probably about the same, and so are the costs of maintenance and depreciation. Starting is usually more certain with a compression ignition engine than with a petrol or vaporizing oil engine, but, because of the high compression ratio of the engine, a larger electric starting motor and battery are needed. More care is needed in storing Diesel fuel than petrol or vaporizing oil as far as the need for cleanliness and the importance of getting rid of sediments are concerned, but good bulk storage tanks look after this, and it is to be noticed that Diesel and vaporizing oil are much safer than petrol to store. One more point to be considered is that vaporizing oil engines are more prone to dilution of the lubricating oil than are petrol and Diesel engines, and less frequent changing of the oil in the sump is called for.

The cost and the rate of consumption of the fuel itself is, however, the greatest factor to be considered, and the answer will depend chiefly on the comparative cost of the fuels. In making the calculation it is well to determine a figure first for the number of hours per year the tractor is likely to run, and the number of years it is expected to be used before it is scrapped or sold secondhand. On farms where the tractor is the only one, or on farms where several tractor drivers are employed, 1,000 hours of work per year is a reasonable figure to take. For the life of the tractor we will reckon seven years, though of course many tractors continue to give good service for ten or 15 years. Then, reckoning fuel consumption per hour to be in the proportions of one for petrol and vaporizing oil and three-quarters for Diesel oil, we can arrive at the relative fuel cost for the seven years. For fairly heavy three-furrow ploughing we can reckon 1½ gallons of petrol or vaporizing oil per hour, and 1⅛ gallons of Diesel oil. However, we must remember that much of the 1,000 hours we are taking to be a year's work will be lighter work, such as harvesting and other transport, and the consumption per hour will be smaller, and it is to be noted that at small loads a Diesel engine is usually more efficient than a spark ignition engine.

In the case of the vaporizing oil tractor we must adjust the figure for fuel costs to allow for the fact that the engine has to be started on petrol and run on this lighter fuel until it is warm. About one gallon of petrol to every 20 gallons of vaporizing oil must be allowed.

The calculated differences in fuel costs for the seven years must then be compared with the differences in capital cost of the tractors designed to use those fuels. It is usually considered that when depreciation and interest on capital are taken into account, the additional cost of a tractor which burns a cheaper or more efficient fuel must be recouped by fuel cost saving within five years, otherwise its purchase is not justified.

One more factor to be considered in choosing what fuel shall be used is the possibility of standardization on one kind of fuel, with the advantage of simplification of storage arrangements. Many oil-fired grain driers and other plants with heaters use Diesel oil as fuel, and many combine harvesters have Diesel engines. Servicing of engines and the transport of fuel to the fields is all made simpler by standardizing as far as possible on one type and grade of fuel.

DRIVING THE TRACTOR

IT is not often that a new tractor has to be put into the hands of a man who has never driven before. Most boys have had a turn at driving during haymaking or harvest when work was so rushed that the farmer took the risk of having the tractor damaged by inexpert driving to release the regular tractor driver to do some other job. Nevertheless, it is well to let the agent who delivers a new tractor give tuition to some person or other on the farm, so that at least one member shall know any unusual features the particular make of tractor possesses.

Moreover, even at rush times it is not very wise to put a boy to driving without some short instruction. It is worth while to teach him before the emergency need for an extra driver arises. Therefore, when the regular tractor driver on a small farm has learnt the tricks of a new tractor he should pass on the knowledge to any people likely to be called upon to drive.

The first part of driving instruction ought to consist in letting the pupil become familiar with the functions of the various components of the tractor, and with the purpose of the various controls. This is best done when the engine is not running. The pupil should sit on the tractor seat and pull the governor control lever in and out, practise pushing out the clutch until he knows where to find the pedal without looking for it, and move the gear lever to and fro within the play of its neutral position. While he is handling the control levers he should be told what the movement of them does, and a brief explanation should be given of how the power from the engine is transmitted first through the clutch and then through whichever of the gears is selected.

The instructor should then start the engine. It is well at this stage for the pupil to be shown where the engine cooling fan is. When it is spinning round, the fan is nearly invisible, and on some tractors it is easy to touch it. The governor should be set low so that the engine will run slowly, and then let the pupil throw out the clutch, engage the low gear, and let in the clutch again and drive round and round some open space until he has got the 'feel' of the steering. It is well for the instructor to ride on the back of the platform of the cab of the tractor so that he can take over control if the pupil gets into trouble.

When the pupil has become confident in setting the tractor in motion and steering it and stopping it, he should be shown how to start the engine. If the tractor is a small one the pupil should learn how to start the engine by hand as well as by electric starter, but he must not try to swing it, and he must keep his thumb on the same side of the handle as his fingers, so that in the event of a backfire the handle will be wrenched out of his grasp before it can injure his wrist or arm.

Next the pupil should be shown how to back the tractor towards an implement and how to hitch the implement to the drawbar. The following is the best routine for attaching a three-point linkage plough to the tractor to teach a beginner.

Back the tractor so that it is square with the plough cross shaft, and the tractor and plough top-link connections are in line. The tractor must be correctly positioned before any attempt is made to attach lower links. If an error is made, it is easier to drive ahead and back up again than to attempt to man-handle the plough into position. Before dismounting from the tractor, wind the levelling lever in a clockwise direction, lowering the right lower link until three threads are showing on the lift rod. Dismount and attach the left lower link to the cross-shaft; secure with linch pin. Attach right lower link in a similar manner, using the levelling lever if necessary to align the pin, and secure with linch pin. Mount the tractor, start the engine and place forward end of top link in tractor top-link connection, moving the tractor slightly backwards or forward until the top-link pin can be entered; secure with linch pin.

Raise the plough on hydraulic lift and it is ready for transport to the field. When driving to or from work, it is advisable to raise the right lower link fully, by turning the levelling lever in an anticlockwise direction. This reduces the amount of slack in the check chains and stops the implement from swinging.

To detach the plough, level the plough with the ground, then lower carefully to avoid damaging the rear share. Disconnect plough in reverse order to attachment; that is to say, the sequence should be first the top link, then right lower link, and then the left lower link.

When the pupil has mastered the method of attaching the implement he can practise dropping it into work, and soon he will be ready to learn how to set the implement and how to carry out cultivations.

The operator must get the feel of the machine, and he must never use force until all attempts by gentle methods have failed. For instance, suppose the bottom gear of the tractor refuses to engage when the driver wants to set the tractor in motion. It can, of course, be forced to engage by the driver pushing the gear lever with all his might: but a most distressing noise of grating will come from the protesting gearbox. If this brute force method is used many times, the protests from the gearbox may become the heralds of very serious trouble indeed, so we must try to find a gentler way of persuading the gears to mesh with each other.

The first thing to do is to set the engine to tick over as slowly as it will run without danger of stopping. The second thing is to make sure that the clutch pedal is held down at the full limit of its travel. Before any attempt is made to move the gear lever, the clutch pedal should be held in this out-of-engagement position for several seconds, to allow the spinning plates and gear wheels to come to rest. If the gear still will not engage, the clutch pedal should be allowed to come back up to the top of its stroke and then be pressed down again. This usually causes the gear wheels to move into a better position for the teeth of one wheel to slide between the teeth of its neighbour wheel.

If gentleness is the watchword, detailed knowledge of the construction of a machine is not necessary for sympathetic operation. Nevertheless, the more the operator learns about the tractor, the more pleasure he will get from his work and the less likely he will be to develop bad habits in methods of operation.

For instance, instinct will tell him that it is better to let the clutch engage firmly and progressively than in a series of grips and slips, but until he understands the construction of the clutch he will not know how bad it is to drive with his foot resting on the clutch pedal. The pressure of his foot causes the clutch withdrawal to be partly in operation all the time. This mechanism is designed only for intermittent use. It is supposed to work only while the clutch is held out of engagement. Continuous pressure on the withdrawal bearings will make them wear out very quickly.

Again, in hot weather, a fellow-feeling may cause the driver to keep the engine of the tractor as cool as possible when it is working hard. In the case of a paraffin engine this will be unkind instead of kind. Paraffin fuel needs a high temperature to vaporize it. Unvaporized paraffin in the engine washes away the oil from the cylinder walls and the bearings, and the engine would rather be hot and oily than cold and unlubricated. Therefore the radiator ought to be kept covered with enough blind to keep the water nearly boiling.

Nevertheless, the circumstances in which one must be cruel to be kind are few: and during the period the driver is learning the construction and principles of the tractor, his best course is to do unto the tractor as he would have it do unto him.

Having learnt how to drive the tractor, and hitch it to implements, the next step is to learn how actual tractor cultivations should be carried out.

ACCURATE STEERING

Straight furrows and rows not only please the eye; they help cultivation. Crooked plough furrows prevent a parallel finishing strip, and the time and fuel spent in veering out short ends spoil the efficiency of a big piece of ploughing. In row-crop work a steering error can cut off several yards of plants before the outfit can be brought back on to its course.

Bad steering at the beginning of an operation is difficult to put right later on. A 'dog's hind leg' in the first furrow is usually reproduced in all the furrows drawn alongside it, while in row-crops a bend in the drilling of the seed causes trouble right through the hoeing season and the harvest.

Three-point Sighting

It is well to use three-point sighting when setting a plough rig, or drawing the first ridging furrow, or drilling the first round of seed. One point is the landmark being aimed at. The other two can conveniently be the radiator cap and the fuel tank filler cap of the tractor. As long as these three points are kept in line the tractor is bound to be running straight.

When a rigidly-attached row-crop implement is being used on the tractor, correcting the direction of travel must be done with slow movements of the steering wheel. It must not be done suddenly, for when the front of the tractor is steered towards, say, the right-hand side, the tools at the back will temporarily swing out to the left and cut the plants. Therefore all corrections must be made gently.

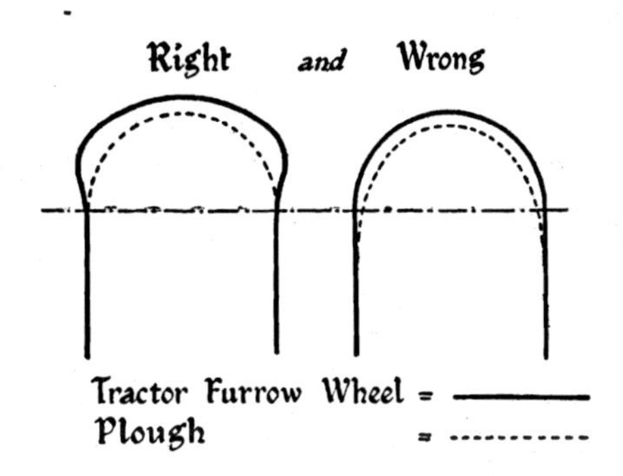

How a tractor is best handled at the headland, when it is turned ready to start work in another bout. As soon as the implement leaves work the tractor steering wheel should be turned momentarily in the opposite direction to which it is intended to make the sweep round back into work.

MAKING AND SPLITTING RIDGES

Planting ridges can easily be cut straight by a tractor with directly-attached ridging-plough bodies, but covering the sets by splitting the ridges is more difficult, for the wheels of the tractor then have to run along the tops of the ridges. Once the wheels begin to slip down into the valley it is extremely difficult to regain the heights before some seed has been damaged.

Splitting can be greatly helped by making the original ridges flat at the top. This can be done by setting the ridging plough shallow in relation to width and to angle of mouldboard. When the plough is set full depth for any given distance between the rows, the peaks of the ridges will be sharp-edged and it will be nearly impossible to keep the tractor along their tops. When the drills are drawn shallower a flat platform is left at the top, and the tractor can be driven along this path.

For easy steering when implements are directly attached the implements should be symmetrically placed along the tool bar, and should all be working at the same depth. It is always tiring, and sometimes impossible, to steer against side draught.

DRIVING BACKWARDS

Reversing with a trailer or towed implement often has to be done. For instance, trailers and ploughs and cultivators and seed drills ought to be put away with their backs against the wall and their drawbars facing outwards ready for a quick and easy start next time out.

This means backing the implement into place, which is a difficult operation at first. The implement seems always to go in the opposite direction to what we intended. Indeed, in reversing a trailer or any two-wheeled implement, the best rule is to steer contrary to instinct. For example, if a straight path is intended, and the trailer begins to turn towards the right-hand side, the tractor must be steered to the right. Instinct warns the driver that this will increase the deviation from the straight path, but, in fact, it will not do so. What it will do is to push the drawbar end of the trailer towards the right. The trailer will swivel on its wheels so that the back of the trailer, which is now the front of the outfit, will move towards the left and bring the outfit back on to its straight path.

All manoeuvring ought to be done at the slowest possible speed. Sudden fast starts must be avoided, for they often bring the outfit into some unwanted position or angle before the driver has time to stop the tractor or correct the steering. The governor control lever should therefore be set low.

When we are accustomed to driving, say, a motor cycle or car with direct throttle control, we are apt to speed up the engine of the tractor lest it stop when engaging the clutch. The throttle of a tractor engine, however, is controlled by the governor, and the governor increases the supply of fuel to the engine as soon as the engine begins to feel the load.

Steering gears must not be subjected to unnecessary strain. In any but the softest soils, the steering should not be turned except when the tractor is moving. To force the front wheels round on hard soil can damage the steering gear or bend the rod or connections. When a tractor has to be manoeuvred round in a confined space by driving backwards and forwards, strain on the steering wheel will be avoided and the turn will be accomplished much more quickly if just before the end of each movement the steering wheel is turned quickly in the direction in which the tractor needs to start off in the next part of the manoeuvre.

CHAPTER SIX

THE TRACTOR IN COLD AND WET WEATHER

WINTER brings several problems to the tractor operator. One of them is difficult starting in a spark ignition engine. This may be due to the spark at the plugs being too weak through condensed moisture affecting the insulation of the ignition system, or the inlet manifold may be too cold for ready vaporization.

Stiffness of the lubricating oil at low temperature can be helped by using a thinner oil, but care must be taken that it is an oil which retains enough body when it reaches the higher temperatures at which the engine runs. Even so, any cold oil is necessarily fairly thick, and the remedy is to ensure that the ignition and carburation are in such good order that only a few turns are necessary before the engine fires.

It is often found in the morning that moisture has condensed as a dew over the surface of the tractor, and one can 'write' on the tank with a finger. This layer of moisture is also on the porcelain insulators of the sparking plugs, and will give the high-tension current an easier path than will the gap at the sparking plug points. The porcelains, then, should be wiped with a warm dry rag. It is often worth while, too, to take off the cover of the distributor and wipe it inside.

COLD STARTING

On cold mornings one should make even more sure than usual that it is all petrol and no paraffin in the float chamber of the carburettor.

Vaporization can be helped by putting a cloth on the inlet manifold and pouring boiling water on to the cloth. If the engine is still stubborn the sparking plugs should be taken out and dried in an oven. Sometimes, if thick oil has been used to lubricate the impulse starting mechanism of the magneto, the trip sticks on a cold morning, and the familiar 'click, click' will be missing. The trip can be freed by injecting a few drops of fine, thin machine oil.

Ordinary low temperatures and high humidities have little effect on the starting of a Diesel engine. The coldness of the oil makes the engine more difficult to turn, but if the injectors are in good order and the compression is firm the engine will start as soon as it is turned.

Part of the resistance to turning a cold engine comes from the thick oil in the gearbox. This part of the resistance can be done away with by propping the clutch pedal in its out-of-engagement position.

FROST TROUBLE

Another cold weather trouble is the fear that an overnight frost will

H

freeze the water in the cooling system and damage the radiator or the cylinder block. Anti-freeze mixtures are good but expensive in the quantity needed for the large radiator of a tractor. Moreover, they are apt to be lost by boiling out of the overflow pipe if the tractor has a vaporizing oil engine, because in these engines the water must be kept near boiling point if the tractor is to run efficiently.

Petrol and Diesel tractors need not run so hot, and if the radiator and other parts of the cooling system are in good order it pays to use an anti-freeze solution. Indeed, in tractors with thermostatic control of water temperature, an anti-freeze is essential if the tractor is to be operated on days of low air temperature. The action of the thermostat can isolate a pocket of water from the water actually being warmed in the water jacket of the cylinder head, and this pocket can freeze even when the engine is hot.

Anti-freeze solutions seem to seek out any weak places and potential leaks in the water system. A radiator which was all right with water will sometimes start to leak as soon as anti-freeze is used and it is well to mix a proprietary leak-stopping compound with the anti-freeze compound. Before putting the mixture into the radiator, the cylinder head bolts should be tightened down to prevent the possibility of any getting into the cylinders or crankcase, where it might cause damage.

When obtainable, a good proprietary brand of anti-freezing solution may be used, but a solution of glycerine is an acceptable substitute. The following table shows the degree of protection afforded against freezing for various strengths of solution.

Per cent of Glycerine in Solution in Water	*Freezing Point*
20	26° F
30	22° F
50	12° F

If commercial alcohol is available as well, a good mixture is 60 per cent water, 10 per cent glycerine and 30 per cent alcohol. This freezes at 8° F. Fresh alcohol must be added frequently to make up for loss due to evaporation.

Do not top up the radiator with water if anti-freeze solution is in use, otherwise the solution will be progressively weakened and freezing may occur.

Top up with an anti-freeze solution when the engine is hot. This will prevent the loss of the solution through the overflow pipe which would occur if the radiator were filled when the engine was cold. When the engine gets hot the water expands and overflows if there is not room for it.

In countries with extremely hot summers and extremely cold winters, the anti-freeze solution ought to be taken out in the summer and saved ready to be put back in the winter. Water containing large quantities of glycerine has a lower specific gravity than pure water, and therefore it cannot cool the cylinder heads and valve ports as efficiently as pure water. This may cause overheating.

To find the difference between the heat capacities of water and of anti-freeze solution, take two small, large-mouthed bottles, each holding a thermometer, and fill them with hot water from the same container. Place one in a quart jar holding water and the other in a quart jar filled to the same level with the anti-freeze solution being used. The temperature of the anti-freeze solution and water should be the same at the beginning of the test, and it will be seen that the bottle cooled by the water will drop in temperature much more rapidly than the bottle cooled by the anti-freeze.

In the case of a vaporizing oil tractor considered too large or old for the use of anti-freeze to be economic, all-winter vigilance is essential if a frozen radiator and cylinder block are to be avoided. The safest thing is to let the water out of the radiator on any evening when there is the slightest risk of frost. Wrapping sacks and straw round the radiator and engine, without letting the water out, may help the engine to start easily after a chilly night, but it is not a reliable precaution against freezing on a really frosty night. Also, it is worth mentioning here that it is dangerous to pack straw near the exhaust pipe of an engine that has only lately stopped work. The straw may catch fire.

EMPTYING RADIATOR

It is well to make sure that the draining plug used to empty the water is really the lowest point of the cooling system. If the tractor is not standing level it may be that a pocket of water will be left below the drain plug. Some tractors have two drain plugs, one at the bottom of the radiator and one on the engine; in this case both should be opened.

We have seen that on some petrol-driven tractors the filling cap of the radiator is an airtight fit. This cap must be removed before the drain taps are opened, otherwise only a little of the water will drain out because no air can enter to replace it.

When radiators are being emptied, the temptation is to leave the tractor as soon as the taps have been opened, and let the water flow away without attention. A danger here is that a flake of scale or some other solid matter may get lodged in the outlet and block the flow, and leave enough water in the cooling system to freeze and damage the cylinder block. If the driver waits by the tractor until all the water has drained out, he can move any blockages of this sort by poking a wire into the opening of the tap.

When the tractor has to be left in the fields, far from a water supply, a container wide enough to hold all the water and low enough to go under the radiator must be provided so that the water can be saved ready for refilling the radiator in the morning. Although covering this container with sacks may protect the water a little against freezing overnight, the water will usually freeze into a solid lump on a hard night, and there is no way of filling the radiator other than carrying water to the tractor. This cannot be avoided. At least there is the consolation that as the water has frozen in the can, it is almost certain that it would have caused damage had it been left in the engine.

When the tractor can be left overnight at a place near a supply of hot water, starting in the morning can be made easier if the emptied radiator is filled with hot water. When this has been done, the engine should be left for some minutes before attempts are made to start it, so that the heat can have time to pass through the metal of the cylinder blocks to warm the inside of the engine. Hot water will not damage the cold cylinder block. The water, even if originally boiling, will only have a temperature of about 200° F when it is actually poured into the radiator, and this is not likely to cause undue stress in the metal.

There are two reasons against this emptying of the radiator every night. One is that fresh hardness is deposited each time from the fresh water, and the other is that the coating of lime salts inside the pipes and cylinder block become harder when they are exposed to air than they do if they are kept in contact with water.

COLD-WEATHER RUNNING

Once the engine has been started on a cold day, it must be made hot quickly to loosen the oil so that the bearings and cylinders are soon lubricated, and it must be kept hot so that the fuel is completely vaporized and burnt. A tin cowl fitted round the carburettor and inlet manifold of a vaporizing oil engine helps to prevent cold winds from causing the paraffin to condense on its way to the cylinders.

In work, the engine is most likely to get cold when it is doing light jobs, or is idling during the loading of trailers. It is well worth while to cover the radiator with a sack or an old coat when the work is light; and if it is inevitable for the tractor to be left standing with its engine running for any length of time, it should be drawn up in such a position that its radiator is pointing away from the wind.

In very cold weather tractor lamps should not be switched on until the engine has been running for a minute or so, otherwise the lamp bulbs may blow out.

In the case of petrol and Diesel engines the water in the radiator need not, and indeed should not, be kept very hot. In fact it is worth observing that if the water tends to boil even when the radiator is not blanked off, then there is something wrong with the tractor, or it is being worked too hard.

Many tractors, Diesel, petrol and vaporizing oil models, are equipped with a radiator thermometer. In the case of a vaporizing oil tractor, this thermometer is extremely useful as a guide to the time at which the fuel can safely be switched over from petrol to vaporizing oil after starting. The change should be made when the thermometer shows 190° F.

DRIVING IN WET WEATHER

The operational difficulties in wet-weather tractor work are chiefly slipping of the driving wheels and sliding of the steering wheels.

Driving wheels lose their grip on wet soil because the soil is soft and offers little resistance to 'shear', and because the space lugs of the wheels, or the tread bars of the tyres, become clogged with mud and so present only a smooth surface to the soil. Nothing can be done to change the soil, but the lugs or tread bars can be cleaned to make them penetrate well and take advantage of what little resistance to shear the soil does possess.

With spade-lugged steel wheels the spaces between the lugs must be cleaned out from time to time with a paddle. This cleaning out is made easier if the middle part of the rim can be kept free of mud by a scraper. The scraper can be made by fixing a tine rigidly to the frame of the tractor, with its point passing between the two rows of wheel lugs and nearly touching the rim.

Such a scraper cannot be used with open-type skeleton wheels, but most of these open wheels are peculiarly well suited to working in wet soil. There is little lodgment for the mud and therefore no scraper or cleaning is needed.

Steering

In some circumstances pneumatic-tyred front wheels do not steer as well as skid-ringed steel wheels, but a new deeply ribbed pneumatic tyre can be very good. In any case, the steering of a wheeled tractor depends upon the grip of the front wheels on the soil in much the same way as the forward movement depends upon the grip of the back wheels.

However wet the soil may be, a tractor running free, pulling no load, will answer to the steering all right. When a load is being pulled at the drawbar the distribution of weight on the tractor is upset; the weight on the back axle is increased and the weight on the front axle is reduced. This helps the grip of the rear wheels but hinders the steering. When the front wheels are turned for a corner they merely slide along the wet soil instead of changing the direction of travel of the tractor.

This tendency can be checked by tying weights on to the front of the tractor, but a simpler way is to plan the work so that there will be no need to try to turn the tractor while it is pulling a heavy load. For instance, in ploughing there will be no difficulty while a straight furrow is being pulled, and the outfit will steer round the headland all right once the plough has lifted completely out of work. This, however, needs a wider headland than in dry weather; for, although a ploughman does not turn the steering wheel until after he has pulled the trip cord of the plough, there is a yard or two in which the plough is still rising out of work and is still exerting enough drawbar pull to spoil the steering in wet weather.

This will leave more headland ploughing to be done, which is itself difficult in wet weather with a tractor. It is nevertheless worth while to go on ploughing the main, middle parts of the field and leave the headlands for a drier day or for horses to do. Similarly, finishing furrows are difficult to make neatly with a tractor outfit on wet soil; the tractor slides sideways when all its wheels are on ploughed land. It may well pay to leave the finishes to the horse plough.

Depth of Ploughing

In wet soft soil an even depth of ploughing with a tractor outfit is sometimes difficult to maintain. If the land-side driving wheel of the tractor should begin to slip, its track will become depressed and it will leave a trough of loose soil. With a three-furrow trailed plough the land wheel of the plough usually follows immediately behind the land wheel of the tractor, and therefore it will run in the trough made by the slipping tractor wheel.

The depth of ploughing, therefore, increases, because the height of the land wheel of the plough determines the height of the plough bottoms. The deeper ploughing causes a higher drawbar load and this in turn causes more slip of the tractor wheel, and very soon the tractor becomes bogged. This trouble can be lessened by fitting a wide extension rim to the land wheel of the plough. The extension rim may ride outside the trough made by the tractor wheel, and in any event the larger area of contact of the widened rim will hold up the plough with less sinking in.

Ignition System

The ignition system of a tractor is well protected against moisture. The only vulnerable parts are the sparking plugs. Plugs will misfire if their insulators are coated with water. Rain will not settle on the insulators while the tractor is running and the engine is warm, but if the tractor is left standing idle in the rain the engine should be covered up lest the rain drive in sideways on to the plugs.

If the insulators do get wet they must be dried with a cloth, for the high-tension current will prefer to travel along the coating of water rather than take its proper path across the sparking plug points.

Diesel engines are not sensitive to rain, as long as it does not get into the fuel. When any tractor tanks have to be filled in rainy weather care should be taken to protect the opening.

If a tractor has a vertical exhaust pipe it is worth inverting a small can over the outlet of the pipe when the tractor has to be left idle in the rain for long periods. This will prevent any possibility of water getting into the engine through open exhaust valves.

CARE OF TYRES

THE care of front tyres is much the same as the care of motor car tyres, but rear tyres, the driving wheel tyres, must have some special attention if they are to remain sound in the walls during all the time that their tread remains serviceable, which on most soils is seven years of hard use.

In particular, the tyres must be protected from oil. Oil causes rubber to swell and then rot. As soon as a smear of oil is seen on a tyre it should be removed by rubbing with a rag. Also, any oil should be wiped off the rim or any other parts of the wheel near enough for there to be any risk of the oil creeping on to the tyre. Tractors sometimes leak oil out of the wheel bearings of the back axle casing and if the leak is bad it may pay, for the sake of the pneumatic tyres, to renew the oil seals of the bearings.

TYRE PRESSURES

Tyres wear rapidly if the rubber, or the fabric upon which the rubber is built, is flexed excessively. If the tyre is not kept pumped up hard enough, the walls of the tyre become unduly bent at the part of the wheel in contact with the ground, and may crack.

The deflected part of an excessively soft rear tyre, on a tractor that is pulling a heavy load, is in a complex curve, a kind of ripple. This is because the tractive effort is being taken through the deflected part of the tyre. It may be that a pull of as much as 2,000 lb in each tyre is trying to drag the outer part of the tyre tangentially away from the inner part. The softer the tyre the more pronounced is this ripple, and the more liable is the tyre to be damaged. The complex deformation of the tyre takes place once in every revolution of the wheel, and the wheel goes round about a thousand times in an hour.

When ballast is loaded on to the tractor to help the grip of the driving wheels, the tyres should be pumped up harder to withstand the added weight.

Besides prolonging the life of the cover, the correct inflation protects the tube. If the tyre is not pumped hard enough, the inner tube will tend to creep round inside the cover. The valve is clamped in the rim, and so the creeping tube tugs at the valve until in the end it is wrenched out of its seating.

Hand pumps can be used, or motor-driven compressors, or a simple pump which takes the place of one of the engine's sparking plugs. With this, the engine is allowed to run, slowly, on the remaining cylinders.

The pressure and suction generated in the 'sparkless' cylinder cause the pump to draw air through a valve in its barrel, and force the air through the flexible connection into the tyre.

When the pump is being put on, or taken off, care should be taken that neither the pump end nor the tyre connection end fall into the soil and become gritty. Grit on the pump will get into the engine cylinder and will injure the cylinder bore; and grit in the connection will be blown into the tyre where it may lodge between the valve and its seating and cause an air leak.

The engine should be running on petrol, not paraffin, for this job, since if any paraffin vapour is drawn into the cylinder it will not be exploded and it may wash away the lubricating oil. It is not likely,

Inflating a tractor tyre with a portable self-contained pump unit with miniature petrol engine.

however, that fuel will ordinarily pass into the cylinder. The operation takes place as follows. On the suction stroke of the piston, the engine exhaust valve is closed and the engine intake valve is open. This valve condition would normally cause petrol to be sucked into the cylinder by the downward-moving piston. As the suction and exhaust periods of one cylinder of an engine usually overlap the same periods of another cylinder, there is always a suction or vacuum in the intake manifold. With the spark plug pump installed, the pumping piston does not have an opportunity to create a vacuum stronger than exists in the intake manifold. The opening of the pump inlet valve permits air at atmospheric pressure to neutralize the suction of the downward motion of the engine piston. Therefore, no fuel is likely to be sucked in the cylinder and forced into the tyre.

PUNCTURES

Care should be taken when the pneumatic-tyred tractor is driven about

Darvill strakes extended to help wheel grip. (*See Chapter* 7.)

Donaldson Skidmaster girdle. (*See Chapter* 7.)

A Fordson track-laying tractor ploughing, with both tracks riding on the unploughed land. (*See Chapter* 8.)

A Robot post-hole digger mounted on a David Brown tractor. The auger is operated through gearing from the power take-off shaft of the tractor. (*See Chapter* 8.)

Ground-crop sprayer mounted on Allis-Chalmers tractor. The booms are lifted for transport. (*See Chapter* 8.)

The Bamford hedge cutter is an example of a machine attached to a tractor but operated by a separate engine. This machine is being used by two men, whereas some mechanical hedge cutters can be manipulated by the tractor driver alone. The accuracy of steering with a second operator usually compensates, however, for the additional labour involved. (*See Chapter* 8.)

The length of the top link of this three-point linkage system can be adjusted by turning the sleeve which has a right-hand thread one end and a left-hand thread the other end. Altering the length of the top link changes the pitch of the implement. (*See Chapter* 8.)

the yard not to run the tyres over kerb stones and threshold stones. The shock may fracture a few strands of the fabric of the tyre casing. Such fractures are often the origin of a bad burst. For this same reason of guarding against bursts, flints ought to be picked out of the tyres before they work their way through the strands of the fabric.

If even a slow puncture should occur in a tube, it is well worth while to repair it by removing it from the cover and patching it. If the tube is first completely deflated it will be found that even the rear tyres can be levered off and on the rims much more easily than would be expected from their large size. Moreover, on most makes of tractor, one side of the rim can be detached from the wheel, and then the tyre and tube can be lifted straight off. The half-rim is kept in place by a spring ring or by a series of nuts or studs.

If the patch is being put on to the tube of a rear wheel, it should be stuck on very firmly. It needs to be put on more securely than a patch on a motor car tyre tube. In a motor car wheel the tube is at comparatively high pressure and therefore presses the patch firmly between itself and the inside of the cover. Also, motor car tyres get hot when the car is running. The pressure and the heat cause the repair solution to vulcanize the tube and the patch into perfect adhesion. A tractor driving-tyre tube, however, does not get hot because the tractor never travels fast enough, but the tyre is used at much lower pressure than a motor car tyre. Since the rear tyres have to transmit the drawbar pull exerted by the tractor, and since the pressure of air in the tube is low, there is considerable flexing, with, consequently, a possibility of movement between the tube and the inner surface of the cover. Patches must, therefore, be put on to the tube very firmly.

Vulcanization by artificial heat is, of course, always the best way of repairing the tube. Small vulcanizers, which use either electricity or a pellet of solid fuel to provide the heat, can be bought quite cheaply and they are simple to use.

If ordinary patching has to be used, the part of the tube near the puncture should be cleaned with petrol and then roughed up with a wire brush. Sand paper or emery paper should not be used, because in some papers the material used to hold the abrasive to the paper interferes with the properties of the solution; and it is difficult to clear away all traces of the powder from the tube without soiling the surface again. If a wire brush has been used it does not matter if a little powdered rubber is left on the surface of the tube.

The solution should be spread on to the tube evenly, and allowed to set past the 'tacky' stage; indeed, it should be left until it is nearly dry before the patch is put on.

Leaking valves are, however, more frequent than punctures. The leak can be detected by spreading a film of moisture over the opening of the stem of the valve. If the moisture forms into a bubble, the valve is leaking. Valve plunger cores can be cleaned and re-used, but it is better to renew the cores.

STEEL-WHEEL JOBS

Pneumatic tractor tyres are invaluable for such jobs as hay-making and harvest, and for road travel. On some soils, however, they must be changed for steel wheels when a long stretch of ploughing or cultivating is to be done. When a tractor goes to work along the road its steel wheels and a jack can be towed behind on a trailer. Then the wheels can be changed as soon as the outfit gets inside the field.

WHEEL CHANGING

Front wheels need not be changed as often as the back wheels, and with detachable skid rings steel front wheels can be left on permanently. Changing front wheels is not a field job, since, in most makes of tractor, it entails dismantling the ball bearing and is best done on a firm, level floor, with a piece of cloth spread out under the wheel to receive the ball bearing and other parts of the bearing.

Rear wheels on some light models are fixed on by a six-pin fitting in much the same way as are motor car wheels. On some other tractors the hub of the wheel is keyed on to the tapered axle shaft end and held by a single large castellated nut retained by a cotter pin. On others, a split hub clamps on to splines cut in the shaft.

Sometimes when the wheel has not been removed for a long time, a wheel puller has to be used to free the hub from the shaft. When the wheel has been loosened, however, the actual taking off and rolling away can be very hard work. But there is a knack in doing this job, and when that has been learned, two men working together can take off the wheel, roll it away and replace it by the alternative wheel without bearing the full weight of a wheel.

A good, finely adjustable jack is needed. The tractor should be jacked up so that the wheel is right off the ground. Then the main nut or other clamping device should be removed and the wheel loosened from the shaft, either by rocking it to and fro, or if necessary, by a wheel puller. Next, the man working the jack must release it gently until the wheel is touching the ground and is just bearing its own weight; the weight of the tractor is still resting on the jack. The top of the wheel should then be pulled away from the tractor until the inside of the wheel hub is only about 1 in along from the end of the axle shaft. While the wheel is held in this tilted position the second man must jack up the tractor higher until the wheel is clear of the ground. Then the man pushes the top of the wheel towards the tractor and his partner lets down the jack until the wheel is resting on the ground again. Pulling the wheel upright will now free it from the tractor and it can be rolled away. The alternative wheel can be put on in the same way.

The best method of getting the keyways, or splines, into line, is to crank the tractor engine, with the magneto switched off and with bottom gear engaged. This will turn the axle slowly round. The second

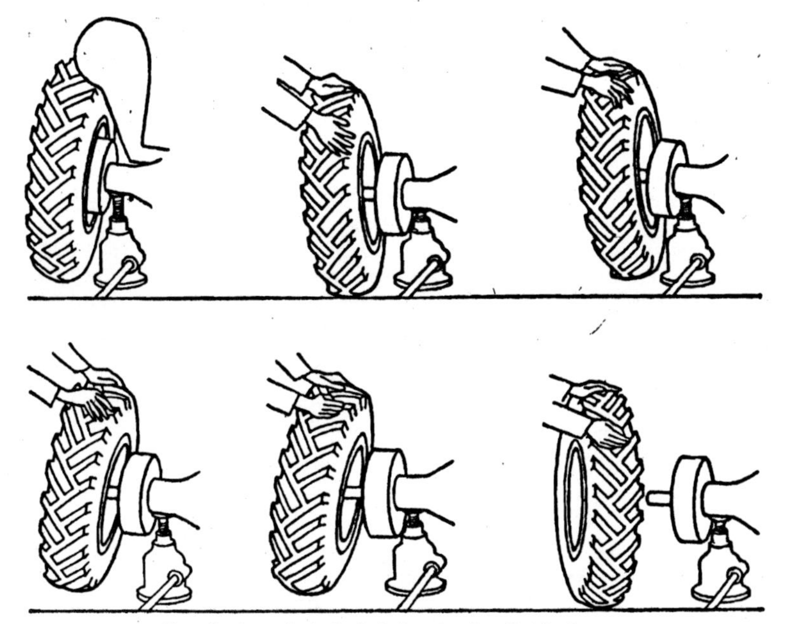

Steps in the method of wheel changing described in the text.

man can watch from the side of the tractor while he holds the wheel, to see when the key is in line with the keyway.

For this method of turning the axle shaft the tractor brake has to be off, and, therefore, the wheel not being worked upon must be chocked. It is as well, in any case, not to rely upon the brake but to wedge firm blocks in front of and behind the other wheels lest the tractor move and fall off the jack during the operation.

When the wheel has a six-nut fitting, it is likely to be easier to get the wheel into position; and there will be no need to turn the axle through many degrees. Indeed, the brake can be kept on while the wheel is refitted. For these wheels, the following removal and replacement routine is best.

Apply the brakes and put a jack underneath the rear axle shaft housing. Before raising the wheel fully from the ground, loosen the six wheel nuts. Raise the wheel just sufficiently, and then remove the nuts. The wheel can then be removed from the studs. Take care not to damage the threads by allowing the weight of the wheel to be taken on the studs.

Before refitting the wheel make sure that the flange of the axle shaft and the mating flange of the wheel are clean. Place the wheel in position so that the register on the flange of the shaft enters the counterbore.

Take care not to damage the wheel stud threads when the wheel is being put into position. Apply grease to the threads of the studs and replace the nuts. Do not tighten these consecutively round the circle, tighten a little at a time diagonally across the circle. This will ensure that the wheel is correctly seated and will minimize any tendency for the nuts to work loose when the tractor is working. Lower the jack and take it away. The nuts should be rechecked for final tightness after the tractor has been in use for a short time.

There is a special precaution to be observed when dual rear wheels are to be removed. When tractors equipped with dual rear wheels have been used in heavy soil conditions the space between the dual tyres may become clogged with soil which will become packed quite solid. To avoid the possibility of injury to the operator, wheels in this condition must not be detached before either the soil is removed or the tyres are deflated. If the soil is left packed solid between the inflated tyres when the wheels are removed, the pressure can strip the last few threads of the wheel studs, and cause the outer wheel to fly outwards.

BALLASTING TYRES

The tread bars of a pneumatic-tyred driving wheel will clean themselves to some extent by the flexing of the tyre. The amount of flexing depends in part upon the amount of deflection, or 'flatness' of the tyre, but the tyre must not be used too flat or it will be damaged. Also, it is better to gain the increase in deflection by adding ballast to the tractor rather than by letting air out of the tyre. The extra weight will help the grip.

Air and Water Ballast

A cheap and convenient way of adding weight is to fill the tubes of the rear tyres partly with air and partly with water. For inserting the water a special valve is used in place of the ordinary air valve. This special valve has a manual release to allow air to escape when the added water has compressed the air enough to cause a back pressure which would impede the entry of more water.

To this special valve can be connected either an ordinary water hose from a mains supply or from an elevated tank, or a stirrup pump hose. If the water filling is done when the tractor is jacked up and the wheel has been turned until the valve is at its highest position, about two-thirds of the volume of the tube can be filled with water, and this amount of water weighs about 2 cwt for a tyre measuring 11·25 in × 28 in.

When as much water as possible has been put in, the special valve is taken out and the ordinary air valve is replaced. Then the remaining capacity of the tube can be filled with air by pumping up the tyre in the usual way. The weight of water added has the same effect on the relation of tyre deflection to air pressure as would an equal weight of

FILLING TYRE BY MEANS OF HAND-PUMP

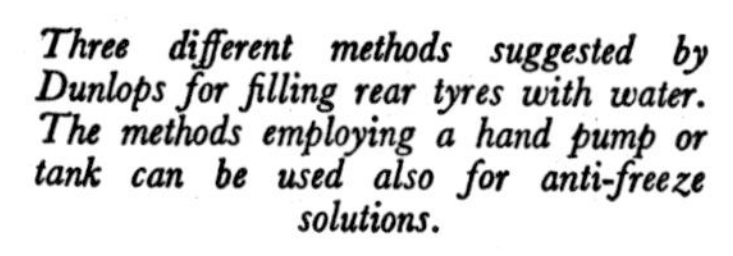

Three different methods suggested by Dunlops for filling rear tyres with water. The methods employing a hand pump or tank can be used also for anti-freeze solutions.

FILLING TYRE FROM MAIN WATER LINE

FILLING TYRE FROM TANK

iron bolted to the wheel. To avoid excessive deflection the tyre will have to be pumped to a higher pressure than before the weight was added.

100 *Per Cent Liquid Ballast*

The methods of water filling so far described fill only about two-thirds of the volume of the tube. The remaining third is air. This air space is inflated just as a tube with no water in it would be inflated, and the pressure in this cushion of air determines how much the tyre will deflect at the part of its circumference which is in contact with the ground. By using a type of air-water valve which has an extension piece of plastic piping fitted to the air outlet, a tube can be filled to as much as 90 per cent of water, because the pipe can extend upwards inside the tube, nearly reaching its uppermost boundary. This provides a greater weight of ballast, and the remaining 10 per cent or so of space is the air cushion. It is possible, however, to fill the tube completely with water, giving considerably more weight. Other advantages also are claimed, such as prevention of loss of pressure and the need for periodical inflation, since the water does not leak as readily as air does, for even rubber is slightly porous to air.

This 100 per cent water filling needs, however, a power-driven water pump and air extractor, and the help of an implement agent who has the necessary equipment must be sought. Usually the tractor to be treated will have to be taken to the agent's premises, but some dealers have portable outfits which can be brought to the farm.

Calcium Chloride Anti-freeze

Water in the tubes of tyres will freeze on a hard winter night. The freezing is not likely to damage the tube or tyre, but the tractor must not be driven while ice remains in the tubes, lest the edges of the ice cut the rubber or cord. The freezing can be prevented by adding calcium chloride to the water: 2 lb of flaked calcium chloride to each gallon of water gives ample protection. The mixing is best done outside the tyre.

Considerable heat is developed by the mixing, and the calcium chloride must be added to the water, and not the water to the calcium chloride. It will be found convenient to make all the calcium chloride into a concentrated solution, say about 5 gallons of solution. Then when some 10 gallons or so of water has been put into the tube, the solution can be put in, then more water added until the tube is filled. (By the way, calcium chloride must not be used as a radiator anti-freeze; it corrodes metal.) Solutions of commercial calcium chloride are slightly acid: therefore 1 lb of slaked lime should be stirred in with each 100 lb of calcium chloride, to neutralise acidity.

Ballast for Ploughing

There are many occasions in ploughing when ballast may be very necessary, but only one of the wheels needs the added weight. In Chapter 2 we considered a load of, say, 14 cwt on each back wheel when the tractor is on an even keel. If, however, the tractor is tilted so that one side is 8 in lower than the other, as it is when ploughing, the weights on the two back wheels change to about 11 and 17 cwt respectively. The lower, or furrow-side, wheel, therefore, has usually sufficient weight upon it to enable the tyre to grip the soil firmly enough to pull the plough, but the land-side wheel may be too light, and may slip. Hence, the best way usually to add ballast for ploughing is to bolt iron weights on to the wheel, or put water into the tube of the tyre, on the land wheel side only. Ballast added to the furrow side of the tractor will not add weight to the land wheel side.

Chains or Strakes

In conditions where ballasting does not give the pneumatic-tyred driving wheels sufficient grip, chains or strakes can be used to improve traction.

Chains help, but in some circumstances they do not provide enough extra grip, and also they are apt to chafe the walls and tread of the tyre. Much better than chains are the patented girdles made up of spiral strakes connected by chain. These girdles are not difficult to fit. The girdle is laid on the ground; the tractor wheel is driven on to it;

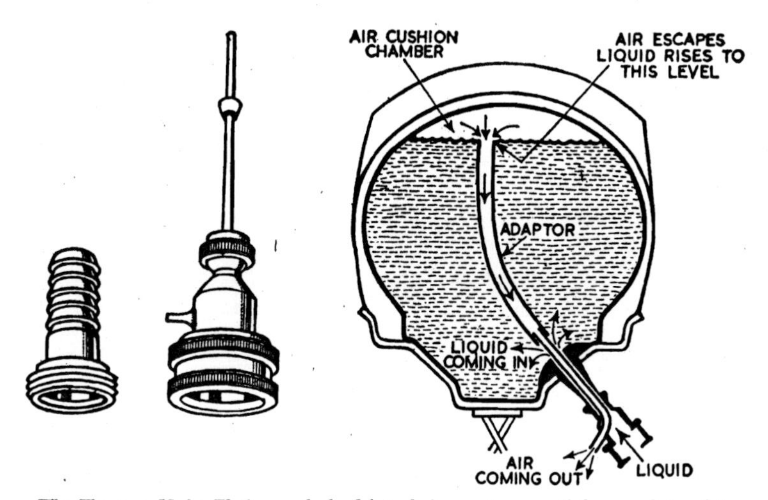

The Firestone Hydro-Flation method of introducing water or anti-freeze solution into tractor tyres. Left, the valve parts and flexible tubular adaptors which have to be chosen according to the size of the tyre being filled. Water pressure hose is attached to the Hydro-Flator valve.

WATER BALLAST FOR TYRES

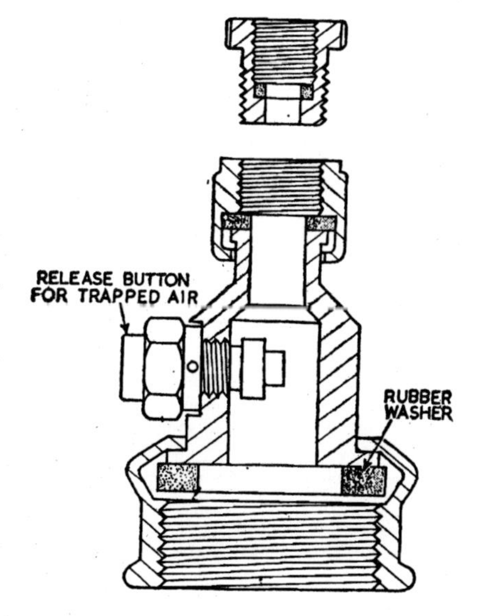

A Schrader water-filling connection with a manual release valve for allowing the air to escape when it has been replaced by water or anti-freeze solution.

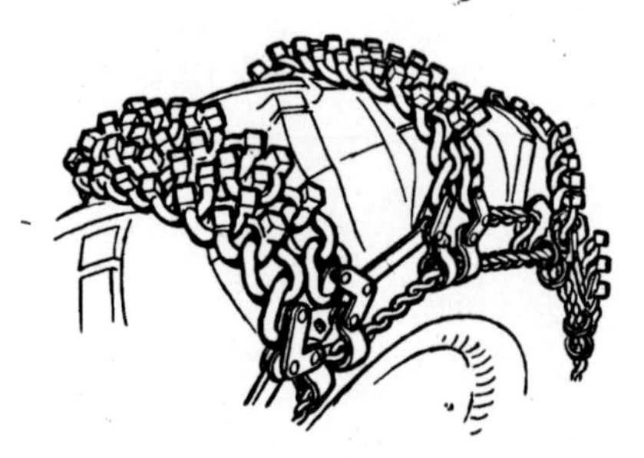

These chain strakes, used in France, were designed specially for use on tractors in log hauling where timber has to be brought out of woods where the soil is soft. The links of the chains have cube-shaped studs on them to help the wheel grip.

and then the ends of the girdle are lifted over the tyre and hooked together by a coupling.

Other non-skid attachments bolt to the wheels and carry iron lugs that project beyond the tread of the tyre. Some types are designed to be fastened on to the wheel easily, and to be quickly and completely detached. Others remain on all the time, but the lugs are retractable so that when they are not needed they can be drawn back towards the centre of the wheel and the pneumatic tyre can be used in the ordinary way; the weight of the folded-up strake helps the tyre to grip the soil.

It is important that the design of the strake should not allow any parts of it to touch the tyre even when the walls are deflected by contact of the tread with the ground.

CHAPTER EIGHT

IMPLEMENT WORK

THE use of the tractor as a portable power unit for belt work in driving threshing machines and other food preparing machinery, and estate equipment such as saw benches, is dealt with in the chapter on belt work, but most of the work of the tractor will be in cultivations, in harvesting and in transport. The implements for these field operations are either directly mounted on to the tractor or are drawn through the drawbar plate and choice of these two methods has to be made.

In the case of a machine with moving parts, we may have to decide also whether the mechanism shall be driven through a power take-off shaft or through land wheels. For rotary cultivators, power drive is essential. Mowers towed through the drawbar plate can have land-wheel drive, or mounted engine drive, or can be operated through a power take-off shaft from the tractor. Coupled mowers are all driven by the tractor power take-off. Some are in the form of a cutter-bar assembly mounted directly on to the tractor; others, semi-mounted, have a framework and castor wheel or wheels. The principle of the cutter-bar mechanism is the same in all types, and most of the following notes on operation and overhaul apply to all kinds of mower. Drive by V-belt is popular in power-operated machines, and some makers take advantage of this form of drive to provide adjustable knife speed.

Land-wheel-drive trailer mowers are easily hitched to and uncoupled from the tractor, and are therefore convenient for intermittent use; they can be used with any make of tractor. Knife speed depends solely on the forward speed and cannot be altered to suit crops or conditions. Wheel slip may occur in very difficult conditions but most modern tractor mowers are heavily built in order to minimize this trouble.

Power take-off drive trailer mowers are preferable to land-wheel drive for large areas, where the little extra trouble of connecting up is worth while. They can deal with all crops, including heavy leys and linseed, and can also be used for topping of pastures and thistle-cutting at high speed, because knife speed can be kept low relative to forward speed.

Direct-mounted mowers are attached either to the rear or to the side of the tractor. They are simple and easily controlled when working. However, mounting may take longer with this type than with the trailed types. Direct-mounted mowers can be attached only to the type of tractor for which they were intended, and this may restrict their usefulness on a farm employing several different makes of tractor. A side-mounted mower does not entail the removal of the tractor drawbar;

it can be left in place at times when the tractor is doing other work.

Most reaper-binders and some combine harvesters employed with tractors are usually power take-off drive, though many combine harvesters have a separate engine and are therefore independent of the tractor as far as the speed of drive to the mechanism is concerned.

As an example of an implement which is more independent of the kind of tractor, we can consider the fertilizer distributor. Nearly all these machines are trailed and in nearly all of them the mechanism is driven from the land wheels of the machine. In this case, then, thought need be given only to the type of machine, and to its sowing width in relation to the power of the tractor. Few normal-sized distributors give a full load even for tractors, and often two or three can be hitched together so as to increase the rate of work. A multiple hitch should be made in such a way that the distributors can be easily unhitched and used separately when required. It should also be made so that the distributors neither miss strips nor give a double dressing, and do not foul each other when turning. The use of a multiple hitch is only practicable where fields are large and fairly level.

The advantages of mounted ploughs and cultivators, in a size and weight within the range of the lifting and holding capacity of the hydraulic mechanism, are beyond doubt. The outfit is more compact than a tractor and trailed plough, and it can be driven backwards easily, and it is readily transported as a unit from field to field.

Directly attached ploughs can be worked with a narrower headland than trailed ploughs since no implement wheels need be on the soil when the outfit is out of work, and therefore there can be no strain on the implement caused by a sharp turn. There is, however, little disadvantage in a wide headland for ploughing, because the field can be finished by round and round ploughing at the edges.

In row-crop operations, there is great advantage in being able to work right up to the edge of the field leaving only a track for a cart or a headland wide enough for any harvesting machines that will be used, and directly-mounted implements allow this to be done easily. In market garden work in small plots it is often worth while to reverse the tractor into position at the edge of the field, then drop the implement into work.

The following are some notes on methods of connecting implements to the tractor and on the use of those connecting parts which transmit power to work the implement and convey control-impulses. It will be useful first to deal with any adjustment which has to be made to the tractor when particular implements are attached to it.

ATTACHMENT OF IMPLEMENTS

When a directly attached mower is fitted to the tractor the adjustment of the wheel track must be such that the wheels will run in the space between the swaths. When a trailer is to be used the track width of the

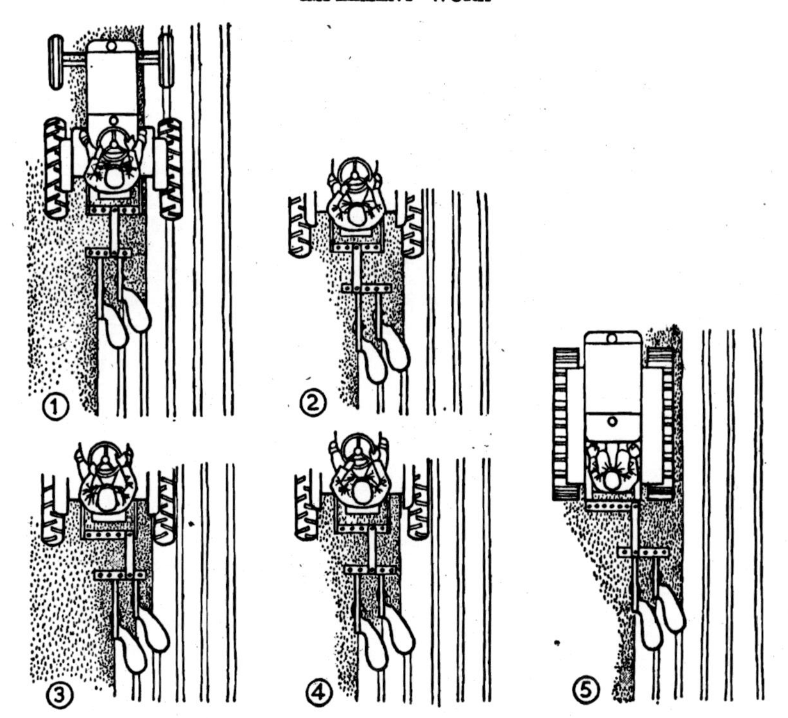

Side draught is inevitable if width of ploughing is narrower than the track width of the tractor. To let a two-furrow plough trail squarely behind a tractor, the tractor would have to run on the ploughed land as in (1). In (2) the hitch has been moved over on the plough hake, towards the land side, in order to let the furrow wheel of the tractor run in the open trough of the previous bout of work. But this puts all the side draught upon the plough and will cause it to run crabwise and cut too wide a front furrow. In (3) the hitch has been put out of centre on the tractor instead of on the plough. The plough will now do good work, but the tractor will be difficult to steer. The furrow-side driving wheel has to bear more than its fair share of draught. The furrow-side wheel lags behind and makes the tractor tend to turn towards the furrow side. (4) shows the best compromise; the side draught has been shared between the plough and the tractor. (5) shows a tracklayer with a two-furrow plough; both tracks must run on land, therefore side draught is very great, but usually a tracklayer pulls a wide width of work and side draught is small.

tractor should be made as nearly as possible equal to the track width of the trailer; this allows the trailer wheels to ride on firm soil in the depression made by the tractor wheels, and in soft ground this can make all the difference between the tractor pulling the trailer easily and not pulling it at all.

Another example of the importance of wheel track adjustment to match the implement is in the case of a directly attached plough. If the tractor is to be used with its wheels running in the furrow and in contact with the furrow wall, then the width of furrow for a single plough, or the width of the first furrow in a multi-furrow plough, will be determined by the setting of the wheels.

In ploughing, adjustments have sometimes to be made to the top link of the tractor. Shortening of the top link pitches the plough forward and increases penetration, but this adjustment should not be abused. With a correctly set plough and with shares in good condition, penetration ought to be almost automatic and certainly the plough ought not to need excessive tilting.

Attachment of three-point implements requires little muscular effort if the tractor is backed into position in such a way that only a little man-handling is necessary to bring the attachment points on the implement into their sockets on the tractor. This is true only if the implement is of a kind which can stand level on the ground when it is detached from the tractor. Most implements, including multi-furrow

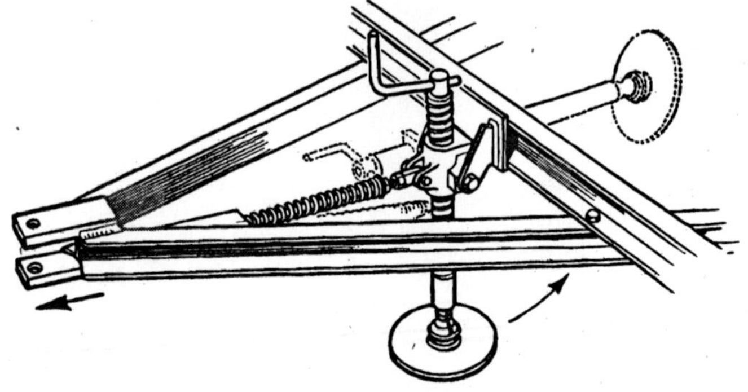

A jack to support a two-wheeled implement when the tractor is taken away. This jack, supplied on the drawbar of a Bamford trailer spreader for farmyard manure, folds away against a spring when it is not in use, as shown by the curved arrow. When the jack is to be screwed down into position to support the trailer, the spring acting in the direction shown by the other arrow holds the jack in place.

mould-board ploughs, will stand perfectly firm and level. A disc plough, however, usually does not stand level on its own and so a jack is needed. In some makes a suitable jack stand is included in the design of the plough. The stand is always carried on the plough and is merely dropped into position when the plough is to be detached from the tractor.

Two-wheeled trailers also need a jack of some kind to maintain them at the right level for easy attachment to the tractor. Some trailers, however, have automatic hitches. One make of trailer has a hook assembly which is attached permanently to the tractor. The drawbar has a telescopic member, and the following is the routine for connecting tractor to trailer. Lower linkage and as the drawbar falls, the hook will disengage and hang below the tractor. Reverse the tractor squarely to the trailer until the hook is directly beneath the eye on the tractor hitch beam. This is simplified by sighting along the telescopic member of the drawbar. Disengage the clutch and apply the tractor brakes the moment the hook strikes the hitch beam. Notice that the drawbar rides up the trailer hitch beam before hook and eye come into line. When the hook and eye are correctly in line, put the tractor into neutral gear and engage the clutch. Move the hydraulic control lever to the lift position so that the hook engages the eye and lifts the front of the trailer. Continue lifting until the automatic latch on the telescopic member operates, then move the hydraulic control lever to the drop position.

Trailed implements which are lifted out of work by a slave hydraulic cylinder need a flexible coupling from tractor to implement of the same kind as the one mentioned above for trailers. Trailed implements which are controlled through the lift arms of the tractor need no connection by hose. A good example of these implements is a trailer-hitched tandem disc harrow in which the angle of the gangs can be adjusted by raising or lowering the three-point hitch arms.

By moving the tractor hydraulic control lever to the lift position the gangs will be immediately straightened. This should be done at all times when turning. It permits short turns and avoids the formation of ridges and hollows at the headlands. Forward movement of the hydraulic control lever allows the tractor lower links to fall and the gangs to angle.

The degree of angle is controlled by the distance that the links are allowed to fall and is selected before operation by placing a yoke in the requisite notch on the top link rack. Maximum angle is obtained when the yoke is placed in the notch furthest from the operator, and minimum angle in the most forward position. The distance that the hydraulic control lever is moved forward has no effect on the angle of the discs.

There are kinds of directly attached implements which are not truly field implements but are really nearer to barn or estate machines. Examples of this kind of equipment are directly attached saw benches, hammer mills and concrete mixers. The three-point attachment mechanism is used to pick up the implement and transport it to another situation of work and it serves also to position the machine in relation to the tractor belt pulley or power take-off shaft while the machine is being operated.

For a machine of this kind we will consider a wood saw driven by flat belt from the tractor belt pulley. The method of attaching and operating such a machine is as follows. Remove the power take-off cap, chain anchor straps and left hand chain. Fit pulley to the left. Replace the two top bolts from the belt pulley housing with studs provided, fitting hitch assembly bracket. Lower linkage and adjust with levelling lever until lower links are parallel. Connect assembly at lower links and assembly bracket. Start engine and raise saw very slowly. If it fouls the tractor mudguard, adjust levelling lever until clear.

If belt adjustment is necessary disengage pulley drive by means of the power take-off lever on the left side of transmission housing, lower saw frame to the ground and switch off engine. Slacken the four holding bolts and turn the belt-adjusting screws until the belt has approximately correct tension. Tighten bolts and adjusting screws evenly. This adjustment controls the centring of the belt on the pulleys. Revolve pulleys by hand to see that the belt centres correctly. If not, reset the adjusting screws as required. When the correct tension and belt position are secured retighten the four holding bolts very tightly.

To operate the saw, lift the hinged section of the blade guard. Start the engine slowly. Engage the belt pulley drive and gradually open throttle until the desired speed is attained. About three-quarters governor opening should produce the speed required for most sawing operations.

Drawbars

Often the drawbar pull is taken from a horizontal plate attached rigidly to the back axle casing or from a bar of triangulated structure attached to the three-point linkage of the tractor. In this plate or bar a series of holes is arranged transversely to the direction of travel of the tractor. This enables the hake of the plough or other implement to be attached either in the mid-point of the plate, or to one or the other side. The chief purpose of being able to offset the implement is to make it possible to combat side draught when the tractor is pulling a plough.

In some other makes of tractor the pull is taken through a swinging drawbar which is pivoted at a point between the back and front axles, and is free to move within a small arc horizontally. This swinging drawbar allows the tractor to pull the implement round through a smaller circle at the headland than is possible when a fixed drawbar is used. Swinging drawbars can, however, be fixed at any point in their travel, so that implements can be offset.

Some tractors have a clevis, or jaw, at the drawbar of the tractor. A drawbar pin holds the implement to the tractor. The pin passes through one hole in the drawbar and two holes in the clevis. More usually, as we have seen, the drawbar is a single plate and therefore the implement to be used with it should have a clevis. Many tractors have a convertible drawbar which can act either as a plate or a clevis.

Several countries have standards which specify the position of the drawbar in relation to the ground and to the rearmost point of the back wheels, and also the thickness of the drawbar plate and the size and position of the holes in it to receive the drawbar pin. Conformity with a standard on these points enables any type of tractor to be hitched to any type of trailed implement. The British Standard covering these details is BS 1495: 1948, and its provisions agree in all important points with the American Standard on the same subject prepared jointly by the Society of Automotive Engineers of America and the American Society of Agricultural Engineers.

It is important that drawbars shall be fixed very firmly to the tractor and shall be capable of carrying considerable loads. In the case of a four-wheeled trailer the vertical load on the tractor drawbar is very small, but in the case of two-wheeled trailers and other two-wheeled implements it can be very high. When a single-axle trailer, with its axle set towards the rear, is heavily loaded in front, the vertical load on the drawbar often exceeds half the weight of the loaded trailer, and this load is increased in certain circumstances of braking and gradient. These loads not only demand a strong drawbar, they also impair the stability of the tractor by causing the front wheels to tend to lift. This effect on stability is smallest when the drawbar is fixed as low as possible and as near to the axle of the back wheels as possible. The British and American standards recognize this point, and their specifications are really a help towards safety as well as interchangeability.

CONTROL OF IMPLEMENTS

Hydraulic Control

Directly attached implements are usually lifted out of work by hydraulic force transmitted from a pump driven by the engine of the tractor, and so are some of the larger trailed implements. Many other systems of lifting have, however, been tried in the past. At first implements were lifted out of work by hand-operated levers. The levers were arranged to give the greatest lifting advantage and often the work of lifting the implement was helped by springs. The next step was to use mechanical lifts driven through gearing from the engine of the tractor. This method was fairly satisfactory but the movement was too inflexible for this duty of lifting an implement smoothly out of the soil. Then pneumatic lifts were employed. An air compressor, mounted usually at the front of the tractor where it could be driven by belt from the engine, provided compressed air which was conducted to a cylinder and piston arranged so that the air pressure would lift the implement out of work. This again was a fairly satisfactory arrangement but the compressibility of air was a disadvantage. Air confined in a closed space such as a pneumatic tyre will squeeze or compress into a smaller space under increased pressure. Air so confined possesses bounce. Increased pressure will squeeze it into a still smaller space, and it expands under lessened pressure. This is one reason why air is not now used in tractor control systems. The implement would bounce up and down on a springy cushion of air.

Liquids, however, will not compress, and that is one reason why hydraulic lifts have become universal. For all practical purposes we can say that pressure applied upon a liquid in a container is transmitted in all directions within that container and acts at right-angles to the surface of the vessel. Because of the equal intensity of applied pressure, a liquid is a very convenient medium for instantly and efficiently transferring a force from the point where it is exerted to where it is required to act. Moreover, a liquid allows a small force exerted through a long distance to move a large resistance when exerted through a shorter distance.

One illustration of this principle is the hand-operated hydraulic jack which consists of two cylinders and pistons. The piston of the small cylinder is worked by a pump handle, and that of the larger cylinder, called the ram piston, lifts the weight. Because the area of the pumping cylinder is much smaller than that of the cylinder which houses the ram piston, many long easy strokes of the pump handle produce a small but powerful lift of the ram. There are automatic suction and delivery valves, also a hand-controlled release valve which lowers the jack by allowing the oil in the ram cylinder to return to a reservoir.

In a tractor, the oil pressure is generated not by hand pump but by a reciprocating or rotary pump driven from the engine, the flow of the oil being regulated by a hand lever which is within easy reach of the driver.

If the pump is connected to a hose, and oil is pumped into the cylinder, piston and load will rise. Pump pressure per square inch must be higher than cylinder pressure to lift the load. In some tractor systems, a control lever slides a simple spool valve to one side or other, uncovering ports which direct oil to and from the cylinder. When a hydraulic system is operating in neutral, the piston in the cylinder is held stationary by oil trapped on either side of the piston. The valves are closed, and oil is by-passed to the sump.

On some makes, when the lever is held between neutral and full forward or full back with slight sustained pressure of the hand, action is slow. The farther the lever is pushed forward, the faster the action. When the driver lets go the lever, it returns to neutral automatically and the action stops. Pushing the control lever full forward slides the spool valve to one side. This opens the oil passage, and oil is pumped into the rear part of the cylinder. The piston travels outwards and raises the implement. Oil in front of the piston is released by a ball valve, and flows back to the sump.

The driver need not keep his hand on the lever. The lever stays full forward until the action is complete, and the spool valve then returns to neutral automatically. When the control lever is pushed all the way back, the spool valve slides to the other side, opening another ball valve and uncovering another oil passage, and this allows the implement to sink to a pre-set limit.

Hydraulic Control of Trailed Implements

A hydraulic jacking-cylinder connected to a tractor by means of a flexible pipe and mounted with quickly detachable pin joints on a trailed implement can entirely replace the mechanical self-lift and hand-operated devices used in trailed ploughs and cultivators, and its

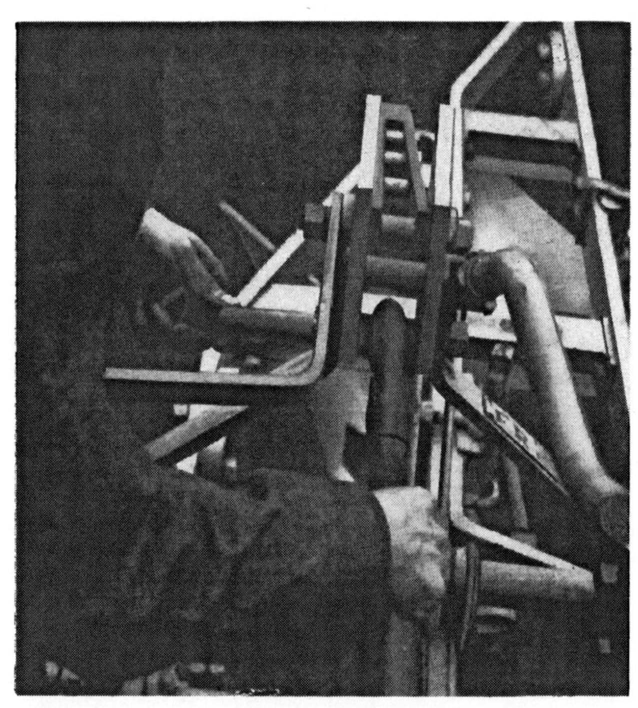

The top link of a three-point linkage system is the last to be fitted when an implement is being attached. It has a detachable pin and a retaining cotter. (*See Chapter* 8.)

The links of a tractor with three-point linkage are attached to the implement through a ball joint and are secured by a linch pin. The nearside lower link usually has no separate adjustment for height and therefore is the first to be attached when the implement is being connected up. The offside link can be adjusted to bring it opposite its attachment point. (*See Chapter* 8.)

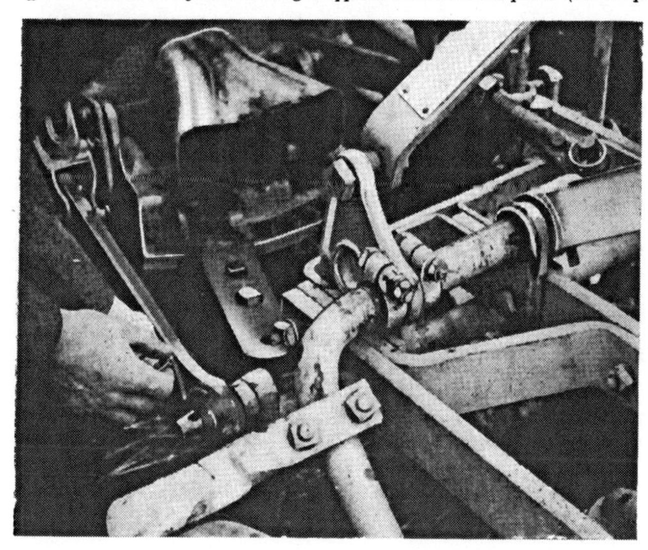

Making a finish in ploughing. The driver is adjusting the height of the offside link on a Fordson tractor. (*See Chapter* 8.)

TWO DRAWBAR ATTACHMENTS

A form of drawbar attachment to allow use of trailed implements with a Ferguson tractor having three-point linkage. (*See Chapter* 8.)

A form of drawbar attachment to allow use of trailed implements with a David Brown tractor having three-point linkage. (*See Chapter* 8.)

A Sutherland Power link connecting the hydraulic three-point linkage of a tractor to the trailer body in such a way that operation of the lift mechanism transfers weight to the tractor and improves wheel grip. (*See Chapter* 8.)

Automatic hitch on Ferguson trailers. In the left-hand picture the trailer has been backed until the eye of the trailer drawbar is over the hook of the tractor drawbar. In the right-hand picture the tractor link has been lifted by the hydraulic system so that the hook can engage in the eye. (*See Chapter* 8.)

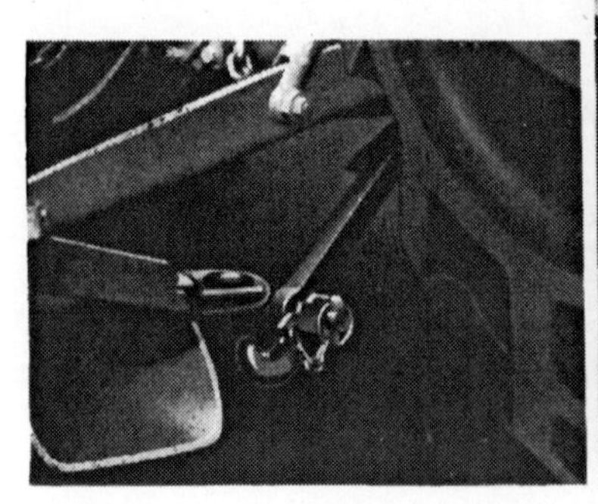

A Lockheed-Avery Breakaway Coupling which allows the hydraulic line from tractor to trailer or implement to be connected or disconnected without loss of hydraulic fluid and without entry of dirt. The unit illustrated is a double connection, for both pressure and return lines. (*See Chapter* 8.)

External hydraulic cylinder and ram used for lifting the boom on an Allman powder duster. Hydraulic power is generated by the tractor's hydraulic system. (*See Chapter* 8.)

External hydraulic pressure pipes connecting the hydraulic system of a tractor to a lifting cylinder. (*See Chapter* 8.)

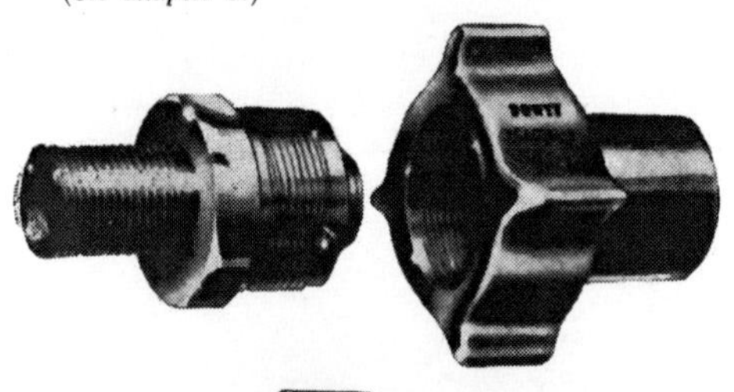

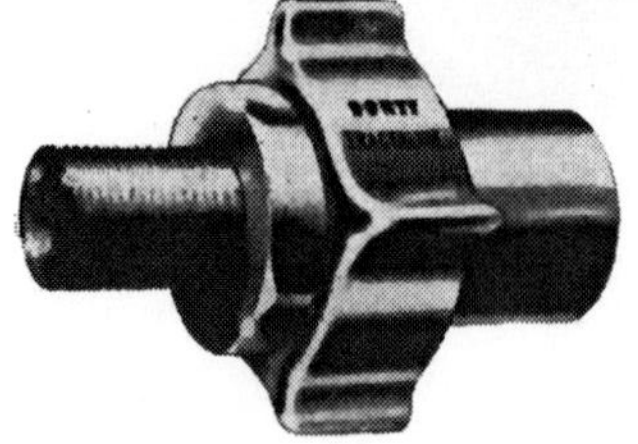

A self-sealing coupling for an external hydraulic pipe line from tractor to implement. The pipe can be disconnected without loss of hydraulic fluid. (*See Chapter* 8.)

Rockford Over-Centre clutch which can be kept in the engaged or disengaged position without thrust, useful for power take-off control. (*See Chapter* 8.)

Splined shaft of the power take-off transmission on a Fordson Major tractor. The shaft has a safety guard over it. (*See Chapter* 8.)

An adaptor made by Lawrence Edwards and Co., Kidderminster, for reducing a **1⅜-inch** *power take-off shaft to 1⅛-inch power take-off shaft.* (*See Chapter* 8.)

The adaptor sleeve made by Lawrence Edwards and Co. can be used to convert a 1⅛-inch splined shaft so that it can be used with power take-off implement designed to work with a 1⅜-inch splined shaft. One of the splines on the sleeve is grooved to take a pinch bolt or set pin with which the implement coupling is held in position. (*See Chapter* 8.)

advantages are many. Mechanical lifts sometimes become unreliable and difficult to engage when the machinery becomes choked with grit; and, being driven from the land-wheels, they may fail to operate when the soil is slippery. Hydraulic-control mechanism is independent of land-wheel drive, and is safely sealed against dirt. The benefit of hydraulic control over hand-lift is even greater. Raising a heavy implement by hand can be very hard work indeed, even when some spring-assisted mechanism is incorporated in the lift. With hydraulic control the movement of the hand-lever needs no effort at all.

The advent of hydraulic control brought one possible disadvantage, the loss of flexibility in the interchange of tractors. Whereas a trailed implement which is mechanically lifted can be hauled by any one of many types of tractor, the introduction of an hydraulic connection could restrict the use of a particular tractor to a limited range of implements. There is, however, a British Standard for hydraulic lifts for trailed implements, and implement manufacturers design their implements to use a standard portable cylinder. One cylinder will fit many different implements of various makes.

The cylinder is quickly removed from one implement simply by pulling out two pins; no tools are needed. A latch holds the implement in the raised position. The cylinder is just as easily put on another implement or machine. The hose lines going to the cylinder are extremely simple to connect and disconnect. Self-sealing breakaway couplings separate with a 12 lb pull, and valves in the couplings instantly close, spilling barely a thimbleful of oil. If the connections are broken with the implement raised, trapped oil holds it in raised position. If the plough strikes a rock and the hitch trips, the operator need only back up, push the hose connectors together, raise the plough over the rock with the hydraulic control and resume work.

The pressure in the connecting pipe is controlled by the valves operated by the tractor driver and the position of the ram piston in the cylinder on the implement responds to fluctuations in the pressure. Because of this dependence of the implement cylinder upon the hydraulic pump and control valves, it is often called a slave cylinder.

The British Standard for Hydraulic Lifts for Trailed Implements, BS.1773:1951, defines the space that the detachable cylinder can occupy and specifies that the cylinder shall be of the double-acting type, and that the system shall be capable of doing work to the extent of 50,000 to 55,000 inch-pounds in two seconds. The reason for specifying double-acting cylinders is that they give more positive location of the piston when the control valves are closed, and that the rate of descent is less affected by changes in oil viscosity due to changes in temperature. There may be conditions in which a single-acting cylinder is adequate, but if such cyliners are made it is desirable that they should be mechanically interchangeable with the double-acting cylinders which are the subject of the Standard.

To make it possible for one hydraulic jack to be interchanged with

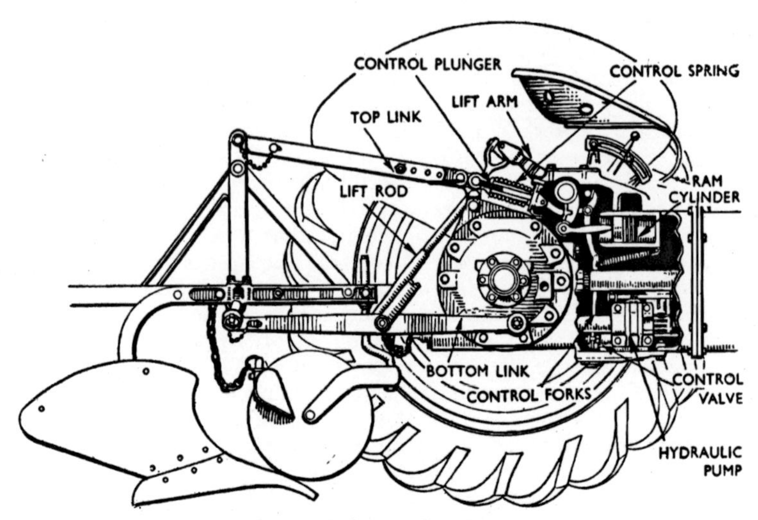

Ferguson hydraulic lift mechanism.

all the trailed implements using this kind of lift, the Standard lays down that the jack, without hose connections, shall lie wholly within a specified space. The tappings on both the tractor and the jack cylinder are to be standardized pipe threads, and a means of control is to be provided to enable the piston to be extended and stopped in any desired position within its stroke, which shall be of a uniform length.

The point of view adopted in the Standard is that the hydraulic jacking-cylinder is an extension of the tractor, and not an integral part of the implement, and that when one implement is changed for another the jacking-cylinder remains coupled to the tractor. Nevertheless, the incorporation of self-sealing, quick-release couplings in the hydraulic hose is recommended so that the user may, if he wishes, leave the jacking-cylinders attached to the implements and uncouple the hydraulic hose from the tractor. Differences in the length of flexible hose required by different implements can also be accommodated by using these couplings. There is no specification for the quick-release coupling itself, but the standardization of the screwed connections at the tractor and the jacking-cylinder allows latitude for the exchange of hose and coupling assemblies by the user when fitting his various implements to the tractor.

The tractor manufacturer decides the effective diameter which will allow the cylinder to give the specified force and stroke. The mechanical interchangeability between jacking-cylinders of different makes is sufficient to assure the implement maker that his product will be operated satisfactorily by all jacking-cylinders that comply with the Standard.

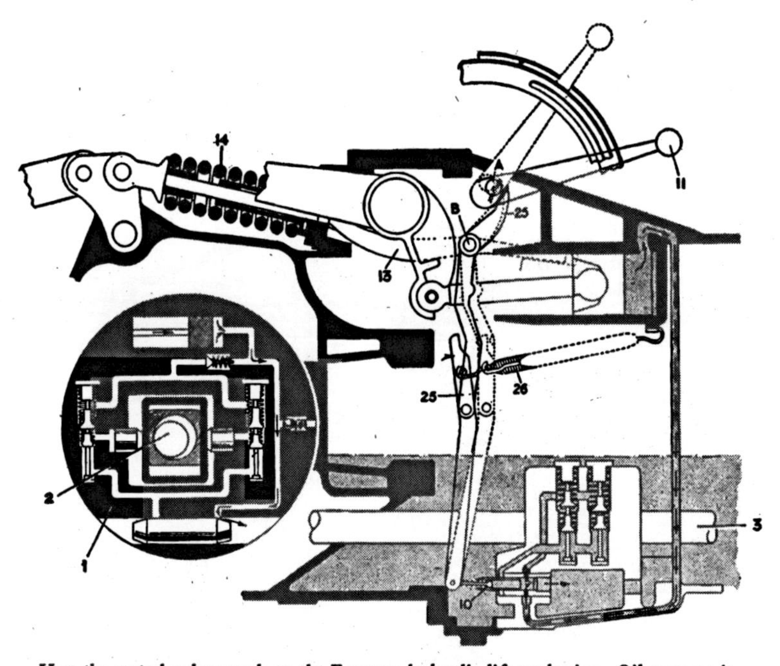

How the control valves work on the Ferguson hydraulic lift mechanism. Oil pressure is generated by the pump, operated by the action of cams on the power take-off shaft 2. *The pump has a check valve which prevents back pressure reaching the pump and a pressure lifting valve which operates when an attempt is made to lift a greater load than the mechanism can well manage. When it is desired to lower the implement into work, the control valve* 10 *is actuated by a hand control fork assembly* 25 *operated by the hand control lever* 11. *Forward movement of the lever causes the fork to pivot about connection B on the control spring fork* 13 *and to withdraw the control valve. Withdrawing the control valve allows the oil to flow away from the cylinder when the implement weight forces the piston forward, so lowering the implement. The amount the valve can be withdrawn is limited by the lower ends of the fork contacting the centre axle housing, from which point further movement of the control lever causes the lower end of the fork to pivot against the tension of the fork retracting spring* 26. *With the implement in the ground, forward movement of the tractor causes a forward pressure, set up at the top of the implement, against the control spring* 14 *pivoting the hand control fork about the control lever shaft at point A. The implement will continue to penetrate until the pressure against the control spring moves the control spring fork and, with it, the hand control fork and valve sufficiently to bring the valve to the central position. While this state of balance is maintained, the implement position remains constant. Implement depth is dependent on the setting of the hand lever.*

Cylinders for operating trailed implements, called slave cylinders, are separate from the tractor and are connected with the oil pump valve through a flexible pipe. For directly-mounted implements it is much easier to make the cylinder a part of the tractor. The designer can arrange for the cylinder to be enclosed within the body of the tractor,

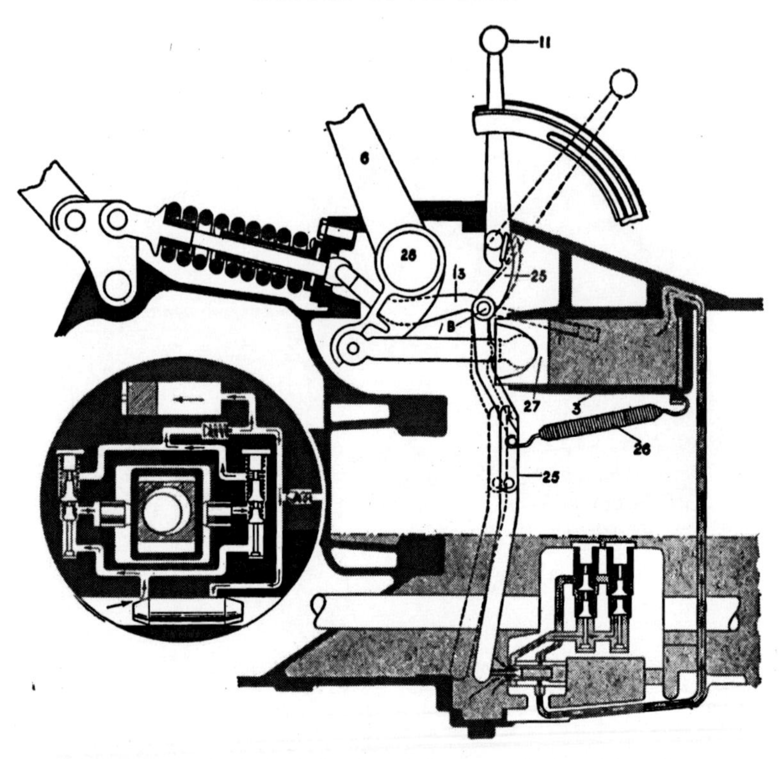

How the implement is raised out of work by the Ferguson mechanism. Rearward movement of hand control lever 11 allows hand control fork 25 to pivot about point B on control spring fork 13 under the action of fork retracting spring 26, so pushing the control valve to the 'lift' position. Oil is pumped into hydraulic cylinder 3, moving piston 27 along cylinder and turning lift shaft 28 with lift arms 6, thus raising the implement. Pump shut-off is obtained when the skirt of the piston protrudes far enough out of the cylinder to bear against lugs on the hand control fork, pivoting it about point B until the control valve returns to the central position.

It will be seen that in this particular form of hydraulic unit, the pivoting action of the lower end of the hand control fork is an important feature in the design of the automatic depth control and setting of the implement. The hand control fork linkage is so designed that a small movement of the control spring fork can produce a comparatively large movement of the control valve. This ensures sensitive operation of the control valve. If the hand control fork were a rigid structure the control lever range would be limited by the travel of the control valve. The pivoting action of the lower end of the fork when the control valve is withdrawn to its limit (that is, when the fork end butts against the centre axle housing) ensures a wider range of control lever movement, besides preventing damage to the fork mechanism when the overload release is in operation.

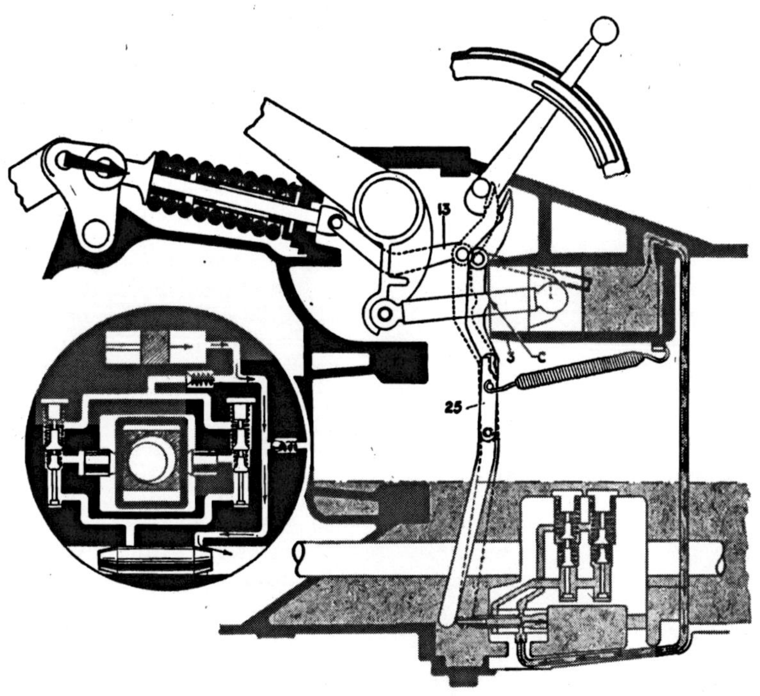

A safety device is incorporated in the Ferguson system. Excessive forward movement of the control spring fork 13, *which occurs if the implement meets an obstruction, causes the lugs on the hand control fork* 25 *to strike the skirt of the hydraulic cylinder* 3 *and pivot about point C. Thus the reverse movement at the lower end of the hand control fork moves the control valve out to the 'drop' position. This relieves the effective weight of the implement from the tractor rear wheels which thereby lose traction and the tractor stops with rear wheels spinning, without damage to the implement.*

and the piston can operate lifting arms which pivot in bearings housed in brackets which are also part of the body of the tractor. This is the basic construction of the three-point linkage used on many tractors. Two lifting arms raise the two lower links; the top link, a compression link, is not power-lifted but moves in sympathy with the lower links.

Much the same kind of control valve can be used in hydraulic systems for directly attached implements as the one which has been described for the control of slave cylinders. It is more usual, however, for a ram type of piston to be used rather than a piston with a connecting rod with big-end and little-end bearings. The ram piston rod has no little-end gudgeon pin; it ends in a semi-spherical projection which fits into the head of the piston in the manner of a ball and socket joint. Contact is made when the piston withdraws from the rod when the weight of the implement is not acting in such a direction as to keep the rod pushed against the piston. It will be seen, therefore, that this arrangement is a lift only. The implement is not pushed into work by hydraulic pressure; it falls into work by its own weight, and is only lifted by hydraulic power. Moreover, in most makes the mechanism is a simple lift with only two positions, one in which the implement is held lifted right out of work, and the other in which it is free to rest on the soil. The depth of work has to be controlled by an adjustable land wheel in the same way as with a trailed implement, by a cranked axle and regulating screw.

In one system of hydraulic attachment of implements, the depth at which the implement works in the soil is controlled hydraulically as well as the lifting. A slide valve to control the flow of oil is operated by a small hand lever, but its movement is also controlled in part by a spring-loaded link in the hitch. The purpose of this is to maintain the depth of work automatically at the pre-set level. If the implement tends to go too deep, the additional draught caused by that extra depth moves the slide valve sufficiently to raise the implement enough to counteract the tendency.

The control valve is actuated by a fork assembly which is operated by the hand control lever. Forward movement of the lever causes the fork to pivot and to withdraw the control valve. Withdrawing the control valve allows the oil to flow away from the cylinder when the implement weight forces the piston forward, so lowering the implement. The amount the valve can be withdrawn is limited by the lower ends of the fork touching the centre axle housing, from which point further movement of the control lever causes the lower end of the fork to pivot against the tension of the fork retracting spring. With the implement in the ground, forward movement of the tractor causes a forward pressure, set up at the top of the implement, against the control spring which surrounds the connecting bar of the top link, and this pivots the hand control fork about the control lever shaft. The implement will continue to penetrate until the pressure against the control spring moves the control spring fork and, with it, the hand control fork and valve sufficiently to bring the valve to the central position. While this state of balance is maintained, the implement position remains constant.

Implement depth depends on the setting of the hand lever. Rearward movement of the hand control lever allows the control fork to pivot on the control spring fork under the action of a retracting spring so pushing the control valve to the lift position. Oil is pumped into hydraulic cylinder, moving the piston along the cylinder and turning the lift shaft with lift arms, thus raising the implement. Pump shut-off is obtained when the skirt of the piston protrudes far enough out of the cylinder to bear against lugs on the hand control fork, pivoting it until the control valve returns to the central position.

The hand control fork linkage is designed so that a small movement of the control spring fork produces a large movement of the control valve. The pivoting action of the lower end of the fork when the control

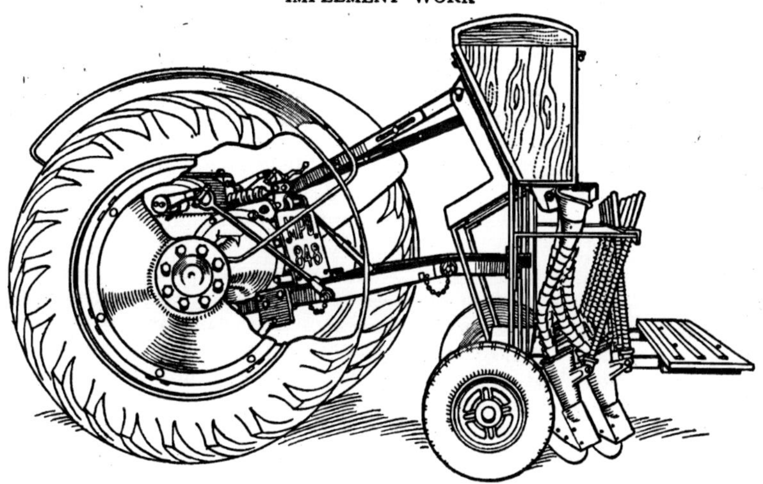

A corn drill made by Rayne Foundry for mounting on the three lifting arm points of a hydraulic system. The implement has its own depth wheels. The seed mechanism is force-feed and it is driven by a belt.

valve is withdrawn to its limit and the fork end butts against the centre axle housing ensures a wide range of control lever movement, and also prevents damage to the fork mechanism when the overload release is in operation.

A safety device is incorporated in the system. Excessive forward movement of the control spring fork, which occurs if the implement meets an obstruction, causes the lugs on the hand control fork to strike the skirt of the hydraulic cylinder and pivot. Thus the reverse movement at the lower end of the hand control fork moves the control valve out to the drop position. This relieves the effective weight of the implement from the tractor rear wheels which therefore lose traction and the tractor stops with the rear wheels spinning.

When implements attachable by three-point linkage are to be used interchangeably between various tractors, conformity to a standard of dimensions for the hitch-points becomes even more necessary than when trailed implements are concerned. The British Standard that deals with three-point linkage, B.S. 1841: 1951, Attachment of Mounted Implements to Agricultural Tractors, specifies two different sizes of hitch, one for light-medium wheeled tractors and one for medium-heavy wheeled tractors. It defines the position of the hitch points and the dimensions of the ball joints which form the connections. It also gives figures and limits for the lift range and the power stroke of the lift, and a minimum angle of inclination for the adjustability of one or other of the lower links by the levelling lever. Oil pressures and weight-lifting capacities are not specified.

It is to be remembered, however, that different implements demand different zones of clearance to enable them to be lifted fully out of work without fouling any projection on the tractor. Therefore, if an implement being considered for purchase seems to be very bulky, and particularly if it seems to project far forward towards the seat of the tractor, it should not be bought until it has been seen mounted on the make of tractor on which it will be used. This is necessary even when both tractor and implement conform with the specifications in B.S. 1841: 1951.

In some designs of tractor it is difficult to make a towing drawbar that conforms with the specification in B.S. 1495: 1948, and yet gives clearance for an implement mounted on a three-point linkage that conforms with B.S. 1841: 1951. Therefore, some makers make the towing drawbar detachable, and others make a shallow drawbar plate which can be extended by an adaptor to bring it into the specified position for trailed implements.

There is a limit to the weight and size of implements that can be attached to the tractor in this direct way, and can be carried to and from the field, lifted out of work. In a long implement, say a four-furrow plough with plenty of spacing between the mouldboards to allow rubbish to pass, the centre of gravity is very far back from the points of attachment to the tractor. This means that any swinging up-and-down movement of the plough mass when the outfit is bouncing over rough ground with the implement raised out of work is operating at the end of a very long arm or lever, and the moment the weight exerts may be sufficiently large to break the lifting arms or other parts of the raising gear.

It is better for these large implements to be mounted on their own wheeled frame, and for the lifting to be managed by a slave cylinder with hose pipe from the tractor's hydraulic system.

On some tractors a pre-set linkage control can be used in conjunction with the hydraulic lift to control the position of a plough in work by pre-determining the amount the lift arms fall as the plough is lowered into the soil. A cam is fitted to one of the lift arms, and a manual control lever is used to pre-select working depth. An adjustable telescopic upper link is fitted in place of the standard upper link.

When the plough and tractor are in operation on level ground the lower links pull the plough at the pre-set depth and the telescopic upper link is in compression, the stop pin in the link being hard against the rear of the slot in the link sleeve.

If the tractor wheels pass over a bump, or the front wheels go into a hollow, the telescopic upper link extends and permits the plough to maintain its pre-set depth due to the soil forces operating on the plough. Conversely when the tractor rear wheels encounter a hollow, or the front wheels pass over a bump, the lift arms are free to move in an upward direction, maintaining the telescopic upper link in compression and the plough at the pre-set depth.

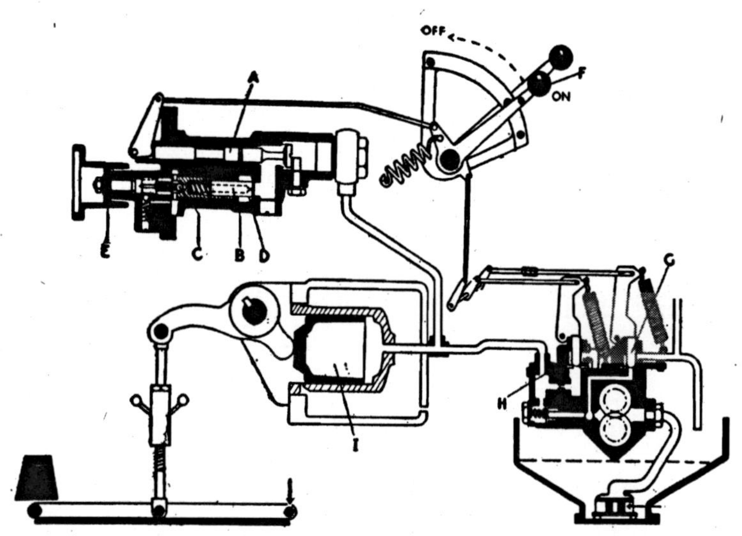

David Brown hydraulic lift system, with Traction Control Unit, in which the proportion of the weight of the implement carried by the rear wheels of the tractor can be controlled by the driver. Shut-off valves at A are controlled by a lever F situated alongside the main implement lift lever. The main lever varies the pressure control valve H. There is a by-pass valve in the hydraulic pump at G. B, C and D are parts of a pressure and flow control valve which has a regulating screw at E. I is the ram cylinder piston. The oil inlet to the system is marked by an arrow. When the levers are arranged so that the flow of oil begins to tend to lift the implement out of the ground, part of the weight of the implement is taken by the rear of the tractor, and this improves traction.

It has been seen that when directly mounted wheel-less ploughs are at work, part of the weight of the implement, and indeed part of the weight of the soil on a cultivation implement, is carried by the rear of the tractor and therefore helps the grip. On one tractor which uses depth wheel controlled implements, the hydraulic system can be adjusted while the tractor is in use to provide for weight transfer. This is used when the driving wheels slip and show signs of needing ballast. The downward pressure on the rear wheels can be increased by as much as 35 per cent when the control lever is set to give the maximum transfer of weight from implement wheels to tractor wheels.

Details of Hydraulic Systems

Tractors with built-in hydraulic systems for directly attached implements usually have a tapping from the system for a flexible hose pipe to conduct the oil to a slave cylinder on an implement or on a self-tipping trailer, but another method of operating a trailed implement is to connect the operating mechanism of the trailed implement with the tractor's hydraulic lifting arms through a series of levers. In some

K

American tractors the pump and control valve are the only built-in components. The lifting is accomplished by external cylinders even when directly attached implements are used. Some such tractors have provision for individual control of articulated tool bars. It is possible to use four slave cylinders on the tractor, one to operate one half of a front mounted bar, another cylinder to operate the other half of the front bar, a third cylinder to operate half of the rear bar, and a fourth cylinder to operate the remaining half of the rear bar. The control lever within reach of the driver can lift any of these four parts at will. One tractor designed on this principle can also take three-point linkage implements; external cylinders and linkage operate the lifting arms of the linkage.

In all hydraulic systems the speed of drop of the implement depends on the viscosity of the hydraulic fluid. This means inevitably that the implements drop more slowly when the tractor is first put to work and the fluid is cold than they do when the tractor has been running for some time and the hydraulic fluid has become warm and thinner.

When ordinary engine or transmission lubricating oil is used as the fluid, all rubber hose and washers have to be made of synthetic oil-resisting rubber. On some few systems, however, a liquid of the brake fluid type is employed. It is extremely important that the correct type of fluid should be used and also that the viscosity should be as near as possible to that recommended by the manufacturers of the tractor. In very cold weather it may be permissible to drain the hydraulic system and refill it with a fluid of slightly lower viscosity in order to make implement operation more speedy.

On built-in lifts where a gravity return is used, the hydraulic pump is sometimes driven from behind the clutch, though this means that when the clutch is out the pump does not work. This is not particularly inconvenient for built-in lifts, but where slave cylinders are employed it is some disadvantage, and for this purpose it is better for the pump to be run direct from the engine at a point in front of the main clutch so that whenever the engine is working the hydraulic system is in operation.

There are two types of control valve, the piston type and the poppet or plunger type. The piston valve seals by a close diametral fit which is usually helped by a ring seal. The poppet or plunger valve is satisfactory except for the fact that the spring may pick up vibrations and cause the valve to hammer on its seat.

The types of pump mainly used are the gear type and the multi-cylinder reciprocating type.

The rotors of gear pumps are hydraulically balanced endways by relieving the covers at the ends or by drilling a hole through the shaft with holes coming out behind the rotor faces. This balancing is necessary to prevent heavy end loading throwing the rotors against one face of the housing.

The period of operation under full pressure is kept down by employing a hold position on the lift which allows the pump to operate with a free or low-pressure discharge. The only work done by the pump is useful work, operating the lift.

A multi-cylinder reciprocating pump need not have as great a capacity as a gear-type pump. The slip on a gear pump is considerable, whereas it is very small on a reciprocating pump. Consequently, the heat generated in a reciprocating pump in the oil is smaller.

Some lift systems can operate at pressures between 1,000 and 2,000 lb/sq.in, and others work at as low as 400 lb/sq.in. They do not normally work at their maximum pressures, but these are the pressures at which the safety relief valve blows off.

The high-pressure systems need only a small amount of oil to do the work, and the action of the lift in either direction is smooth and fast, and is not very much affected by small changes in viscosity of the oil. The benefit of needing only a small quantity of oil is seen when a tipping trailer or front-mounted loader is in use. If a large ram is used with low-pressure oil the bucket or trailer will take some time to return because so much oil has to be forced out of the ram.

More elaborate sealing arrangements are necessary on the pipe joints and valve facings in a high-pressure system than in a low-pressure one. If the sealing points are within the transmission casing or the housing of the lift this does not matter, but if the pump is mounted outside the transmission casing very good sealing glands have to be arranged at points where the driving shaft passes through the casing.

On the hydraulic ram, one type of seal used is a cup seal, usually moulded in synthetic rubber. The oil pressure behind the ram opens out the seal and causes it to press against the cylinder wall and prevent the passage of oil. This kind of seal cannot pass over a port in the cylinder wall. If it did, it would be damaged on the return stroke. If ports are used the ram has to be sealed by means of piston rings.

A system in which the hydraulic equipment uses the same oil as the transmission gears is compact, and it is possible to have the whole system within the body of the tractor. It is protected from outside, but it may pick up abrasive metal from the gears, so a suction filter and a magnetic plug are used to keep back any particles of this kind.

The separate oil system allows the power lift to be built as an extra, without being part of the machine. The correct grade of hydraulic oil can be used without reference to what any other part of the tractor may need for lubrication, and the system can be sealed off completely enough to do away with the need for elaborate filters.

Power Take-off Implements

Many implements have arrangements for using the power of the engine to operate their mechanism directly through a shaft. Reapers-and-binders, for instance, can have the cutting, sheafing and tying mechanism driven direct from the power take-off instead of from the ground

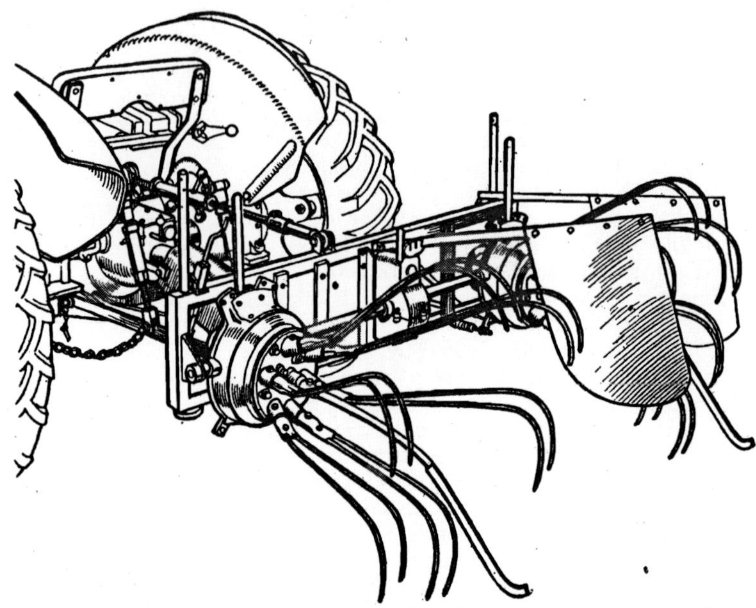

A David Brown-Albion hay swath turner for direct tractor attachment and power take-off drive. The position of the rotors is adjustable so that different widths of row can be turned.

wheel of the binder. This helps where the crop is thick and the ground slippery. If necessary, the tractor can travel in bottom gear, and yet the cutting mechanism will be driven at full speed. By fitting an additional clutch in the main transmission system of a tractor it is possible to arrange for a live independent power take-off. The power shaft can revolve whenever the engine is running, even though the tractor is stationary.

The driving shaft is square and it fits into a square sleeve to make a telescopic connection between the tractor and implement. The extensibility of this connection and the inclusion of suitable universal joints enable the tractor and implement to turn corners without strain to the mechanism, and provide some flexibility in fitting one implement to tractors of not quite the same design. This flexibility is, however, not sufficient to allow interchangeability if the various tractors have the power take-off shaft situated in very different positions in relation to the drawbar and to the centre of the tractor. Therefore, the Standard B.S. 1495: 1948, Agricultural Tractor Details, includes a specification for the position of the power take-off shaft. It also defines the diameter of the shaft and the size and shape of the splines on it, and gives a normal speed of rotation of the shaft. Conversion sleeves can be bought to make out-of-standard splines match up, and also secondary drives to change the position of the power take-off shaft.

For easy turning of the tractor and implement, the hitch point should lie midway between the universal joints of the shaft. When the two front universal joints are located with the hitch point on the drawbar midway between them, the angle is divided equally between the universal joints when the outfit turns a corner. The yokes of the two universal joints have to be in the same plane. That is to say, if the shaft is lying on the floor, both yokes should be flat on the floor and not one flat and the other standing up. If the shaft is not assembled in this way it will vibrate.

Correct positioning of the shaft and provision for flexibility are particularly important when an implement mounted on three-point linkage is power-driven. Although there may be no need for the implement to be driven when it is in the lifted position at the headland, the shaft will continue to rotate, and the shaft must neither come out of its sleeve nor press solidly against the far end of the inside of the sleeve. Mounted potato spinners are one type of implement in which flexibility of the drive must be watched carefully.

BELT WORK

DRIVING stationary machines such as feed mills and circular saws is a useful secondary occupation for a tractor even though this duty may not provide a truly economical load for the engine. It is wrong to buy stationary engines for belt work if a tractor, already purchased, can be spared to do the work. Moreover, the tractor is able to propel itself to the job, whereas a stationary engine has to be transported. Often the tractor can also tow the machine to the site where it is going to operate, as it does with threshing tackle.

Tractor belt pulleys are of greater diameter than those on small stationary engines, so it may not be possible to set the tractor engine to run slowly enough to provide a belt speed suitable to the machine that is being driven. If this is so, the pulley must be removed from the machine and a larger one must be fitted in its place. The size of pulley needed can be calculated by multiplying the diameter of the tractor belt pulley by the speed of the tractor belt pulley and dividing by the desired speed of the machine.

Suppose the diameter of the tractor belt pulley is 12 in. Its speed of rotation at normal settings of the engine governor can be found in the tractor instruction book. Let us suppose that it is 540 r.p.m., this figure being within the limits specified by British and American standards. Suppose the machine is to be driven at 600 r.p.m. Then the theoretical diameter of the pulley to be fitted on to the machine is $\frac{12 \times 540}{600} =$ 10·8 in.

This would be the right-size pulley if no slip at all occurred. In practice there is always at least 5 per cent belt slip so that the machine would actually run slower than the required 600 revolutions per minute which our calculation was designed to give. Therefore it is well to reduce the diameter of the driven pulley by roughly 5 per cent, and in this instance we should fit a 10 in pulley. Very few machines need the speed to be exactly correct, and in any case pulleys are not obtainable in odd sizes, but in a range of sizes with fairly wide steps between each dimension.

It may be that the tractor instruction book refers to belt speeds instead of pulley speeds. This is the travel of the belt in feet per minute (ft/min).

Belt speed (ft/min) = Diameter of pulley (ft) × its r.p.m.

The circumference of the pulley can be found by putting a string around the pulley and then measuring the length of the string. For example, suppose the circumference of the pulley is 2·65 ft and the pulley runs at 1,000 r.p.m., the tractor would have a belt speed of

2,650 ft/min. Most tractors have a belt speed of about 2,650 ft/min when the engine is running at one or other of its normal governed speeds, but some which have a clutch in the belt pulley have a much higher belt speed.

If the instruction book cannot be found there are several ways of measuring pulley speed, either with the tractor stationary or with it running over a field. It can, of course, be measured directly by means of a revolution counter or a speed indicator; but if such an instrument is not available there is a simple method for estimating the speed of a tractor engine. When it is doing belt work, locate some part of the belt-driven machine which is turning slowly enough for it to be counted, disconnect the machine from the tractor, and turn the main pulley of the driven machine by hand and count the number of revolutions needed to turn the slow moving part one revolution. This is the gear ratio or reduction ratio. By counting the revolutions per minute of the slower-turning part and multiplying by the reduction ratio, the revolutions per minute of the main drive pulley can be determined. To determine the number of revolutions per minute of the tractor engine measure the circumference of the belt pulley on the tractor and the circumference of the belt pulley on the driven machine.

The governed engine speed can be estimated also when the tractor is travelling. The revolutions per minute of the rear wheel can be counted, and this number multiplied by the gear ratio will give the speed of the tractor engine in r.p.m. If the gear ratio is not known, it can be determined by jacking up one rear wheel, grounding the engine so it will not run, and then turning the crank by hand and counting the number of revolutions necessary to turn the engine over to give one revolution on the rear wheel. This number must be multiplied by two to take account of the action of the differential gear. When only one wheel is allowed to turn, the free wheel turns twice as fast as it would normally.

In the case of a vaporizing oil tractor, if the machine provides only a small load for the engine, the radiator blind must be kept closed, or an old coat must be hung over the radiator to keep the engine warm enough the vaporize the paraffin fuel.

A tractor with spade-lugged wheels usually needs no chocking when doing light belt work. A pneumatic-tyred tractor, however, sometimes creeps along the ground towards the machine it is driving, even when the brake is firmly on. This slackens the driving belt.

The best way to scotch a pneumatic-tyred tractor is to place a thick, heavy plank of wood resting on the top of the front tyre and wedged between the ground and the lower part of the rear tyre. When it is chocked in this way the tractor can still move slightly backwards and forwards through the elasticity of the tyre; but this is an advantage rather than a disadvantage, for it gives the drive a flexibility that often helps when the load is intermittent, as in mills and pulpers.

With a rubber-tyred tractor on belt work it is well to hang a chain from the tractor to the ground to provide an easy path to earth for any electricity that may be generated by the movement of the belt.

If, as on some tractors, there is little room between the edge of the belt and the inside of the front wheel, and if the belt whips sideways, it may rub the edge of the wheel. While this does not matter much if the wheel is iron, some sort of protector ought to be fitted between the tyre and the path of the belt if pneumatics are fitted. If the plank method of scotching the tractor is used, the plank can be allowed to overlap the tyre just enough to take the belt rub.

Positioning Belting

The pulley of the driven machine must be exactly in line with the tractor pulley. Whenever possible the belt should run horizontally, so that the weight of the sag of the belt can help it to hug the pulleys tightly. The weight of a vertical belt tends to pull it away from contact with the lower pulley. If a vertical belt has to be used, the drive should be as short as possible.

A horizontal belt should be as long as possible; and, if circumstances permit, the direction of rotation of the pulley should be such that the pulling side of the belt is at the bottom and the slack side is at the top. The droop in the upper run of the belt increases the arc of contact round the pulleys.

Where the pulley of the driven machine has to rotate in the opposite direction to the driving pulley on the tractor the belt must be crossed. Provided the run of the belt is fairly long, this crossing has no bad effect. In fact, a crossed belt sometimes grips better than a single-looped belt, because the crossing increases the arc of contact between the belt and each pulley.

When a crossed belt is being used it is important to see that the fastener does not project beyond the width of the belt. If it does, it will jar every time it comes round to the place where the belt crosses. The jarring and scraping may wear the belt or weaken the fastening.

Care of Belting

Once a year, Balata cotton belts and leather belts should have castor oil rubbed into the outside face. The oil makes the belt more supple so that it can conform better to the shape of the pulleys and grip them more firmly. Rubber cotton belts, however, must not be oiled.

Sometimes a belt gets saturated with engine oil through a leak in the bearing housing of the machine, and it loses its gripping power. Usually the oil can be absorbed by ground chalk but if the belt still slips it must be taken off and scraped and left packed in sawdust for a few days. Resin should never be used as an anti-slip. It does cause the belt to grip better temporarily, but it glazes the surface of the belt and in the end makes the slip worse.

Sometimes a belt slips merely because it has become hot and has expanded. The more it slips the hotter it will get. Often this slip can be cured by pouring cold water on to the outside of the belt.

When the pegs in patent fasteners become worn into steps they should be renewed. If the belt has become frayed where the teeth of the fastener grip, the belt should be shortened enough to get rid of the torn edges and a new fastener should be used. Usually the belt will have stretched somewhat, so that the slight shortening does not matter, but if the tractor cannot be brought near enough to the machine, it is better to put in an extra piece of belt and an extra fastener rather than rely on the fastener gripping the frayed end of the belt. Similarly, when a laced belt breaks, the old holes should never be used again. The length of the belt should be sacrificed to give the fastener a fair chance.

It is most important that the end of the belt should be cut quite square to the length. If the ends are not square the belt will always be pulling unevenly. It will wobble on its travel, and one side of the fastener will be taking most of the driving strain and will soon give way.

Used in the right place in the right way and given regular overhauling, transmission of power by belt is extremely efficient and free from trouble; but in addition to the periodic overhaul there is a daily attention which will much prolong the non-slip life of the belt. Before the machines are set to work, the face of the belt should be cleaned free from surface dirt. For Balata cotton belts and leather belts a cloth damped in paraffin is best for removing the dirt. For rubber belts a cloth damped in water should be used.

Paraffin or lubricating oil which has leaked or splashed on to a rubber cotton belt must be cleaned off at once, otherwise the rubber will swell and crack and the belt will become defective.

TRANSPORT WORK

MUCH tractor work is really transport. Manures have to be carted to the fields and crops have to be carted away from them to the farmstead or even to railway station or customer. On large farms some of the transport work can be done by lorry or motor cart, but tractor trailers do most of the work, at any rate the part of the transport work that lies within the boundaries of the farm.

Lift boxes which fit on to the three-point linkage of the hydraulic lift mechanism of a tractor are useful for short jobs such as taking milk churns to the road. For long journeys their capacity is not great enough to warrant the tractor being out of commission for other uses. Front-mounted hay sweeps with horizontal tines are used to collect hay from windrows and carry it to a stack or stationary baler, and buck-rakes fitted to the three-point linkage at the back of the tractor are used to pick up newly cut grass for silage. The buck-rake, which has its tines closer together than a hay sweep, is driven backwards along the swath of grass, with the rake lowered. When the load has been collected on the tines the rake is lifted by the three-point linkage so that the load is well off the ground and the tractor can be driven away forwards at quite a high speed. This method of cartage can be ued also for corn sheaves and for kale, but it is hardly economic for journeys of more than ¼ mile, and the trailer is the chief farm transport vehicle for longer journeys.

It is well to have at least two trailers, so that one can be a four-wheeler and one a two-wheeler, because each kind has special advantages. Four-wheeled trailers are easy to hitch to the tractor, because only the weight of the hinged drawbar has to be lifted to bring the clevis up to the height of the tractor drawbar plate. Another advantage is that they do not pitch so much as long two-wheelers when they traverse ridge-and-furrow land. Four-wheelers will stand on their own, and can be loaded or unloaded while the tractor is away on another job, then they can be easily hitched to the tractor when it returns. Hitching a loaded two-wheeled cart-type trailer on to an ordinary tractor drawbar is a difficult task for one man, unless the load is stowed in such a way that the trailer almost balances on its axle. Once a two-wheeled trailer is attached, it has, however, the advantage of being compact and easily reversible. A close-coupled two-wheeled trailer, attached to its tractor, becomes virtually a six-wheeled lorry, and if the axle is well to the rear of the trailer, the weight of the load is taken largely on the rear of the tractor where it helps the tractor's wheel grip. An alternative method of improving grip is to provide for the trailer wheels to be driven by the tractor engine through a transmission-speed power take-off shaft. This arrangement converts the outfit into a six-wheeled four-wheel-drive vehicle. Some trailers for tractors with hydraulically operated lifting arms can be automatically hitched and raised off the ground even when they are already loaded.

Tipping trailers are useful for roots and farmyard manure, and for discharging combine-harvested grain into the receiving pits of grain-drying installations. Most tippers are two-wheeled, although four-wheeled tipping trailers are made. Some trailers are made to tip to the side as well as to the rear, and this is a great convenience in grain handling. Use is made of hydraulic mechanism to operate the tip. An extension pipe and flexible hose fitted to the hydraulic system of the tractor conduct the fluid pressure to work the tipping mechanism of the trailer. Even with trailers towed by tractors not having hydraulic power, the hydraulic method of tipping can be used. A hand pump mounted on the drawbar supplies the oil pressure needed to work the lifting cylinder.

Many kinds of trailer bodies are available. Some are merely a platform, with perhaps harvest ladders at each end; others have deep sides, close-fitting enough to hold grain. When the particular needs of a farm cannot be met by a ready-made trailer, a chassis can be purchased, and a platform or body can be built on to it on the farm. If several different bodies are needed for short periods during the year, a trailer which can be changed from one type to another by use of conversion sets may prove economical, and these convertible trailers are becoming popular.

A power-driven moving floor can be operated to load the trailer evenly and also to discharge the load automatically. These trailers are used particularly for green crops for silage and grass drying. When a short grass loader has deposited sufficient material into the rear of the trailer, the grass can be moved forward by operating the floor. This allows space at the rear for more grass. For unloading, the floor moves in the reverse direction and deposits the grass on the ground.

Another kind of self-emptying trailer has a hinged floor.

On a firm level surface the capacity of even a small tractor for pulling loaded trailers is great. Gradients, however, and soft soil conditions cause the trailer to resist being pulled along. A little planning of routes within the farm can give the tractor an easier task. If a route is being laid out for haulage work and a short but steep gradient along it provides an obstacle which will reduce the possible load over the route, it is as well to see whether this obstacle cannot be by-passed or levelled away with a bulldozer.

A route used continuously can deteriorate quickly if it is not properly maintained, and it is for this reason that serious thought should be given to the making of some hard roads along routes which will be much traversed by vehicles. These roads need not be continuous; it is surprising what an improvement can be obtained by paving quite a small area round a field gate. In the main part of the field, the cart tracks rarely intersect, and the once-over journey on the soil does little harm; but all tracks converge at the gate, and very soon the soil there becomes so churned up that it causes vehicles to offer high rolling resistance.

The final stage in farm milk transport is the milk churn platform which is the link between the farm vehicle and the milk distributor's lorry. The platform should be sited so that the milk lorry can be manoeuvred right up alongside it without having to traverse ground which may get soft in winter, and the height should be gauged carefully according to the height of the trailer and also the height of the lorry. Sometimes a two-level arrangement can be contrived by making use of sloping land so that the trailer can drive up to the platform one side and the lorry the other side. Then the trailer body should be higher than the platform and the lorry should be lower. This device can save some lifting when the churns are moved from trailer to platform, and also when they are moved to the bottom layer of the lorry. The platform should be strongly made and well creosoted.

Tractor mounted front-end loaders are useful aids to transport work. They are cranes which are attached to the body of the tractor and yet leave the drawbar free for trailers to be hitched and towed. The derrick carrying the bucket or grab is lifted and lowered hydraulically by power transmitted through a pipe from the tractor's hydraulic system. Front loaders are used chiefly for picking up farmyard manure from heaps, but they have many other applications.

Bulldozers and angle dozers, important to the making and upkeep of farm roads, also are attached directly to the body of the tractor and are controlled hydraulically, but some earth scoops and graders are attached to the rear of the tractor through the three-point linkage.

OVERHAUL

WHEN a tractor has run for about 1,000 hours, the accumulation of soot and hard carbon in the cylinder ought to be cleaned out. This decarbonizing can be merely the removing of the cylinder head and the scraping away of the carbon from inside the combustion chambers and from the top of the piston. It is, however, better to make decarbonizing the occasion for a more general overhaul of the engine. The valves should be ground in and the piston rings should be loosened and the grooves cleaned free from carbon. In addition to the necessary readjustment of the valve tappets any little jobs that can be done more easily when the engine is dismantled than when it is together should be done at the same time as decarbonizing.

REMOVING CYLINDER HEAD

Before any parts are removed it is well to place a tarpaulin sheet under the tractor to catch any nuts and bolts or other components that may fall to the ground.

If new gaskets for the cylinder-head joint and inlet and exhaust manifolds are not available, special care must be taken when the head and manifolds are taken off. Whenever it is possible to obtain them, new gaskets should be used.

When the bolts holding down the cylinder head have been removed it should be taken off gently. It should not be prised off with a screwdriver or the machined faces may be damaged. If it is stuck, the head should be tapped sideways with a wooden mallet to break the seal. In some makes of tractor it is necessary to swing the radiator forward slightly, so that when the cylinder head has to be lifted it can be lifted straight up, and so not damage the gasket.

The carbon deposit from inside the cylinder head should be scraped off with an old screwdriver or some such blunt tool.

Before work on the cylinder block is started, all openings leading into the sump or other inaccessible places should be stopped up, each with a piece of non-fluffy rag.

VALVES

The valves should next be removed. If the engine is a side-valve model a valve-lifting tool must be used to compress the spring for the cotter to be removed. This tool takes its purchase from the top of the cylinder block and provides a fulcrum for leverage. If the engine is an overhead-

Using pieces of cloth to protect tappet holes and other apertures difficult to clean out. Carbon deposit is being scraped off two piston heads, while the other two cylinders are protected by rag.

valve type, the best way to hold the valves in position while the cotters are being removed or replaced is to use a small block of wood as spacer between the bench and the valve head.

Before the valves get mixed up, each one should be marked so that it can go back to its original place. Alternatively, since it is difficult to make punch marks on the hard valve metal without risk of bending the stem or distorting the head, it is better to place the valves in holes drilled in a wooden block until the time has come for them to be put back into the guides to be ground in.

The valve stems and guides must be cleaned with paraffin until the gummy oil is all gone and they work easily with no sticking. Then they should be ground into their seatings with fine emery paste. Some valve heads have a slot for a screwdriver blade, and the best way to turn these valves to grind them is to use a screwdriver bit in a brace. Some valves, however, have smooth-topped heads, and must be turned by a suction grinder, which is a pencil-shaped piece of wood having a rubber suction cup at the end, like the projectiles that stick on to a wall when they are shot from a toy gun.

Valve grinding can be made easier if a light spring of suitable size is threaded over the valve to engage against the guide and hold the valve off its seating whenever the pressure is released from the screw-driver brace or other grinding tool being used.

Coarse grinding paste should be spread on the contact surface of the valves, and the brace should be rotated backwards and forwards, the valve being lifted at frequent intervals and given a quarter turn to bring it to a fresh position. This should be carried on until an even and continuous seating is obtained on both the valve and the seat. The coarse grinding paste should then be wiped off and some fine paste

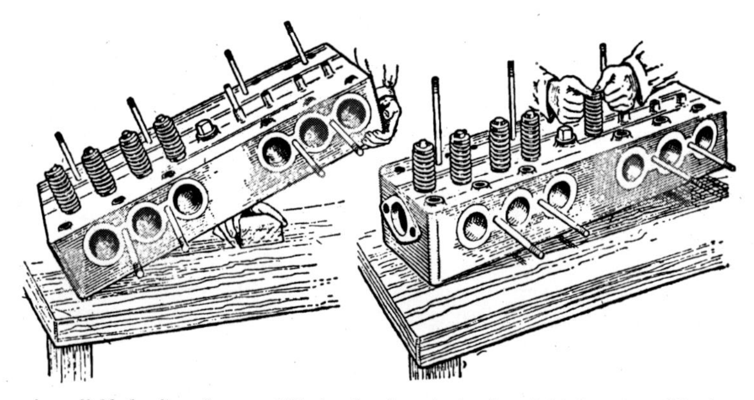

A small block of wood can usefully be placed on the bench to hold the valve while the cotters are being removed or replaced. The valve spring can be compressed and the cotter is easily dealt with.

substituted so that the final grinding shall give a finished surface. Valve seats should not be pared away any more than is necessary to restore the contact surface and produce a gastight seal. If the seating becomes too deeply recessed, the path of the gases is obstructed. The valves and valve seats should be wiped with a clean rag, and before assembly the valves, seats and guides should be washed with petrol to remove all traces of grinding paste. If the valve guides are much worn they should be replaced.

Attention must also be paid to the condition of the valve springs, since these deteriorate with use. They can be examined by standing them on a flat surface to see whether they are all the same length. If possible, they should be compared for length with a new spring or with the maker's measurement for a new spring. Springs that have lost their resilience will be short, and they must be replaced.

How the carbon deposit is scraped from the combustion chamber, and how a brace with a driver bit can be used to grind in a valve.

PISTONS

Usually, to get at the pistons the connecting rods have to be detached at the big-end bearings and drawn through the open sump. This is laborious, but it provides the opportunity for testing the big-end bearings for play and for tightening the bearings by removing shims if play is detected. Also, the removing of the sump gives opportunity for cleaning all the oil filter gauzes thoroughly by washing them in petrol.

The safest way to remove the piston rings from the piston is to get four strips of tin about $\frac{1}{4}$ in × 4 in and use them as slide rails for the rings. The top ring should be attacked first. One end of the ring should be picked out with a screwdriver and strip of tin No. 1 should be slipped between the ring and the piston. Then the other end of the ring should be treated in the same way and strip No. 2 should be inserted there. Now strip No. 1 can be moved round the ring three-quarters of the circumference. This will make it possible to insert strips Nos. 3 and 4 which should be spaced so that the four strips are at equal intervals round the piston. Then the ring can be made to slide up towards the top of the piston and it can be slipped off with no fear of breaking it. The other rings can be removed in the same way.

The piston itself should be cleaned. The carbon around the extreme edge of the top of the piston should be left undisturbed. Its presence helps to keep back the pumping of oil from the underside of the piston to the upper side into the cylinder where it will be burnt away. The ring grooves should be scraped, and then the piston should be washed in paraffin. After this the piston rings should be scraped and cleaned with a rag and replaced on the piston. The strips of tin can be used in this refitting. Care should be taken that the piston ring gaps are not in line or there will be a compression leak. The rings should be arranged so that the gaps are staggered round the piston. Oil should be spread over the piston before it is put back into the cylinder.

GASKETS

The exhaust manifold should be cleaned free from carbon. All gaskets should be greased before they are put on. Wherever possible new gaskets

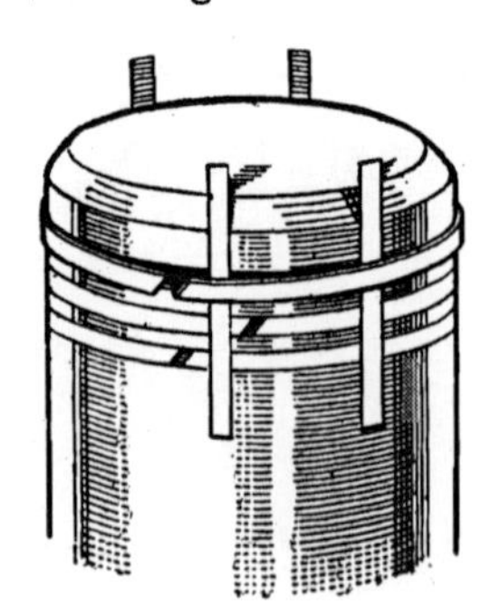

Strips of thin sheet iron being used to help detach piston rings from the piston. When the rings are re-fitted they should be put on in such a way that the gaps are staggered around the circumference of the piston. If the gaps are too nearly in line with each other gas may escape through them. It is important to check their position just before the piston is led back into the cylinder.

A Tasker four-wheeled trailer chassis which will take several kinds of body. (*See Chapter* 10.)

A Ferguson trailer with hydraulic tipping. (*See Chapter* 10.)

Hard surfacing renovated cleats on a crawler track. (See Chapter 11.)

(Below, left) A half of a metal bearing housing in which marks can be seen in the surface of the bearing metal, but the scores are not sufficiently deep for renewal of the bearing to be necessary. (See Chapter 11.) (Below, right) A taper roller bearing. The outer race has been detached and the rollers can be seen in their cage. They are in contact with the inner race. Taper roller bearings are slightly adjustable for wear since the further the cage and inner race are inserted into the tapering outer race the smaller clearance there is between the rollers and the races. (See Chapter 11.)

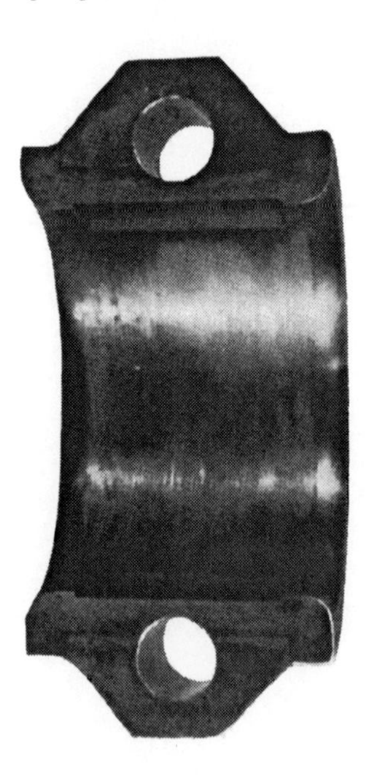

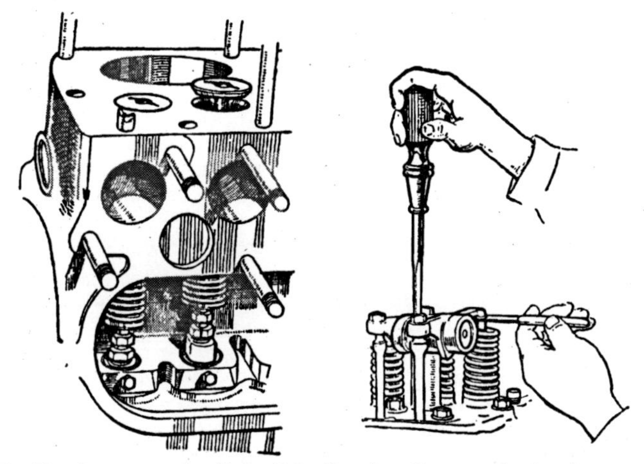

Left, side-valve tappets of a kind which allow for adjustment of valve clearance. The lower of the two hexagon faces on each tappet is a lock-nut; the higher one, the actual tappet head, can be screwed up or down when the lock nut has been loosened. Adjustment must be continued until a feeler gauge of the thickness recommended by the tractor manufacturer will just pass under the foot of the valve stem, when the tappet is at its lowest position, off the cam. Right, a usual type of adjustment on overhead-valve tractor engines.

should be used. Once the asbestos and copper have been compressed, much of the sponginess of the gasket has gone. It is this sponginess which helps the gasket to fit into uneven places and make a good seal.

REPLACING CYLINDER HEAD

When the cylinder head has been put back into position and the nuts of the holding-down bolts have been put on finger tight, each nut in turn should be tightened a little by spanner. If any one nut is tightened hard in advance of the others the cylinder head may be distorted by the uneven strain. After the engine has been running a little and is warm it will be possible to give all the nuts another half turn.

ADJUSTING TAPPET CLEARANCE

The tappet clearances will need to be readjusted, since the valves have been ground in. Moreover, in an overhead-valve engine, the relative position of the cylinder head and camshaft is bound to have been changed during the removing and refitting of the cylinder head.

In some side-valve engines the clearance between the valve stem and the push rod can be increased only by filing away the end of the stem. Since this is a one-way adjustment, and the clearance can be made larger but not smaller, it is essential to carry out the filing in small stages, measuring the clearance after every stage. The clearance should be not more than 0·02 in. It is also important that the end of the stem

L

should be filed perfectly level, otherwise only part of its face will be in contact with the push rod. Both too little valve clearance and too much clearance can bring poor engine performance. Too little clearance can also cause local overheating of the valves and seatings. If the valve never closes there will be a leak of gas under compression. This will reduce power output, and the gas escaping under high pressure will generate heat at the seating. If the valve never opens fully, the gas cannot pass in or out of the cylinder freely and quickly enough, and the power output of the engine will be smaller than it would be if the valve was opening to its full intended extent.

When adjusting valve tappets, make sure that the piston is in the firing position for the cylinder on which the tappets are being adjusted. To find the firing position, remove all the spark plug leads so that the engine cannot run, then take the lead of one cylinder and hold it about ¼ in from the engine block and turn the engine until a spark is noted at the gap. When a spark is noted stop turning the engine immediately and adjust the tappets on that cylinder.

To adjust the tappets on the other cylinders, give the crank a half turn if it is a four-cylinder engine, or a third of a turn for a six-cylinder engine, and adjust the pair of tappets on the cylinder in the firing position. The order of adjusting will be the same as the firing order.

If some of the tappets appear to have a very wide gap before adjusting, the valves may be sticking in the guide and not closing. If this is suspected, run the engine and pour some paraffin or flushing oil round the valve stem to free it, before making final adjustment. Valves must have ample clearance or they will be held open and burn. Stainless-steel valves need more clearance than the other steel valves so if the tractor has had new valves of the improved type installed, give them about half as much again clearance as was right with the older-type valves.

On most tractors, the oil for lubricating the valves is carried to the rocker arm shaft under pressure through a small pipe. Make sure that this pipe is open. To check its condition, have the engine running, remove the valve cover, and see if oil is flowing around each of the rocker arms.

BEARING ADJUSTMENT

Some big-end bearings have detachable shims for wear to be taken up. Others can be adjusted only by removing metal from the faces of the housing. Only a very little metal must be filed away, or the two halves of the housing will fail to make a true circle. If the bearing is much worn it should be re-lined.

The following is a good routine to follow when adjusting main bearings. Remove crankcase. Remove wire through the front main bearing screwheads. Undo the screws. Remove the cap. Examine the bearing cap metal to be sure that the bearing is making perfect contact over all its area. A number of shims, about 0·005 in and 0·0025 in thick, are usually found between the bearing cap and the cylinder block. Remove shims of equal thickness from each side of the crankshaft.

Replace the cap, do up the screws and test the crankshaft for tightness by turning it with the starting handle. The drag of the bearing should be just felt. It may be necessary to undo the bearing cap and add or remove shims several times before the adjustment is correct.

When the front main bearing is correctly adjusted, undo its bolts again, and repeat the adjusting process first on the centre bearing, and then on the rear main bearing. When all these are satisfactorily adjusted, do all the screws up tightly and re-wire the heads. Then make sure that the bearings are not so tight that it is impossible to turn the crankshaft with the handle. The crankshaft must have end float of between 0·002 in to 0·006 in when finally adjusted.

If the big-end bearings are adjusted at the same time, see that when all bearings are tightened it is possible to turn the crankshaft with the handle.

After adjusting main or big-end bearings it is advisable to run the engine light for some little time, keeping it well supplied with water and oil before running it under load.

Bearings with replaceable linings need careful fitting. Each half of the split bearing is made about 1/1,000 in longer than half of the circumference of the bearing, so that when the two halves are put into place the ends project slightly above the faces of the housing. When the nuts of the bearing housing are screwed down the linings become crushed and this forces them into close contact with the housing so that it is impossible for them to move in it. If the linings and bearings are put together by hand and inspected it may seem that the linings have been made a little too long, and that the projecting ends ought to be filed off. If this is done, the lining may soon become loose.

In replaceable bearings the width across from end to end of the insert part is made greater than the diameter of the bore into which they fit. This is to give springiness to the inserted linings and this holds them in place while the bearing is being put together.

If ball or roller bearings have to be replaced, the new bearing must be pushed into place by pressure on the inner ring only. No pressure must be put on the outer ring of the race. If the bearing is being moved by using a piece of pipe as a drift, the pipe must be of a diameter such that it will bear only on the inner ring.

Support for a bush, or for any other kind of bearing, must reach as nearly as possible all round it, otherwise the bush or bearing will become distorted.

If worn-out thin-walled bushes have to be removed, a hacksaw blade can be used to split the bushing. The split will relieve the surface pressure which has been on the bush ever since it was first pressed into place and so the bush can quite easily be pulled out or pushed out. If a hacksaw blade cannot be got into position it may be possible to use the one corner of a triangular file to split the bush.

Roller and ball bearings, although they are able to withstand such very great pressure and to allow rotation at such very high speed, are vulnerable to quite unexpected sources of damage. For instance, the acid in the sweat from the fingers of an operator fitting a bearing may be the original cause of a complete breakdown in a bearing. When a ball or roller bearing is removed and will be used again it should be washed carefully with paraffin and dried thoroughly, and then a lubricant should be worked into the cage and on to all balls or rollers. The bearing should be wrapped in oiled paper to keep grit out of the lubricant and it should be kept covered until the time comes for it to go back on to the job.

Big-end bearing caps and main-bearing caps should be checked for alignment when they are removed for the bearing surfaces to be replaced, by placing them on a surface plate or piece of plate glass. If they do not rest squarely on the surface but wobble, they have been strained, and it is better to replace them.

ENGINE KNOCK

It is well here to point out that not all knocks are due to worn or ill-adjusted bearings, and the following notes describe some knocks, due to causes other than loose bearings, which can occur both in spark ignition engines and Diesel engines.

Ignition knock in petrol and vaporizing oil engines is caused by running with the ignition lever too far advanced for the particular conditions under which the tractor is working. The cure is to retard the ignition just enough to stop the knock. Ignition knocks and knocks caused by accumulations of carbon sound very much alike. They can be distinguished by retarding the ignition; an ignition knock will vanish but a carbon knock will persist.

A knock due to using the wrong type of fuel resembles a carbon knock, and it may persist even when the ignition is retarded. Along with this knock there is likely to be a tendency for the engine to overheat. The characteristics of certain lamp oils cause this kind of knock in engines not designed specially for them.

Carbon knock occurs only after the tractor has been in service for a considerable time. It is noticeable when the engine is hot and pulling hard.

Piston slap is due to wear in the pistons and cylinders. It is most noticeable when the engine is cold, and diminishes as the engine warms up. This wear is usually due to the tractor air cleaner allowing uncleaned air to enter, due to shortage of oil or excess of dirt in the sediment trap, or with unsuitable or inferior oil in the engine. It also occurs after prolonged service, owing to normal wear. The cure is to have the cylinders rebored and oversize pistons fitted, or rebored and sleeved, with new standard-size pistons.

Some of the above kinds of knock can occur also in Diesel engines,

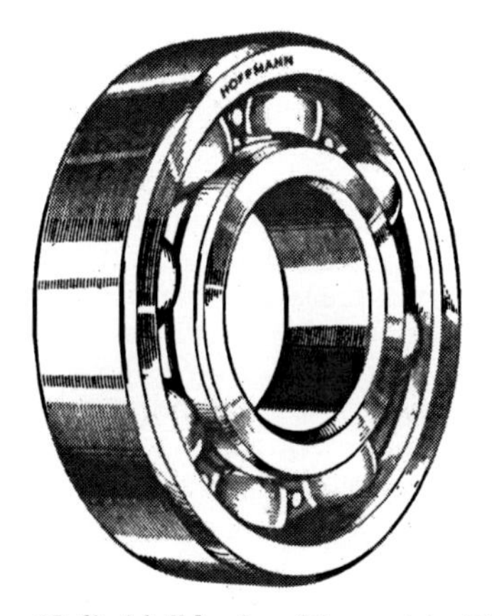

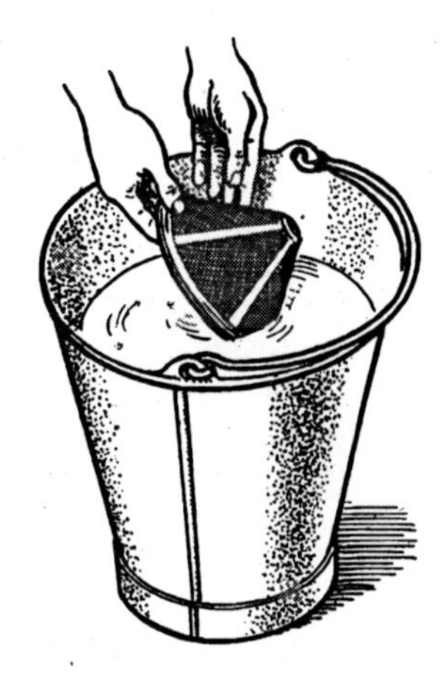

(Left) A ball bearing of the cage type. There is a housing over each ball in the series which separates the inner and outer races. (Right) A conical filter screen, from an engine sump, being washed in paraffin.

but all Diesel engines have a noise which might be described as a knock, even when they are new and correctly adjusted. This is really the noise of the combustion of the fuel under the conditions of high compression which obtain. It is sometimes called 'characteristic Diesel knock', and it is harmless.

When the big-end and main bearings of an engine do have to be replaced, it is likely that the crankshaft journals will be found to have worn oval and will have to be reground. Crankshaft regrinding is certainly a job for an expert, and it is usually carried out at the same time as reboring of the cylinders, though the primary reason for a major overhaul is generally to put right excessive consumption of lubricating oil.

Excessive oil consumption and lack of compression usually call for a rebore or new liners. Wear makes the cylinders oval and tapering, the bore being larger at the top than at the bottom. The excessive clearance between piston rings and cylinder walls brings loss of power, over-consumption of lubricating oil, and even oiling-up of sparking plugs or carboning of Diesel oil injectors.

Most of the lubricating oil present on top of the piston is boiled and burnt by the heat of the explosion, and the vapour so produced may in some small measure condense, although most of it passes out of the exhaust ports. Probably any lubricating oil which does find its way on to plugs and injectors does so in the form of a liquid spray, owing to the turbulence taking place in the cylinder, and it is only when the quantity of oil finding its way past the piston rings is very great indeed that it settles on the plugs or injectors before it can distil away as vapour.

Sometimes the fitting of new piston rings will improve matters; but if calliper readings show that the cylinders are seriously out of round then they must be bored out or re-lined and new oversize pistons must be fitted. A difference of 5/1,000 in between the smallest and largest diameter of the cylinder bore on tractor engines is reckoned to indicate that reboring or relining is needed.

Although reboring of cylinders ought not usually to be carried out on the farm, it is often useful to be able to make tests on an engine to know whether reboring is indeed necessary. A useful measuring device which is not expensive to buy but can be of service when decisions have to be made about engine overhauls is a compression indicator, a pressure gauge which can be inserted into a sparking plug hole or injector hole and held in gastight contact through a rubber washer. A gauge of this sort can give quite a good indication of whether the bore is leaking past the piston rings or whether the valves are leaking.

When the engine has been run long enough for it to become warm, all the sparking plugs or injectors are removed and the compression gauge is inserted into one of the sparking plug orifices or injector orifices. In the case of a petrol or vaporizing oil engine the throttle and choke should be fully opened. The engine is then rotated with the starter motor or other means until the gauge registers a steady reading. The test is repeated on all cylinders and a note is made of each reading, and if the compression pressures recorded do not vary more than 20 lb/sq.in the aggregate effectiveness of the piston seals and the valves may be regarded as equal for all the cylinders. If, however, the variation exceeds this figure, then one or more cylinders may have some mechanical damage which has not affected the others. If the readings for all the cylinders are about equal, comparison with the manufacturer's figure for the reading in a new engine can be made.

Further tests can give some idea on whether any loss of compression is due to escape of gas through the valves or past the piston rings. To perform these tests, one tablespoonful of heavy oil is injected into one cylinder through the sparking plug orifice, by a force-feed oil-can. The oil will form a temporary seal between the piston rings and the cylinder. Immediately after injecting the oil, a pressure test is taken. The maximum pressure obtained is recorded, and the test is repeated for all cylinders. The readings obtained from this wet test are compared with those obtained from the dry test. Any difference represents the loss of compression due to piston ring leakage. If the compression of any cylinder showing a low dry test reading remains low during the wet test, loss of compression due to defective valves or to a faulty cylinder head gasket is indicated.

If there is a serious loss of compression due to leak between the piston rings and the cylinder it may in some cases be rectified without reboring the cylinder, because the trouble may be only worn piston rings. To investigate further, the pistons must be removed, and the cylinder bores inspected for scores and wear. Pistons should be measured immediately below the ring lands, and at right-angles to the gudgeon pin axis. Cylinder bore wear should be measured with a cylinder gauge or callipers for the full length of the bore over which the piston travels, but particularly over the part of the bore just below the highest point of ring travel, because the greatest wear occurs at that part.

Clearance should be measured at the side and front of the cylinder. In engines with cast-iron pistons the total clearance between the skirt of the piston and the cylinder wall should be about 0·005 in. Piston rings should be cleaned and checked for side wear in their grooves and for gap clearance in the cylinder. The top ring will show most wear. Put each ring in its own cylinder and with the piston push the ring more than half-way down so that it is placed squarely in the cylinder. Measure the gap or end clearance with a thickness gauge. If new oversized rings are used, some filing may be needed. For the top piston-ring gap, 0·003 in clearance should be allowed for each inch of cylinder diameter; and for the bottom ring and oil ring, 0·002 in for each inch of diameter.

If standard-sized piston rings are too small, rings with oversized diameters, starting with 0·005 in oversize or larger, can be used. If a cylinder takes more than a 0·015 in oversized ring, it is usually in need of reconditioning.

If oil pumping cannot be stopped with new piston rings, the cylinders usually must be rebored or relined.

If new liners can be fitted it is best to replace the old liners with new ones complete with new factory fitted pistons and piston rings. If the cylinders are not fitted with removable liners, then reboring and refinishing the cylinder walls and fitting with oversize pistons and rings is necessary. Reliable cylinder reboring and refinishing services are available through implement dealers and garages. Work on the farm is probably best confined to deciding when reboring is needed.

In engines having alloy steel valves and valve seat inserts, valve reconditioning is often performed by means of special grinding equipment. The valves themselves are refaced with a special stone refacing machine and the seats are ground with an electrically operated stone grinder equipped with special angled stones for any size and angle of valve seat. If the valves and seats are reconditioned with these stone machines, grinding or lapping valves with a compound may be unnecessary.

After a rebore an engine should be run in as if it were a new engine. Engine oil or upper cylinder lubricant should be added to the fuel, and additives to the sump oil are useful.

ELECTRICAL OVERHAULS

Actual repairs of the electrical equipment on the engines of tractors and other machines are best left to an expert, and, indeed, the most satisfactory treatment is often the fitting of reconditioned replacement parts obtained from the distributors of the equipment. Nevertheless, it is necessary in the first place to trace the particular component causing the trouble. A systematic search is the only way of investigating electrical faults, and a few simple testing instruments will help this orderly investigation. Some of these can be bought cheaply and some can be made up in the farm workshop.

Testing leads are perhaps the most useful and easily constructed pieces of equipment for the farm electrician. A few 3 ft lengths of insulated wire should be fitted with crocodile jaw terminals at either end. These crocodile jaws are made primarily for clipping on to the leaden terminal posts of batteries, but they can be clipped also to most shapes of metal. If a break in a wire or a bad connection is suspected, the part of the circuit which might be at fault can be bridged by one of these clip-ended wires. This is a handier, and more reliable, method than holding the wires in contact by the fingers.

The idea of a wandering, easily connected wire can be extended to make a useful continuity tester. One end of the wire carries a crocodile clip, and the other is connected to the middle terminal of a brass lamp holder of the kind which takes a single-contact bulb. A brass or steel point is soldered on to the outside of the lampholder. To use this tester, a motor car tail-lamp bulb of the same voltage as that of the system being tested is put into the lamp holder and the crocodile clip is clamped on to some well-earthed part of the engine. Then, if the sharp point of the tester is pierced through the insulation of the wire being tested, the lamp will light if the wire is alive. For testing in some parts of a circuit, the tester will have to be provided with its own battery.

On engines having battery-and-coil ignition the tester can be used to examine the wire leading to the low-tension contact on the distributor. It will also serve as an indicator in timing the engine, for if the tester point is held against the distributor contact and the engine is turned slowly the bulb will light when the contact points are open. When the points are closed the bulk of the current will pass across them rather than through the lamp bulb, so that the light will go out. If the lamp does continue to glow when the points are closed, it shows that they are not making good contact. This may be due to the points being dirty or pitted, or to looseness or a fracture in the breaker fittings.

In some situations it is more convenient to have the probing point of the tester on the end of a second flexible wire, rather than attached to the body of the lamp holder. It may thus be worth while making up two sets: one will have the probe on the body of the lamp holder firm and convenient for use in spots which are easy to reach; the other, with the more flexibly attached probe, will be better in awkward places, such as when dealing with the wiring behind the facia board of a tractor.

The field circuits in a generator can be tested for continuity by putting the tester in series with a field coil and a battery. They can also be tested for a possible short circuit to the framework of the generator.

Simple voltmeters also can be useful for detecting faults in cut-outs

Fault	*Cause*	*Repair Needed*
No charge indicated on ammeter even though batteries are not fully charged	Fuse burnt out	Renew fuse and check charging system
	Drive belt slipping	Adjust drive belt
	Brushes on dynamo dirty or worn too short	Renew, clean or reset brushes
	Faulty commutator on dynamo	Clean with fine glass paper
	Faulty windings	Have field or armature re-wound
	Faulty connections	Check wiring and terminal
Rate of charge too low	Dynamo drive slipping	Adjust belt drive
	Faulty connections	Clean or service commutator
	Voltage control faulty	Have regulator unit serviced
Rate of charge excessive	Faulty connections	Check wiring and terminals
	Faulty voltage control	Have regulator points serviced and re-set
Fuse burnt out	Short circuit	Insulate or renew faulty wires
Starter motor inoperative	Faulty connections	Check wiring and terminals
	Switch faulty	Short out starter switch and test circuit. Remove starter motor cover, check wiring and starter terminals, check operation of solenoid switch and service as required
	Starter motor faulty	Have starter motor serviced

and regulators. For example, a voltmeter across fuse terminals will show whether the fuse is intact. The cut-out control can be tested by connecting the voltmeter to terminals across it. If any flow is indicated when the control points are open, the cut-out is faulty. When no current flow is indicated at the fuse terminals, renew the fuse wire and make the following tests.

Connect the voltmeter to the regulator terminals, and if a battery voltage is registered, the regulator is faulty. Run the dynamo at 1,000 r.p.m. and note if any charge rate is shown on the tractor ammeter. Connect the voltmeter to the regulator terminal and the earth terminal and the dynamo voltage should be about 14 volts in a 12-volt system. If no charge is indicated on the ammeter, short-circuit the regulator with a short length of insulated wire and note if a charge is then indicated on the ammeter. If a charge is now shown, this indicates faulty and dirty cut-out points. If no rate of charge is shown, this indicates a faulty regulator or voltage control unit.

If the starter motor fails to operate, the trouble may be due to faulty starter button. To prove this, place an insulated screwdriver in contact with the positive terminal of the starter motor and the solenoid switch terminal. If the solenoid switch moves and closes the contacts, failure of the starter motor to rotate may be due to burnt or faulty solenoid contacts.

Faults such as this, and failure of the dynamo to charge, will have come to light during the regular maintenance inspections, but the table on the previous page may be useful for a special check when the electrical system is being inspected.

TRANSMISSION OVERHAULS

The drive gear and clutches very rarely need overhaul, and when they do the job should be done by the agent, for he has special jigs and tools. Even the replacing of components of such parts as the steering mechanism is usually best done by an agent. For instance, old king-pins and their bushes can often be knocked out with a hammer and drift for replacement, but they come out much more easily and safely if a hydraulic press can be used. The same applies to the pins in the links of crawler tracks, and removal and replacement are best done at an agent's.

The pins and bushes on tracks wear oval, and they can be given a new lease of life by turning the bushes and pins round through half a turn. In the same way it will possibly be found that the sprocket teeth are all a little worn, but only on the driving face side of each tooth. Accordingly, if the sprockets are changed from one side of the tractor to the other, fresh and nearly unworn faces will be presented to the drive.

It will be found usually, however, that most repairs to the tracks themselves, as distinct from adjusting their tension on the sprockets, cannot be undertaken on the farm, and the tracks, when they have been detached, should be sent to an expert repairer who has the heavy press needed for removing pins and bushes. But the importance of frequent examination of the tracks cannot be over-estimated, for its purpose is to guide the operator in deciding the point at which the tracks must be submitted to an expert. Timeliness here may reduce the need for renewal parts. Wear in one part of the machine often leads to quick wear of some other part which works in conjunction with it. It is not at all unusual for the replacing of some small part to postpone the need for replacing some larger or more complicated part.

In the case of rigid girder half-tracks, the tendency of the track is to lose the original curvature provided for, and eventually to become flat. When reconditioned the curvature of the track is restored by building up with welding the interlocking abutments, the surplus metal being machined so as to give the correct radius to the track; the pin holes in the links are rebored and oversize track pins are fitted. This is a job for a dealer. It is essential to have the tracks reconditioned when, after all the adjustment has been taken up, they show signs of becoming flat. To continue running the equipment after this time might mean that the tracks could only be reconditioned once, and in extreme cases the wear may reach a point when reconditioning at all becomes impossible.

There must always be some track slack. The track is to be driven in compression, not tension; unnecessary wear and tear is involved if the track is in tension. After considerable service, however, undue track slack can develop, and it is desirable that it should be taken up a stop on the idler bracket, provided that when this has been done some track slack is still left.

ESTIMATING THE JOB

Before looking over a tractor with a view to deciding what repair or overhaul is necessary, clean the machine thoroughly with steam or a safe solvent and water under pressure.

The cost of reconditioning a tractor can be determined pretty accurately following a thorough inspection to reveal what parts are needed, what adjustments should be made, and any other requirements. It is well to plan for a thorough job, because most parts or assemblies have a close working relationship with other parts, and it is not good to mesh old parts with new ones.

When replacing parts or assemblies, it is well to refer to the manual and not rely too much on memory. Always use the manual that is specifically designed for the make, model, and serial number of the actual tractor concerned. When ordering replacement parts, state the number of each part needed along with the make, model, and serial number of the tractor. When the model of the tractor is not known, it is well to state the date the tractor was purchased new, the tractor horsepower, and other information that will assist in identifying the model.

To ensure having the right spare parts on hand when reconditioning the tractor, it is well to place the order for parts well in advance of the date the overhaul is to be started.

SERVICING AND MAINTENANCE

REGULAR servicing will reduce the number of stoppages in the field, and will do away with the need for many emergency repairs that would otherwise have to be done. There are, however, some adjustments that have to be made from time to time to rectify the effects of wear or damage, and the need for these may come at any time. Therefore, regular servicing does not take away the importance of the driver keeping alert for unusual sounds and for reduction of power output. The chapter on tracing breakdown causes gives a quick guide to discovering the origin of faults of this kind, and they are considered in more detail here. Some of the faults can occur in all kinds of tractor, but some refer specially to vaporizing engines and some to Diesel engines.

Some method is needed to ensure that the intervals at which the various servicing tasks fall due are not exceeded. This means keeping records and it is often said that keeping records does not make the tractor run any cheaper. This is not quite true, because records which are quickly but systematically made can help a great deal in servicing, and this will make the tractor last longer. Records will show how much the tractor is being used and when the various services should be performed. Cost records too can be useful in deciding what kind of tractor to buy next, in fixing charges for any contract work that may be done for neighbours, and, which is their chief purpose, to provide data for planning more economical operations for the future.

RECORDS

The forms to be used are best nailed to the wall of the tractor shed where they will keep fairly clean. The design of the forms must be such that as little writing as possible is needed when an entry is made. It is well to keep a pencil on a string near the forms.

Alongside the form for recording servicing, a list of the services, with code letters, and the periods for performing these services should be pinned up. On the form, space provision is made for ticking off the hours the tractor is used. A suitable ruling for a chart which can be drawn in a few minutes is shown opposite. The square spaces are arranged in vertical rows of ten each. The date should be inserted for the first hour of use and thereafter during that same day simply put a tick for each additional hour the tractor is used. Then, whenever a service is performed on the tractor, insert the code letter for that service.

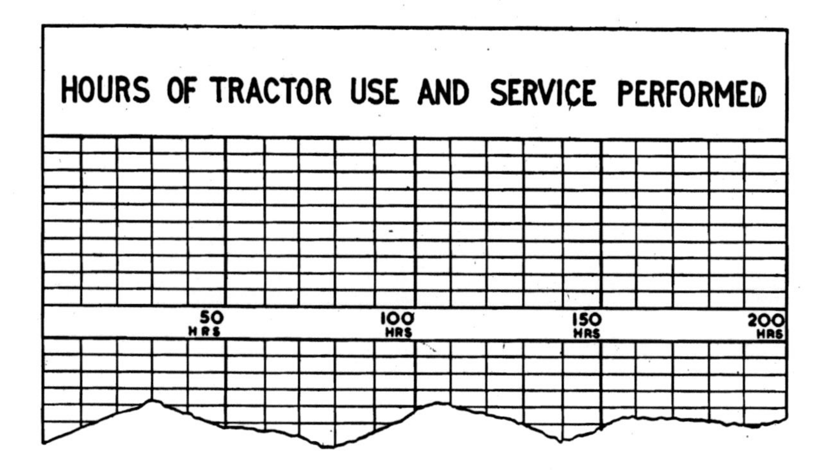

It will be noted that the services posted generally fall for convenience at the top or bottom of the 10-hour columns. Note also that the system provides a record of the hours on an accumulative basis.

In caring for a tractor there are several services which should be performed daily and other services which should be completed at certain regular intervals such as every 50, 100, 200, 400 and 800 hours. Some services are set up on a yearly basis and they should be listed. We can imagine 800 hours of operation in a year. If more than this is accomplished then it will be well to perform the so-called annual jobs at the 800-hour mark rather than wait till the end of the calendar year.

The first step in preparing the services list is to refer to the tractor's instruction manual and pick out the jobs which should be done at stated intervals. If no definite time is listed for a specific service, classify the service where you think it belongs. When all the needed information is assembled, list the jobs or services on the chart to be pinned up alongside the form. The kind of list that would be made for a Diesel-engined tractor is given on p. 175.

All the services which should be completed daily are listed under the code letter 'A', those every 50 hours under 'B', those every 100 hours under 'C', those every 200 hours under "D", those every 400 hours under 'E', those every 800 hours under "F". By selecting periods for 'B, C, D' etc. which are multiples of the 'A' period, all of the shorter-period services that may be due will be completed automatically each time a longer-period service is completed. For example, after 100 hours of tractor usage you would complete the 'C' services, listed under 100 hours, and also all the shorter-period services specified for 'B' and 'A'.

For determining the cost of tractor operation, a form such as shown overleaf would be used.

FUEL, OIL AND REPAIRS					
DATE	TYPE OF WORK	FUEL USED GALLONS	OIL USED QUARTS	HOURS	NOTES ON REPAIRS

To keep a record of the fuel used, a fuel measuring stick must be made unless a storage tank with pump which measures the fuel is used.

IGNITION FAULTS

Faults in the ignition system can waste much power, but their effects are most apparent in making a hard task of starting the engine. Indeed, weak ignition is probably responsible for more tractor starting difficulties than any other cause. A weak spark may ignite the charge in the cylinders satisfactorily while the engine is running and warm, but it may be quite unable to fire the mixture drawn into a cold engine. Also, some of the actual causes of weakness of spark are more to the fore when the engine is cold than when it is running; for example, rain will not settle on the insulators of the sparking plugs while the tractor is running and the engine is warm, but rain driving sideways certainly can settle on to the plugs of an idle engine.

Rain does not often affect the insulation of the magneto or of the wire leads running from the distributor to the sparking plugs, but condensation moisture does sometimes bridge the gaps inside the distributor. Wiping with a dry warm rag will put this right.

Faults in ignition systems can have one of the following effects on a tractor engine:

(1) Difficult starting.
(2) One or more cylinders not firing at all.
(3) Erratic misfiring which cannot be traced to any one particular cylinder.

Sparking Plugs

Difficult starting is often due to the gap at the sparking plug being too wide, a condition brought about by the burning away of the electrodes. Besides making the engine difficult to start, an excessively wide gap at the plug points may strain the magneto, or induction coil, because in order to jump the wide gap the high-tension current has to build up to

Maintenance Tasks on Diesel-engined Wheeled Tractor

A. 10 *hours*
1. Check sump oil level.
2. Check water in radiator.
3. Grease steering links.

B. 50 *hours*
4. Service crankcase breather.
5. Service air cleaner.
6. Add distilled water to battery.
7. Lubricate fan and check tension of fan belt.
8. Lubricate clutch bearings.
9. Lubricate clutch pedal and brake pedal.
10. Lubricate starter and dynamo.
11. Check oil level in gear cases.
12. Clean radiator fins.

C. 100 *hours*
13. Change sump oil.
14. Service oil filter.

D. 200 *hours*
15. Check level of oil in gearbox and differential casing.
16. Clean sediment bowl.

E. 400 *hours*
17. Clean fuel screens.
18. Flush cooling system.
19. Adjust valve tappets.
20. Adjust clutch and brakes.
21. Clean and adjust front wheel bearings.

F. 800 *hours*
22. Change lubricant in gear cases.
23. Tighten all bolts.
24. Clean and test injector.

a very high pressure, at which it may break through insulated parts of the magneto or coil.

The width of gap can be reduced by knocking the side electrode towards the centre electrode. This must be done carefully, for if the electrode is fractured it may fall off when the engine is running, and although it is of fairly soft metal it may damage the valves. It is not advisable, therefore, to hammer the electrode into an acute angle to make it stretch across a badly burnt plug.

The coating of oxidized and corroded metal on the electrodes also increases the electrical resistance. It should be scraped off with a knife. However, too-frequent cleaning and scraping of the electrodes should be avoided. There is no need for the metal to be kept bright.

The points should be set at a gap of 0·025 in. Feeler gauges are available in this thickness, but a postcard can be used instead and is sufficiently accurate. If a corner of the card can be pushed between the plug points and feels to be a firm fit with no play, then the plug is set correctly.

Misfiring also may be due to plug trouble, and the defaulting cylinder can be traced by shorting out each plug in turn while the engine is running. If shorting the plug makes the engine run less evenly, than the plug was doing some good; if it makes no apparent difference then that particular cylinder has not been working. In shorting the plug the screwdriver or hammer used should be connected first with the metal body of the engine and then with the sparking plug top. In this way the operator ought not to receive a shock. Nevertheless, it is well to hold the tool by its wooden handle in case the metal part should slip away from contact with the body of the engine and yet remain touching the sparking plug top.

If the plug of the dead cylinder is taken out and examined, it may be found that the points are bridged by a flake of carbon. Removing the flake with the point of a pocket knife will restore the plug to working order. Possibly, however, the short-circuit is inside the plug. A layer of carbon on the surface of the inner porcelain insulator may allow the current to jump across inside the plug. If the plug is made in two detachable parts the surface of the insulator can be cleaned. It should, however, be wiped only with a cloth as this will move the carbon without scratching the surface of the insulator. Once the surface is scratched it picks up carbon much more quickly and the carbon lies along the scratches and spoils the insulation.

In reassembling a plug of this kind the locking nut should be tightened sufficiently to make a gastight union, but care must be taken not to use too much force lest the insulator should be cracked or the gasket become too compressed.

When a vice is used to hold the barrel of the plug, the vice must not be screwed too tightly against the sides of the body of the plug, because it is possible for the pressure to distort the metal parts sufficiently to fracture the insulator.

An injector has been removed from this Diesel engine for inspection, and a piece of cloth has been pushed into the hole to prevent foreign matter falling in. (*See Chapter* 11.)

Using two spanners to undo a fuel pipe union. This prevents the possibility of twisting the pipe which can occur when only one spanner is used. (*See Chapter* 11.)

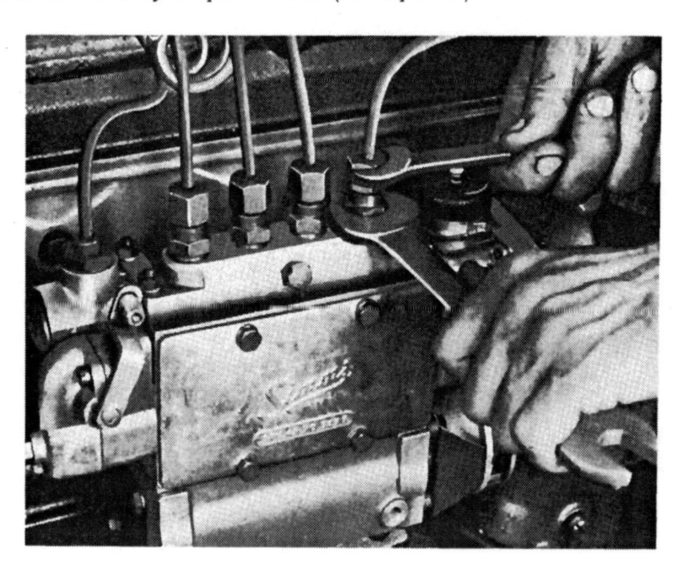

Gauging the gap between the points on a magneto contact breaker. (See Chapter 12.)

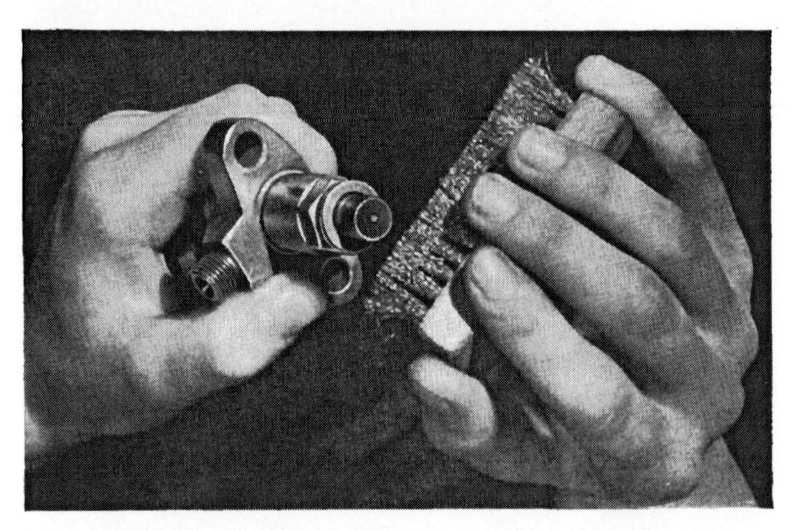

When the injector has been removed, the carbon deposits are removed carefully with a wire brush. (See Chapter 12.)

If the electrode gap has to be adjusted, only the outer electrode should be tapped or bent. Bending the central electrode will break the insulator.

Worn-out sparking plugs should be replaced. Eventually the points become too short to be adjusted to the correct gap, and the insulator material begins to deteriorate. The correct type for replacement must be chosen, for different designs of engine need different kinds of plug. Some types conduct heat away readily and are called 'cold' plugs, others are designed to retain heat and these are called 'hot' plugs.

Unsuitable plugs will be subject to oiling-up, or will quickly deteriorate with burnt electrodes. It is worth noting, however, that continual

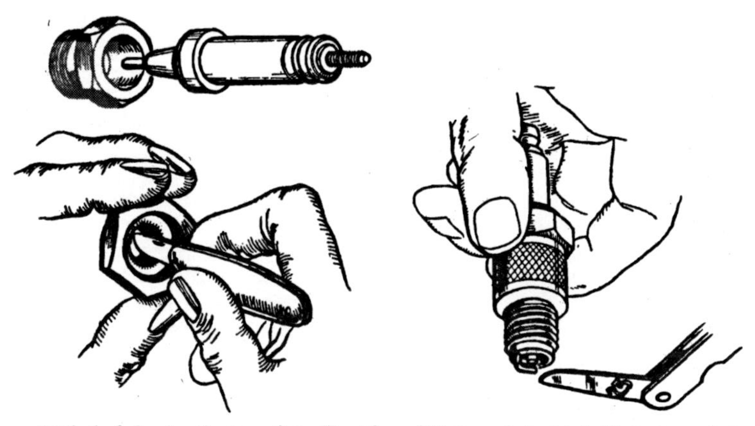

Method of cleaning the type of sparking plug which is made in detachable parts, and of adjusting the gap width between the electrodes. When a vice is used to hold the barrel of the plug, the vice must not be screwed too tightly against the sides of the body of the plug, because it is possible for the pressure to distort the metal parts sufficiently to fracture the insulator.

oiling-up of sparking plugs may be caused not by any fault in the ignition system or by the use of unsuitable plugs, but by worn piston rings or worn cylinder walls.

Gas leakage, either between centre electrode and insulator or between insulator and shell, causes a plug to overheat. A very slight leak between centre electrode and insulator can raise the temperature at the tip of the insulator as much as 40° F.

When plugs are fouled by petrol or vaporizing oil, the insulator shows a dull, black, fluffy carbon deposit. When they are fouled by lubricating oil the insulator more often shows a shiny black hard deposit, and in severe cases a filling of the space between insulator and shell with caked carbon.

If sparking plugs run satisfactorily when the engine is running slowly,

M

but the engine spits back when it speeds up, this is evidence of pre-ignition caused by some part of the combustion chamber becoming hot enough to ignite the mixture before the spark occurs at the sparking plug gap. This hot spot may be carbon, a very hot open valve, or a hot sparking plug. The cause of the extreme heat may be a weak fuel mixture or the water being too low in the radiator, or a stoppage in the water system, a defective pump, or a broken or slipping fan belt.

Should it be necessary to clean a plug in the field, the best way is to fill the sparking plug between the insulator and the shell or body with petrol and then set fire to it, letting the petrol burn off the excess of carbon or oil that has become deposited on the insulator. It is then a good idea to clean the points of the plug and adjust their gap to the correct gap setting, and to replace the sparking plug not in the hole that it came out of but into a cylinder where the plug is already firing satisfactorily. This gives the sparking plug that may be a little weak a chance in a good dry cylinder, and the oily cylinder is getting a dry sparking plug which has been working well in one of the other cylinders.

It is well to wipe the top of the insulator on plugs because grease and dirt which gets blown against them from the fan can cause the spark to jump from the brass terminal to the body of the sparking plug, especially when the tractor has been standing in a damp atmosphere.

In detachable-type sparking plugs care should be taken when replacing the insulator to see that the copper seating washer is down square, before any attempt is made to tighten the gland nut. New washers are worth getting; once a copper washer has been tightened hard to make a joint, the elasticity has gone.

Another important point of gas sealing is the copper washer which goes on the outside of the plug between the plug and the cylinder head. This should be renewed from time to time because this copper washer is largely responsible for transferring the heat from the plug body to the cylinder head and then to the cooling medium. When a plug has been screwed down a number of times on to the copper washer, the washer becomes very hard and consequently does not transfer the heat from the plug body as it was intended to, and slight overheating of the plug takes place.

The colour of deposits on plugs is a good guide to engine condition. If the insulator is ashy white it is a sign that the plug is overheating, and this may be caused by a wrong-type plug, improper installation of plug, shortage of water or shortage of oil, or ignition timing too far retarded. A weak mixture will cause the engine to overheat and give the plug deposit a white colour. If the insulator is a dull black, then that plug has been running too cold. The insulator is not getting hot enough to burn the carbon deposited on the insulator, so the carbon builds up on the insulator until the plug begins to miss. This condition can be brought about by the engine idling for long periods, or by excessive use of the choke. If the insulator has a brownish shade, this is a sign that the plug is running at its correct self-cleaning temperature, and the carburettor mixture is correct.

Of course, failure of one cylinder to fire may not be due to plug trouble. It can be due to a faulty valve. If the plug, placed on the body of the cylinder block, and with its top connected with the high-tension lead, shows a spark when the engine is turned by the starting handle, then the ignition is all right. If the cylinder is 'dead', therefore, there must be some mechanical fault.

Erratic firing is probably caused either by faulty high-tension leads or by a breakdown in the magneto itself. Here again, the trouble may be mechanical or due to some carburation fault such as a dirty carburettor.

High-tension Leads

If the rubber covering the high-tension leads is perished, or swollen from oil, it ought to be replaced. If there seems to be a leak merely in one or two frayed parts of the leads the insulation can be restored by wrapping rubber strip and adhesive tape round the faulty places.

The Contact-breaker

Misfiring may be caused by the contact points of the breaker of the coil or magneto being dirty or out of adjustment. When they are taken out the points can be reconditioned with a fine file. After the points have been brightened they should be wiped with a petrol rag. If they are still pitted after much filing, they should be replaced by new points.

When the breaker parts are back into position the engine should be turned until the breaker points are fully opened. The width of the gap between the points can be set by a feeler gauge to 0·02 in or to whatever the maker's recommendation may be. If the points do not make a snug fit over the faces of the gauge, the locking nut of the tungsten-point holder should be slacked back and the width of the gap adjusted. Here again, if no gauge is available the corner of a postcard will serve well enough as a feeler.

In the case of a tractor having a magneto with impulse coupling, the points can be brought into their fully open position only by turning the engine backwards, by pulling on the fan belt by hand. To make this easy the sparking plugs should be removed to release the compression.

The points should also be examined in their closed position to find whether they are making perfect contact. Sometimes dampness causes swelling in the fibre bush on which the contact arm rocks, and this makes the arm stick in the open position. If this has happened the bush must be eased with a rat-tail file.

Occasionally too, a film of moisture collects inside the distributor cap, and may allow the high-tension current to take the wrong path inside the cap. The inner surface of the cap should be wiped free of water and particles of carbon that may have been shed by the brush.

A puzzling fault which can occur when a distributor cover has been in use for several years occurs when, even after the inside of the cap has been wiped, there is a leak from one electrode to another so that firing becomes erratic. This may be caused by the formation of cracks in the inner surface of the cap. The cracks themselves would not matter, but they fill up with carbon dust and other conducting material which provides a path for their current. Close inspection will reveal these cracks, and a temporary cure is to scrape a trough along the course of the crack until all the carbon has been removed, and then to paint the trough with shellac, but a new distributor cover must be put on as soon as possible.

Another wrong path which high-tension current sometimes takes is along dirt or moisture on the outside of the insulator of a sparking plug. When this is happening the loud 'tick' of the exposed spark can often be heard. Insulators of uncovered plugs should be wiped from time to time. If they get coated with oil, the oil when it hardens becomes sticky and picks up dirt which may be a conductor.

Thimble-shaped insulators called condensators are fitted on some tractors. They cover the upper end of the sparking plugs and reduce surface leakage due to moisture. In fitting them, care must be taken to see they fit squarely over the plug insulator and only make contact with the centre electrode at the top where the high tension lead is attached.

The remaining parts of the induction coil or magneto rarely give trouble, and if they do it is a job for an expert. A magneto may need new bearings and a newly-wound armature. An induction coil may have to be replaced completely. Many troubles, however, in both instruments, are caused by a broken-down condenser. This can cause sparking at the contact-breaker points, pitting them quickly and causing irregular running of the engine and difficult starting. A new condenser is not expensive, but the fitting of it is a job for an expert. Usually a service unit can be supplied by the agent to keep the tractor running while the coil or magneto is being repaired.

Adjustments of this kind are the chief points to watch in ignition systems. In addition, some of the few moving parts must be kept lubricated, but the lubrication must be very sparing indeed because oil can rot insulating material and cause its electrical resistance to break down, or cause the material to swell so much that moving armatures can foul their housing. As an example of how little lubrication is needed, two or three drops of thin machine oil once a week are ample for the bearings of the magneto.

BATTERIES

After every 50 hours of work, distilled water should be added to the battery until the liquid covers the plates. If much water is needed, it is likely that the dynamo has been charging the battery at a high rate.

The charging rate of the dynamo, in equipment not having an automatic voltage regulator, can be governed by moving the third brush to a new position. To increase the rate of charging, the brush must be moved in the direction of rotation of the armature. Moving the brush in the opposite direction will reduce the charging rate; 8 to 10 amperes is the best charging rate for most sets.

The battery connections must be kept tight and clean lest the current from the dynamo finds it easier to pass through the lighting circuit than through the battery. If it does, it may blow the lamp bulbs. The battery acts as a buffer and reservoir, but if the resistance of the battery circuit is too high, the balance is upset. Loose connections are one cause of high resistance, but the internal resistance of the battery itself changes with temperature. In cold weather the resistance is higher. Even when

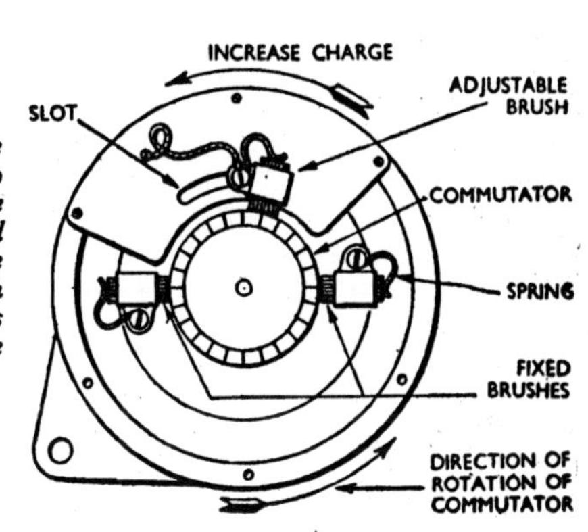

Method of adjusting the position of the third brush on a dynamo in order to regulate the charging rate. To increase the rate of charge the adjustable brush should be moved in the direction in which the armature of the dynamo rotates. Such adjustment is not provided in dynamos used in circuits with automatic voltage control or voltage and current control.

the equipment has automatic voltage control, too much current from the dynamos may for a short while pass into the lamps when they are switched on, and may blow them before the voltage regulator can modulate the pressure.

Ammonia will help to clean away the sulphate on terminals, and petroleum jelly will delay future formation of sulphate. In very cold weather the engine should be run for a minute or so before the lamps are turned on, lest the bulbs should be burnt out while the generator voltage is adjusting itself to the high resistance offered by the cold battery.

It should be noted here that dynamos made for use in sets having automatic voltage control need not be fitted with an adjustable third brush for regulating output.

The voltage adjustment is carried out automatically by the regulator in the control box. The controller itself can be adjusted so that it comes into effect at various different ranges of voltage, but this is an adjustment best made by the implement dealer or an electrical expert.

Overcharging can still occur even when a controlled charging system is in use if there are very low electrolyte levels, or if the battery is reduced in capacity by sulphation.

The chemical action of charge and discharge when a cell works does produce unavoidable wear on the plates and separators. The active material is constantly changing chemically, and this change contracts and expands it, and the gas produced during charging knocks minute particles of active materials off the plates. These particles, together with any material dislodged from the plates by vibration, collect as sediment in the bottom of the battery. The more material the plates have lost the smaller becomes the battery capacity.

Vibration can break plates from their connecting bars and the loose plates may then puncture separators, but although the battery must not be loose in its box, the holding-down bolts must not be too tight, because too much pressure may warp the battery container, or break the partitions between cells, or crack the sealing compound. If there is room between the battery and the base, sides and top of the case which hold it, it is well to put in some strips of rubber sheet cut from old inner tube, to absorb shock.

When the battery is discharging, lead sulphate forms in both positive and negative plates, and the lead sulphate formed from normal discharge is easily removed by charging, but if the plates are allowed to stand with lead sulphate in them, or are operated for long periods in this partly discharged condition, the lead sulphate becomes hard and crystalline, and it expands and warps the plates. Hardened lead sulphate cannot be removed by charging.

Excessive overcharge produces heat, as well as discharging water as vapour, and the heat softens the separators and evaporates more water from the electrolyte.

If the level of the electrolyte is allowed to stay so low that the tops of the plates are exposed, the areas of negative plates exposed will sulphate, and even when water is added the parts of the plates that have been exposed will never fully recover their efficiency. As the level drops still lower the electrolyte becomes more concentrated and can char the separators. Moreover, the paste in the negative plates becomes mushy, and falls from the lead grids. The effects of overcharging on the positive plates is of course to overheat them, due to the rapid chemical action, and this brings buckling and distortion. The plates increase in size and the lead grids shed their active material.

The sulphuric acid solution should show a specific gravity of 1·28 when the battery is fully charged. The solution must not be allowed to be more concentrated than this. Acid can be lost from the battery only by spilling. The vapour driven off in use is water only, and therefore distilled water is all that need be added.

The correct specific gravity of the electrolyte is 1·28 only at 70° F; specific gravity varies with temperature and any reading taken with a hydrometer should therefore be corrected. For every 10° F above 70° F 0·004 should be added and for every 10° F below 70° F 0·004 subtracted. The electrolyte of a discharged battery (specific gravity approx. 1·150) freezes at temperatures below 0° F, and a fully discharged battery (specific gravity approx. 1·110) freezes at 16° F. In cold climates the battery must be kept well enough charged for the specific gravity not to fall below 1·200 at 0° F, 1·245 at − 20° F, and 1·265 at − 30° F. The following table gives specific gravities at various temperatures:

SPECIFIC GRAVITY OF ELECTROLYTE

Fully Charged	*Fully Discharged*
1·264 at 110° F	1·094 at 110° F
1·268 at 100° F	1·098 at 100° F
1·276 at 80° F	1·106 at 80° F
1·280 at 70° F (Normal)	1·110 at 70° F (Normal)
1·284 at 60° F	1·114 at 60° F
1·292 at 40° F	1·122 at 40° F
1·300 at 20° F	1·130 at 20° F
1·308 at 0° F	1·138 at 0° F
1·316 at −20° F	1·146 at −20° F

It is difficult to pour from a jug into the small openings in the top of the battery, and it is worth while to buy or make a pipette to deliver a controlled stream of water which can be stopped as soon as the plates of the battery are covered. A great aid to ease in filling up batteries is the rubber connecting pourer which can be bought for bottles. By an ingenious arrangement of the levels in the rubber connector when the bottle is inverted the flow of water stops when the level of electrolyte is at the required half inch or so above the plates.

If electrolyte has been spilled on the battery it should be wiped off with a rag dipped in ammonia. Dirt should be cleaned off the top of the battery.

To estimate the state of charge of a battery a hydrometer can be used. A usual type of hydrometer for battery work is enclosed in a syringe. The rubber bulb of the syringe is pressed and then released and the solution is drawn up into the syringe so that the hydrometer floats in the solution. Provided that the concentration of acid in the solution first put into the cell was correct, a specific gravity of about 1·280 indicates that the cell is fully charged. When the specific gravity has fallen as low as 1·110 the cell is run down.

A more accurate test of state of charge can be made with a voltmeter to show the voltage drop along a resistance of known value.

Distilled water for batteries should be kept in clean, covered vessels made of material quite insoluble in water. Glass, china, rubber and lead are suitable.

Sulphuric acid at the concentration it falls to in a discharged battery

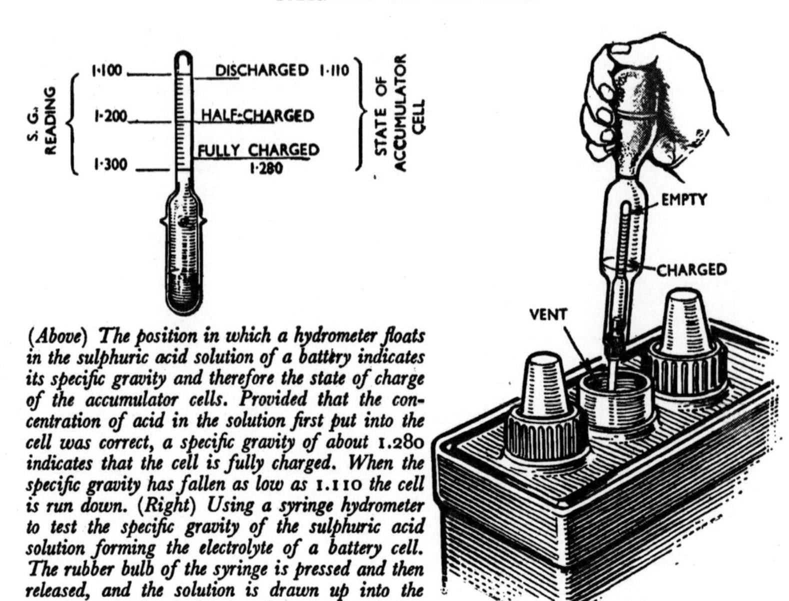

(Above) The position in which a hydrometer floats in the sulphuric acid solution of a battery indicates its specific gravity and therefore the state of charge of the accumulator cells. Provided that the concentration of acid in the solution first put into the cell was correct, a specific gravity of about 1.280 *indicates that the cell is fully charged. When the specific gravity has fallen as low as* 1.110 *the cell is run down. (Right) Using a syringe hydrometer to test the specific gravity of the sulphuric acid solution forming the electrolyte of a battery cell. The rubber bulb of the syringe is pressed and then released, and the solution is drawn up into the syringe so that the hydrometer floats in the solution.*

will freeze at 16° F (16° below freezing point), and the least harm it will do when it freezes is to damage the container; it may also loosen the paste from the plates. But the solution in a battery three-quarters charged, that is sulphuric acid at a concentration giving a specific gravity of 1·245, will not freeze until the temperature has fallen to 50° of frost. Therefore in very cold weather extra care should be taken to keep the battery well charged; and water should be added only when the engine is going to be run very soon. The charging will mix the water and acid before the water, lying in a layer at the top, has had time to freeze.

Continued overcharging is bad for an accumulator, but most battery damage comes from too heavy discharging or from the battery being left for long periods in a low state of charge. We have seen that a discharge too heavy for the size of battery buckles the plates. Starting a cold engine in a tractor or other machine needs a very large current and puts a big strain on the battery. To overcome the inertia of pistons glued to the cylinders by congealed oil, a high amperage has to pass through the electric starting motor. Some of this drain on the battery can be prevented by turning the engine over a few times by hand before the starter is used, to free the oil. This need not be laborious, because it does not matter how slowly the engine is cranked. There is no need to swing it. Holding out the clutch also can ease the work of the starter, because it will then not have to turn the shaft in the gearbox.

There should be little variation in the specific gravity readings from cell to cell on any battery in reasonably good condition, provided the electrolyte in some cells has not been more diluted than that in others. If the variation is greater than 0·025 then it is likely that a positive and negative plate are touching slightly in any cell that shows the low readings. Cells like this will run themselves down in a night. Another, but less frequent, cause of unequal readings is that vibration has shaken some of the active material out of the plates, and that this material was loosened at a time when the plates were partially discharged. The material fell to the bottom of the cells carrying with it a certain amount of sulphuric acid, and then on recharge this acid was not put back into circulation. A cell that shows low gravity for this reason is not subject to running down overnight, but it will have reduced capacity. It may be unable to turn the starting motor, but it will serve fairly well if only low discharge rates are demanded.

Variations in electrolyte gravity are sometimes accompanied by voltage variations from cell to cell when the voltage is tested by a meter with a discharge shunt. If it is found that any one cell is showing a gravity variation of more than 25 points and this same cell shows a voltage difference of 0·2 volts, the only safe thing is to have the battery opened by an expert.

When acid has been spilled or lost by leak, topping up the level with distilled water will lower the specific gravity. This can be corrected after charging the battery by adding a solution of sulphuric acid which has an approximate specific gravity of 1·350 until the specific gravity of the electrolyte is 1·280. Concentrated acid must not be used for this purpose and when the 1·350 solution is being made up, acid must be added to water, and never water to acid, because the dilution of a large quantity of acid all at once causes great heat and spurting.

Before starting to charge batteries from an external source, they should be topped up with distilled water to ½ in above the separators. A usual charge rate for a 12-volt battery is 10 amperes. If the charge is to be complete it should be continued until the specific gravity and cell voltage in each cell show no further rise during five hours of continuous charging and all cells gas freely. The maximum permissible temperature of electrolyte during external charging is 110° F and if this is exceeded the charge should be suspended or reduced to one-half to allow the temperature to fall.

INSULATION

The mechanical condition of the electrical equipment, particularly the bearings of moving parts, must be kept sound, and all insulation must remain effective. Frayed or rotted insulation can cause slow leaks of electricity from the battery to the frame of the tractor. This will run the battery down, and can mean great trouble when the tractor has next to be started; and also it is bad for the battery. If the battery seems to be losing its charge overnight and there is no reason to believe that the battery itself is worn out, then a dealer should be asked to check over the insulation of the electrical equipment with instruments for detecting leaks.

The commutator should be cleaned occasionally by holding a petrol-moistened cloth against it while the armature is being turned. Emery cloth should not be used, lest the dust from it lodge between the commutator segments and short-circuit them, nor should the cloth used be fluffy. The carbon brushes should be looked at to see whether they are much worn, and to find whether they are free to move in their holders. If they are sticking they should be withdrawn and the sides of the brushes should be filed a little. Brushes worn short should be replaced. Irregular generating, and much sparking at the commutator, can be caused by the brushes becoming so short that the springs cannot keep them in firm contact with the commutator.

The bearings of the armatures must be kept sufficiently lubricated, but only a very little oil or grease should be added, if any at all. Some makes are packed with grease when new, with no provision made for renewal of it before the dynamo or motor is taken to pieces for overhaul. Excess lubricant might get on the armature windings and cause the insulation to swell. This swelling might result in the rotating armature fouling the close-fitting field magnets, and so stripping the insulation from the outer wires. The same risk of fouling arises when an armature is expanded by overheating due to too high a rate of current generation, and also when worn bearings allow the armature to wobble.

DIESEL INJECTORS

Of all the causes of bad running in Diesel engines, faulty injectors and nozzles are the most frequent. A knock in one or more cylinders, loss of power, smoky or black exhaust gases, and increased fuel consumption all indicate that the injectors may need attention.

Before removing any of the fuel injectors, the engine and the equipment to be removed must be well cleaned. The high-pressure lines at the fuel pump are then slackened, and the injector unions are disconnected. The injector holding-down nuts are slackened and the individual injectors are lifted out. When they have been removed, each insert should be plugged with soft cloth to prevent dirt entering the cylinders. Each injector should be wrapped in a clean cloth and the set should be placed in a container until they have been serviced or replaced by serviced units.

It is usually best to have all the injectors in a multi-cylinder engine serviced at the same time, but if they have only recently been replaced and if there is reason to think from the beat of the engine that only one injector is out of order, the faulty one can sometimes be traced. When all the injectors have been removed, each should be reconnected to its pipe, but upside down, so that if the engine is turned with the starting handle, the quality of spray from the injectors can be observed. If each injector is in good condition a noise will be heard at the moment of spray. If the nozzle sprays to one side only, or if the spray appears unduly wet, the nozzle should be probed. If there is no spray at all or if the nozzle dribbles, the complete injector should be serviced or replaced.

It is unwise for the injectors to be given much treatment on the farm, but there is no harm in soaking the nozzle valves in petrol to soften the carbon and in scraping the valve seating gently with a brass scraper. The spray holes can be carefully probed with a broach to clear them without enlarging them. If this amount of treatment does not make the injector give the right kind of spray when it is connected with the fuel line, it should be exchanged for a replacement unit.

When the injectors are to be refitted, the plugs should be removed and the injector inserts should be cleaned. Any surplus oil from the injectors should be wiped off before they are put into the inserts. The washers and holding down nuts should be replaced and tightened finger-tight. Then the nuts should be tightened alternately half a turn each until the injector is firmly in position. Each pressure pipe union should then be connected at the injectors but not tightened. Then the pressure pipe unions should be connected at the fuel pump and tightened securely. Finally, the fuel leak-off pipe assembly should be connected. The copper sealing washers must be in their correct places.

When the nozzles are being tested great care must be taken to prevent the direct spray making contact with the operator's hands. The working pressure is high enough to cause the Diesel oil to penetrate the skin of the hand and cause serious injury.

The fuel injection pump should not require attention between engine overhaul periods providing the fuel filters are kept clean by being serviced at the recommended periods. The pump is filled to the correct level with oil when first fitted to the engine and will not require topping up between overhaul periods.

If a Diesel engine appears to have less power than it ought to have, and the compression is all right and the injectors are in good order, it may be that the injection pump timing is not quite correct. Usually there are timing marks on the crankshaft pulley and the front of the engine and the setting can be readily checked.

On both spark ignition engines and Diesel engines, clogged filters in the fuel line can cause erratic running, but a great difference in procedure between servicing Diesel engines and servicing spark ignition engines is that after attending to the fuel line of a Diesel engine it is necessary to take special steps to make sure that the fuel system is full of fuel and contains no air. Bleeding out the air is usually necessary after filters in the fuel pipe line have been cleaned, and sometimes is needed when the fuel tank has been allowed to run dry. The fuel systems of many engines may be bled, after filling the fuel tank, by loosening the fuel lines at the injectors, one at a time, and operating the transfer pump or injection pump unit by hand until the air is forced out and air-free fuel appears at the loosened fitting. As long as bubbles continue to appear the pumping must be carried on.

Usually air release points are also provided on the fuel oil filter and the fuel injection pump and there is often a hand priming lever on the fuel lift pump.

If air has found its way into the fuel line even when there was a good supply of fuel in the tank, checks must be made elsewhere to find where it is gaining access. The first points to check are all the unions along the pipe lines, from the fuel tank to the injector feed pipes. If there is a leaking union on the pressure side of the fuel line, it may be shown by the presence of fuel oil. The pipe between the fuel tank and the fuel lift pump is under suction, and if the union nuts are loose air will be drawn into the system, so these likewise must be tightened to make them completely airtight. Take care, however, not to over-tighten the union nuts lest they split.

DIESEL GOVERNOR FAULTS

Air leaks at the pneumatic governor and venturi unit of a Diesel engine also can cause trouble. An air leak anywhere between the venturi unit on the air manifold and the penumatic governor will affect engine control, and usually causes race at idling control settings. All unions and joints should be checked. If they are in good order, then either the flexible metal pipe or the diaphragm in the pneumatic governor is faulty. A new flexible pipe should be tried out first, but, if the trouble still persists, the diaphragm should be checked for leakage and replaced if necessary.

CLEANING FUEL SYSTEM

Dirt, dust and moisture can contaminate fuel both in Diesel engines and spark ignition engines. Tanks should be examined from time to time for leaks and for corrosion and accumulations of water and dirt. They should be drained and cleaned once a year. Both ends of the fuel line should be disconnected and the pipe blown out with a tyre pump. If the line seems still not to be quite clear, it may be found that the pipe is dented somewhere. When sediment bowls have to be emptied at the 200-hour interval, the bowl and filter should be cleaned and wiped dry with a cloth free from fluff. The gasket must be in good condition and must fit properly.

Mechanical fuel pumps need occasional attention. The constant flexing may cause fuel pump diaphragms to crack, resulting in insufficient fuel supply. Gummy deposits from some fuels may interfere with proper valve action. Water in contact with white metal parts may form whitish, flaky deposits, and dirt particles reaching valve seats may affect the operation of the pump. Leaks at pumps are indicated by fuel on the outside of the pump. If the tractor often suffers from vapour lock, it is worth while to move the fuel lines to a cooler location and shield the pump from engine heat.

Diesel engines are fitted with three or four different types of fuel cleaning unit. The first one is a fuel tank filter located in the fuel tank filler opening and acts to filter out any coarse scale, grit, and other large particles. This screen should be inspected each time the unit is fuelled and should be rinsed in clean Diesel fuel when it is found dirty. Next comes the primary fuel line filter. All of the fuel coming from the tank passes through a primary filter before going to the transfer pump. Where the unit is fitted with a drain cock it should be opened daily to draw off any accumulation of water or sediment. A secondary filter is usually located between the low-pressure or transfer pump and the high-pressure or injection pump to ensure that no trace of dirt is carried to the closely fitting pistons and cylinders of the injection pump. Many engines are equipped with small filters at the injectors. These metal edge or copper filters act as a final barrier to the passage of even microscopically small particles into the injector nozzles. Cleaning of these filters is not frequently necessary and should not be attempted in a farm workshop.

The capacity of the transfer pump is greater than the maximum requirements of injectors, so a positive pressure is always maintained on the inlet side of the injection pump if the pump is working properly. Some types of transfer pump have small screen-type filters which should be frequently removed and cleaned in petrol or fuel oil. The transfer pump usually requires little maintenance attention.

The injection pumps may be unit injectors assembled with the fuel nozzle, installed at each cylinder, or multiple pumps in a single assembly mounted on the side of the engine.

Any high-pressure fuel lines should be examined frequently for leaks by wiping the outside of the fittings with a clean, lint-free cloth and noting if leakage occurs.

CARBURETTORS

Carburettors need regular attention. A float may develop a leak or a needle require grinding into its seating. The only adjustment likely to slip is that of the slow running stop and the governor connections. Any looseness in the carburettor or intake manifold is most likely to lead to weak mixture. It is well to check the manifold occasionally for air leaks due to looseness or damaged gaskets by squirting oil over joining surfaces while the engine is running. If oil is sucked in, a leak is indicated. Leaks can be cured by tightening the studs or replacing the gaskets.

The adjustable needle valve may have a ring worn around it where it has been screwed down on the seat. This wear prevents the proper adjustment of the carburettor. If the point of the needle valve is not in very bad condition, it can be re-pointed. Sometimes both the needle and its seat can be replaced.

The float chamber cut-off shuts the fuel off from the carburettor when the bowl fills to the proper level. Any leakage from the bowl is usually due to dirt under this valve, or the valve may be so badly worn that it will not seat properly. Tapping the carburettor bowl sharply may dislodge the dirt. If the cut-off valve and seat are worn, both parts must be replaced.

The float may be worn at its hinge and where it touches the cut-off. If the cut-off holds properly, replace the worn float parts and then change the thickness of the fibre washer under the cut-off housing to bring the fuel to the proper level. Holes in metal floats can be soldered, but cork floats are more difficult. If the shellac coating becomes cracked, the float can soak up fuel and become heavy. If this happens it is best to fit a new float. If the old one is to be re-used, the shellac must be cleaned off, and the float dried thoroughly, and two coats of shellac applied to it.

It is of course impossible to see the level of fuel in a float chamber, though usually an excessively high level will show as overflowing from the jet. The height can however be measured by using a short piece of rubber hose connected to a glass tube in the form of a manometer. This is attached to the carburettor drain plug hole. The materials needed to make this tester are a glass tube, an 8 in length of 1/8 in rubber tubing, a brass male fitting with 1/8 in pipe thread on one end and a 2 in length of 1/8 in metal tube on the other. The 1/8 in fitting will replace the carburettor drain cock on many makes of tractor. A 6 in ruler or steel scale is used for taking the measurement.

Before attaching the tester, make sure that fuel flows freely from the drain. When measuring the level of the fuel in the carburettor bowl, measure from the top of the chamber where the gasket is located between the carburettor bowl and cover.

The choke should work freely, and it should open up completely in order to prevent excessive fuel consumption. The shafts that hold the butterfly valve and choke valve extend through one side of the carburettor so that the controls can be attached. They must fit the holes snugly or air and dirt will enter the engine. The air will cause hard engine starting and uneven running, and the dirt will cause wear.

ENGINE LUBRICATION

Grease points on the fan and other parts of the engine should be cleaned before the gun is applied to them and so should the end of the gun. All the old crankcase oil should be allowed to drain out of the sump before the plug is refitted when fresh oil is put in. The sump should be emptied while the engine is hot, because if this is done when the engine is cold the oil comes out more slowly, and much of the sludge will hang inside the crankcase and never flow out at all. When the engine has been working the oil is at its thinnest, and the insoluble impurities are well churned into the oil and will flow out with it.

The filter gauzes in the sump should be cleaned when the oil is renewed. This is important, since a clogged filter can either restrict the flow of oil to the bearings, or cause the oil to by-pass and so lose the benefit of being filtered.

If an external filter is fitted it must be serviced regularly so that it shall continue to hold back grit from the oil. Some of these filters have elements that can be restored to effectiveness by being washed in paraffin; others have a renewable cleaning element which must be changed at the intervals recommended by the manufacturer.

CRANKCASE BREATHERS

Crankcase breathers should be kept very clear since, if they are choked, there is no outlet for gases or water vapour that form in the sump. If the water vapour cannot escape it will condense, and then mix with the oil and form sludge.

If the crankcase breather is of the type that uses felt or steel wool to keep grit from getting into the sump and yet allow gases to pass out freely, the felt or steel wool should be washed with paraffin.

TRANSMISSION OIL

It is well to point out here the difference between the duty of transmission oil and that of engine oil, and to explain why eventually the transmission housing should be drained, while the oil is still warm from work, and refilled with new oil.

Transmission oil does not get hot enough for the lubricating value of the oil to break down, and there can be no possibility, as there is in the engine, of fuel mixing with the oil to spoil its property of preventing metal surfaces rubbing together and wearing out. If, however, we picture how the oil may contain little bits of metal that have become chipped off during an unfortunate gear change, we shall see that it is better to get rid of the pieces by changing the oil than to risk a piece becoming jammed between two gear wheels. Also, after many hours of churning, transmission oil contains air bubbles which are injurious to its lubricating property, and which will not completely disperse.

ENGINE OIL PRESSURE

A drop in the reading on the engine's oil pressure gauge after some hundreds of hours of running does not necessarily mean that the engine is deteriorating seriously, nor does it call for a heavier grade of oil to be used. The drop in pressure implies only that less resistance is present in the system. As long as the oil is in good condition and is of a good quality, the reduction in pressure, though it may indicate that the bearings now have more clearance, is in one way a change for the better, for it ensures quicker lubrication of the bearings when the engine is started.

Loss of oil pressure may, however, be serious. It may be due to clogged inlet filters, or leakage from the delivery side of the pump resulting from faulty joints, slack bearings, or badly cut oilways, or it may be due to excessive thinning of the oil because of dilution by fuel. Running the engine with the choke closed, or running the engine too cool, will allow fuel to accumulate in the sump and thin the lubricating oil. A pump with worn parts allowing too much oil to pass between the surfaces can prevent sufficient oil flowing through the bearings. If a given parcel of oil remains in a bearing for too long the oil will become overheated locally.

VALVE STICKING

Some engines develop sticking valves, and this is usually a matter of the oil gumming on the stem of the valve, retarding its movement. Sometimes the gum sets into a pitchy deposit which renders the valve immovable when cold. Because all oils on heating decompose in varying degrees, a heated engine part, such as a valve stem, is bound to accumulate a certain amount of decomposed oil.

Sticky valves usually cause back-firing. Sometimes mis-firing disappears when the engine is warm. Although valve sticking is usually due to gumming, faulty alignment or weak springs also will hamper the movement of the valves.

The remedy for gummed valves is to free the valve guides with a little paraffin or penetrating oil. An ounce of lubricating oil added to each gallon of fuel, or some upper cylinder lubricant, will help to prevent the trouble recurring.

LUBRICATION OF ELECTRICAL SYSTEM

The only parts in the whole tractor that need fine machine oil are the coil ignition distributor, or the magneto and its impulse starter, the dynamo and the starter motor. Two drops in each oil-hole every fifty hours are sufficient, and the best oil to use is the sort sold for the bearings of cream separators or sewing machines.

GREASE GUNS

Most of the other moving parts of the tractor are lubricated by grease through grease-gun nipples. Liberal use of the grease gun not only provides lubrication; it also cleans the bearings. The first few strokes of the gun do not produce any effect that can be seen by the user, but when the space in the bearing housing fills up, grease starts to exude around the edge of the bearing. At first the grease that comes out is hard and dirty, but if the pumping of the gun is kept on long enough all the old grease will be turned out, and new grease will begin to exude. Much of the dirt will have come out with the old grease.

Main fuel filter on Fordson Diesel engine. The filter has a replaceable element to be renewed after each 400 hours of work. The body of the filter must be cleaned thoroughly with a brush and Diesel oil before the new element is put into it. (*See Chapter* 12.)

Working the pump on a Diesel system in order to bleed the pipes and units free of air. A vent plug has been loosened by spanner to let the air escape. (*See Chapter* 12.)

Crankcase filters for the engine oil are sometimes washable but usually have replaceable elements to be renewed at the time of oil change as in this Fordson. It is well also to renew the rubber sealing ring. (*See Chapter* 12.)

In this Fordson engine a sump cover plate with a screen is fitted. When the engine oil is changed the screen is cleaned in paraffin and a new gasket is fitted before the plate is re-fitted. (*See Chapter* 12.)

(Left) When grease is applied to tractor king pin bearings the gun should be pumped until the lubricant appears at the ends of the pin or between the contact faces. This will force out the gritty, used lubricant from the bearing. It is to be noted that roller and other bearings having dust and oil seals should not receive grease in large quantities or at high pressures lest the seals be damaged (see Chapter 12). *(Below) The air inlet and pre-cleaner can be detached for cleaning. (See Chapter* 12.)

PLATE 32

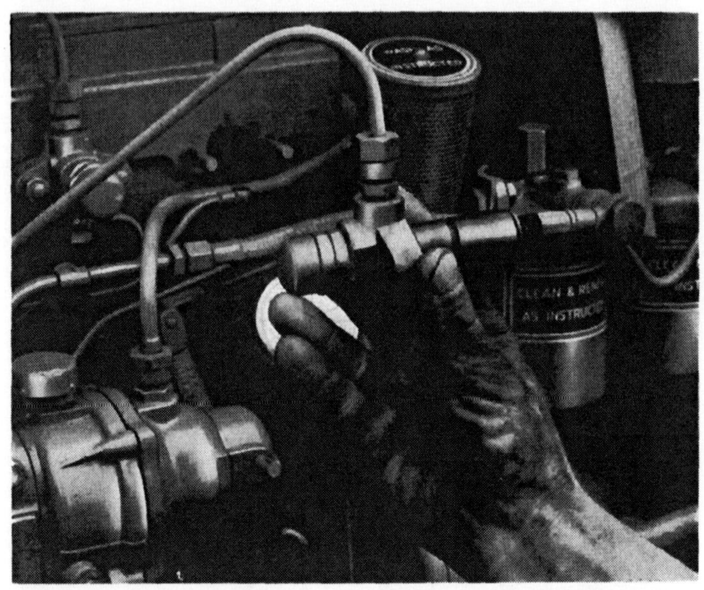

This Diesel injector, removed for test, has been re-coupled to its fuel pump. Carbon can be seen round the hole from which the injector was removed, and this indicates that the joint was not secure, and gas was blowing past it. (See Chapter 13.)

Street plates fitted to the tracks of a County Tractor to comply with regulations in Britain which require a certain area of flat surface contact to protect the highways against damage by cleats. (See Chapter 14.)

A grease gun and details of the nose of the gun. The nozzle has a spring-loaded ball valve.

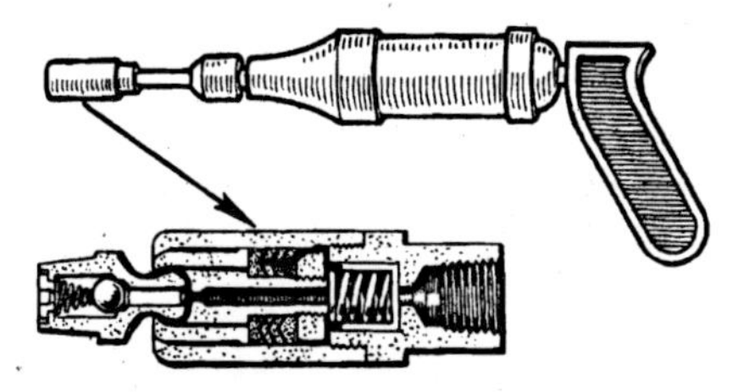

Grease guns have a spring-loaded ball valve at the nozzle. Grease cannot leave the gun until the pressure is sufficient to lift the ball off its seat. The gun must be fitted carefully with clean hands or with a clean piece of stick to act as a spoon, and the main supply of grease must be kept in a closed container and no grit must be allowed to get into it while the gun is being filled. It is usually better to buy grease in 2 lb tins rather than in 7 and 14 lb tins. The large tins remain in use so long before they are finished that there are many opportunities for dirt to enter.

Some tractor manufacturers advise thick oil instead of grease, and this can be poured into the gun.

LUBRICATING GREASES

Lubricating greases are plastic mixtures of lubricating oil and metallic soap, the metallic soap imparting rigidity which prevents the grease from excessive flow. Greases have to be used instead of oils to prevent the entry of dirt in track rollers, at points where, owing to the absence of oil seals, fluid lubricant would not be retained, and for inaccessible bearings permitting only infrequent lubrication. Thick oils, without soap additives, are specified for some bearings which have nipple lubrication, and the oils are applied by a grease gun.

Sodium base greases are very often of somewhat fibrous texture and are not normally very resistant to the action of water. As a general rule they have higher temperature resistance than calcium base greases and can therefore be used where such higher temperatures are expected. Special types of sodium base grease are particularly manufactured for high speed ball and roller bearings, but should not be used in conditions where a considerable amount of water can come into contact with the moving parts.

Aluminium base greases are generally of smooth texture but are very frequently manufactured with high-viscosity lubricating oils and may also contain tackiness agents. They are resistant to water and are particularly suited for track rollers and idlers on crawler-type tractors.

Lithium base greases possess particularly high temperature resistance. They are smooth and buttery in texture and they are water resistant. They may be used for wheel bearings, water pumps and dynamos.

N

GREASING WHEELS

Some front wheels can be re-greased only by removing the wheels. The bearings are packed with a wheel-bearing grease and must be serviced by removing and repacking. In other designs, the wheel hubs are equipped with a pressure gun fitting and must be lubricated daily. If the wheel-bearings are lubricated with the pressure gun, enough grease should be forced in each time so that a very little will work out around the inner dust seal. The grease working out around the inner seal will keep working the dirt out and will help to form a seal, but it is to be noticed that high grease pressure on seals of this kind may damage them.

When wheels and bearings have to be removed for old grease to be cleaned from the wheel hub and bearings, it is advisable to clean the inner bearing while it is still on the axle, and not to remove it unless it is intended to replace the inner seal. After the hub and bearings are thoroughly cleaned by washing in fuel, the inner dust seal should be inspected. If it shows signs of wear it really is worth while to replace it.

Some mechanics prefer to hand pack the front wheel bearings rather than use the grease gun nipples when these are fitted. If this is done a special wheel-bearing grease should be employed. Use about a table-spoon of grease for each bearing and work the grease into the bearing. Do not fill the hub cap with grease, for the pressure of grease can damage the seal if too much is used, but it is to be noticed that overfilling with grease in a tractor front wheel bearing is not as serious as in a motor car front wheel-bearing, which might leak oil on to the brakes. Nevertheless, it is well to hammer over the grease gun nipple so that some-one will not try to lubricate it with a grease gun. The success of hand packing depends upon the use of good dust seals and the proper type of grease.

When replacing the front wheels care should be taken in assembling the wheel so as not to injure the bearings. A good way to tighten the bearings is to turn the wheel slowly while the adjusting nut is being tightened until the pressure causes a slight drag on the wheel, then the nut should be backed off to the nearest cotter key hole or from one-sixth to one-fourth of a turn.

AIR CLEANER

Along with lubrication goes attention to the air cleaner. The oil in it must be drained out regularly at the intervals given in the maker's instruction book, and the sediment emptied out, and fresh oil put in up to the beaded mark on the body of the cleaner. From time to time the whole air cleaner must be cleaned out and washed in paraffin.

These attentions are necessary for two reasons: to maintain the cleaning property of the component, and to keep the path free for the flow of air. If the oil in the cup becomes overcharged with dust, a coating will be formed where the oil splashes on to the inside of the cleaner, and this coating can reduce the rate of flow of air through the cleaner sufficiently to upset the correctness of the mixture of fuel and air entering the cylinders.

The volume of air entering the engine is affected also by the viscosity of the oil used in the cleaner. It was found in some fuel consumption tests that the use of too heavy an oil in the cleaner can cause the engine to use two pints more fuel per hour than it otherwise would need. On this point of the correct viscosity of oil to be used, the tractor maker's instructions should be followed precisely. In some makes of tractor the grade of oil recommended is the same as the grade used in the engine.

For some vaporizing oil tractors disused engine oil from the crank-case is the fluid to be used — that is to say, engine oil slightly diluted by paraffin. In other makes a low-viscosity oil specially bought for the purpose is necessary.

An advantage in using new oil for the purpose is that it is not as likely to change its viscosity during use. If a used oil is placed in the air cleaner cup, it may contain so much diluent that it begins by being too light in body and as the diluent is drawn off the level in the cup is lowered too much. Also when the diluent is being drawn through the separating screen it tends to wash loose any fine dirt that might have collected in the top of the screen and this dirt is carried into the engine. But what is perhaps more important is that after losing its dilution the oil may eventually become much too heavy.

It is important that the oil cup is filled to exactly the proper level as indicated on the cleaner cup. Filling the cup too full may cause oil to be drawn into the engine, taking dirt with it, and may also cause air-intake restriction which would increase fuel consumption.

Fine dirt may collect in the top of the separating screen and occasionally this unit might become clogged with leaves or dirt. Remove the air cleaner, take a white cloth and wipe the inside of the pipe leading to the carburettor to determine if much dirt is passing through the air cleaner.

Wash the separating screen by pouring fuel through it or by moving it up and down in a bowl of paraffin.

Inspect all other possible sources of dirt leakage such as the throttle shaft and inlet gaskets.

Air cleaners need to be attended to at much shorter intervals during summer cultivations than they do in the winter, because the atmosphere contains so much more dust. Tractors being used to drive threshing machines, however, need frequent attention to the air cleaner even in the winter.

TRANSMISSION ADJUSTMENTS

Clutches and brakes are adjustable to take up the effects of wear. The bearings of belt pulleys need correct lubrication. On some makes, the bearings and the mesh of the gears that drive the pulley can be adjusted by shims.

Some power take-off joints have lubricant-sealed bearings and require no lubricant servicing for periods of 800 hours after which the unit must be completely dismantled and lubricated. Some splined connections of universal joints are automatically lubricated, and others have nipples or screw plugs.

CRAWLER TRACK ADJUSTMENT

For most tractors the track adjustment is correct when the track has a sag of 1½ to 2 in at a point half way between idlers and sprocket. This is measured by placing a flat bar along the top of the tracks or by placing the pinch bar on top of the track carrier roller, when it should be possible to lift the track 1½ to 2 in above the roller. If the track is too loose it will come off the sprockets. If it is too tight it will wear the sprockets.

To adjust the tracks, remove the covers at the back of the idlers or in front of the sprocket. Loosen the clamp nuts and adjust the track to the correct tension, drive the tractor back and forward to equalize adjustment, then recheck tension.

If the track becomes longer and longer due to the increasing wear in the pins and bushes the adjusting bolt has to be turned more and more to take up the slack. This cannot, of course, go on for ever and in most instruction books there will be found an indication of how much adjustment can be made before the track must be overhauled so that the slack in the hinges can be taken up by new pins or bushes or by turning the bushes into another position, as described in the chapter on overhauling the tractor.

Although the pins must not be lubricated it is most important that the bearings of the track roller frame should be lubricated regularly with semi-fluid lubricant, at every twenty hours of use. The track carrier rollers and the outer bearings of the roller frame must also have the lubricant replenished every twenty hours. Semi-fluid lubricant is correct in all these places and also for the large bearings which carry the idler wheels.

Wear of the track and sprocket has two effects, both tending to cause the track to ride off the driving sprocket. The first is to make the track or chain too long, and this can be corrected for by tightening as described above. The other is to change the pitch of the chain in relation to the pitch of the sprocket.

The steering clutches and brakes should be examined regularly to see if they need to be taken up. If the pedal for the main clutch has to be pressed right down on the foot plate of the tractor before the clutch disengages fully, the pedal should be adjusted. This can be done by removing a cotter, screwing out the clutch release arm, and then replacing the cotter.

WATER COOLING SYSTEM

At the specified servicing time the water circulating system should be flushed out thoroughly. All that is necessary is to open the drain cocks and pour water into the top tank until such time as the water which comes out of the drain cock is quite clear. In the case of a tube radiator, if tubes become clogged inside, and water does not remove the obstruction, they can be cleaned by means of a thin strip of copper wire or cane, inserted through the filler hole. When the fins and tubes become clogged outside with chaff or seeds they should be cleaned so that the air circulation shall not be retarded. Compressed air is useful for doing this.

A small leak in a radiator may be repaired temporarily by applying brown soap or white lead, but the repair should be made permanent with solder as soon afterwards as possible.

A choked radiator reduces cooling efficiency. The stopping-up of one tube does not seriously interfere with the water circulation but it may prevent proper emptying of the water, and the tube may freeze in cold weather.

Rust and deposits of lime within the system retard the flow of water through the tubes or honeycombs and they also increase the insulation to heat so that the heat cannot pass readily from the water to the metal and into the air. When hard water is used this trouble can become quite serious, particularly when the radiator is emptied each night for fear of frost, and is refilled with fresh water from which a new lot of lime will be deposited in the cooling system.

There are proprietary rust inhibitors which can be used at each filling of the radiator after draining. Some of these preparations are a soluble oil and inhibitors of this kind must be used only in very small quantities, for if too much soluble oil is added to the cooling water it will deteriorate the rubber hose connections and the rubber rings which are used to seal cylinder liners in some makes of tractor.

Lime deposits can sometimes be removed by a solution of washing soda; 1 lb of ordinary washing soda to every 2 gallons of water makes a useful solution for this purpose. The solution should be put into the radiator and the engine should be run, with the filler cap off the radiator, until the solution becomes hot. Then the cooling system should be drained and flushed with clean water.

This is a mild treatment and is safe to use. Something more drastic may be needed, however, particularly for lime-cum-rust deposits, and there are several commercial preparations of acid on the market. If it is desired to make up a solution at home the following are the proportions recommended for the solution; one part of formaldehyde, five parts of commercial hydrochloric acid and 42 parts of water.

This solution is put into the radiator and the engine is then run for two or three hours. It can, of course, be doing work during this period. Then the radiator is drained and the cooling system is flushed with clean water in which some caustic soda has been dissolved. This will neutralize the acid that may be left after the cleaning solution has been drained out. The caustic soda itself, however, must be carefully washed out and so it is well to leave water flowing through the radiator for quite a long time. This can best be done by letting a hose pipe play into the filler opening of the radiator while the drain tap is open. The amount of water flowing through the hose pipe can be adjusted so that the radiator remains nearly full of water. The purpose of the formaldehyde is to slow down the action of the acid on metal parts. It has a selective action; it inhibits the action between hydrochloric acid and metals but interferes only very little with the action between hydrochloric acid and lime.

The fan needs occasional checking to find whether the blades are loose, cracked, or out of line. A damaged fan can lead to damage of the radiator core. If the fan is belt-driven, the belt should be inspected for proper tension; the amount of slack in it midway between the two pulleys should be about ¾ in on most tractors. The fan belt should be replaced when it is worn, oil-soaked or frayed, and it should not be able to ride down in the bottom of the pulley groove; if it does, there will be slippage and rapid wear.

Water sometimes leaks from the pump gland. This may be due to the wrong kind of lubricant. Some tractors have pumps which need a grease that does not saponify with water. If the right grease is used, and the gland still leaks, the cap of the packing gland can be tapped round part of a turn to tighten it. If the leak persists the packing nut should be unscrewed so that the packing can be removed. Emery cloth can be used to clean any rust or corrosion from the shaft, but all emery dust must be removed from the pump. Then the gland can be repacked with asbestos string or other suitable material and the packing nut can be installed. The packing must not be tightened too much or the shaft will bind. Some water pumps have carbon seals which are unaffected by water or oil.

If ever the water system freezes, the engine must not be turned or the key that holds the pump impeller to the shaft may break.

Hose and hose connections may collapse or leak, and reduce circulation or cause loss of coolant and overheating of the engine.

Thermostats sometimes stick, particularly if very hard or dirty water has to be used in the cooling system. If an engine does not warm up within a reasonable length of time after starting it is likely that the thermostat has stuck in the open position. If it sticks in the closed position the water which is trapped in the engine will get hot and boil while the water in the radiator may not get warm at all; indeed, on a cold day it may freeze. A check to find whether the thermostat is stuck in the open position can be carried out by removing the radiator cap when the engine is cold, starting the engine, and observing whether the coolant is being pumped through the radiator. If the coolant is moving through the radiator, it is likely that the thermostat is stuck in the open position.

If the thermostat has to be removed it must be handled carefully. The valve must not be forced to open or close. It can be cleaned in a solution of warm water and washing soda, and rinsed in clean water. The opening temperature of the thermostat can be checked by immersing it in a saucepan of water that can be heated gradually. The thermostat ought to begin to open at about 160° F for a petrol or Diesel engine, and at about 180° F for a vaporizing oil engine.

STEERING GEAR

The steering gear should be adjusted when much backlash has developed. Causes of backlash are wear of the worm gear; wear of the bushes of the pins at the various connecting joints of the steering rods; and even excessive play in the bearings of the front wheel themselves. Each of these possible places should be looked at while the steering wheel is being rocked slowly within the limits of the backlash.

The worm is on the steering-wheel shaft and the worm wheel is on the steering arm which moves the link rod and imparts movement to the stub axle. The steering arm moves through only about a quarter of a circle, and therefore only one quarter-sector of the wheel is ever engaged. Thus the teeth on a quarter of the sector of the wheel become worn while three-quarters get no wear at all.

In some tractors only a sector of wheel is provided; in others a whole wheel is fitted; and an unworn sector can then be brought into use merely by turning the worm wheel to a new place on the spindle of the steering arm. In makes which have only one sector provided, other means have to be used for taking up the effects of wear. Usually the mesh of the teeth can be adjusted by moving the position of an eccentric housing which holds the bearing of the worm wheel. Care, however, should be taken not to overdo this process of adjustment and so leave the gears in tight mesh.

If the steering wheel feels rigid, then the adjusting has been taken too far. Correct adjustment leaves the wheel with about an inch of backlash movement at its rim.

FUELLING THE TRACTOR

It is well to have a fixed routine for refuelling a tractor from a storage tank, and it is better to fill the tractor tank at the end of a day's work rather than at the beginning. The moisture-laden air in the tank is replaced by the fuel and this prevents condensation overnight. Also during the night foreign matter and water present in the fuel has time to settle to the bottom and if the drain cock is opened in the morning the accumulated water and sediment can be drained out of the tank.

TRACING BREAKDOWN CAUSES

HOWEVER well the tractor is maintained, there are mornings when it fails to start, and there are days when it runs irregularly and lacks power. Quite often the cause of these troubles is some simple defect that can easily be put right; but finding what is wrong may take a long time.

In tracing the cause of troubles of this kind, it is important to follow some regular routine, so that we shall not be adjusting first one thing and then the other until we get to a point where we do not know whether the present trouble is the original one or whether it is one we have brought about through some adjustment we have upset during the search for the original trouble.

The following are suggested routines for tracing troubles. The first two drills refer to vaporizing oil engines and the second two cover points peculiar to Diesel engines. In both sequences the purpose of the first drill is to find why the tractor refuses to start at all, and the second one is to find the cause of bad running, or of mysterious stoppages in the field.

VAPORIZING OIL TRACTORS

When the Tractor Fails to Start

(1) Float chamber of carburettor empty, or containing some vaporizing oil instead of petrol. Open drain cock: if no liquid flows out see if petrol tap is turned on and tank contains petrol. In case of vaporizing oil tractor, if liquid flows out, smell it. If paraffin, empty the chamber and refill with petrol.

(2) Setting of choke valve may be wrong. If the engine has been turned many times with choke fully closed, sparking plug points may be coated with wet petrol. If choke too far open the mixture may not be rich enough to fire in cold engine.

(3) Ignition may be switched off, or the switch wire may leak. Possibility of leak can be checked by disconnecting the wire at magneto end. If engine will then start, wire is defective. If switch is controlled by rod, see if rod is bent or disconnected.

(4) No current reaching plugs. Test by taking lead off one plug and holding lead, by insulation, so that its end is about $\frac{1}{4}$ in from some clean metal part of the engine. Turn engine. If no spark occurs, battery may be discharged, or contact breaker or magneto may be out of order, or the high-tension leads may be disconnected or broken.

(5) If current coming through lead, but engine still will not fire, sparking plugs may be defective, or may be wet with water or fuel. Wipe film of moisture off insulators. If this does not cure, take out plugs and wipe points. If points are wide (measure with feeler gauge or postcard), adjust by knocking side electrode towards centre electrode. Plugs can be tested by laying them on their sides on the cylinder block, with leads attached, and turning the engine over. If the plug is all right, spark will occur at electrode points.

(6) If no current coming through leads, contact-breaker may be stuck in open position. Usually caused by fibre bush being swollen. Ease the bush a very little with a rat-tail file.

(7) Film of moisture or particles of carbon may have collected inside distributor-cap and allowed high-tension current to take wrong path inside the cap. Wipe inner surface of cap.

(8) Impulse starter may be stuck. If it is, the familiar click-click will be missing when the engine is turned. Probable cause is congealed oil, and this can be loosened by injecting a few drops of fine machine oil. Less likely cause is a broken spring which must be replaced.

(9) Gasket between carburettor and manifold may be broken and allow air to enter and weaken the mixture so that it will not fire in cold engine. New washer must be fitted. Temporary expedient is to cover break with fire-clay paste.

(10) Poor compression shown by lack of resistance at starting handle, may make engine difficult to start from cold. This can be caused by overchoking leading to oil being washed from cylinder walls, but usually indicates engine needs new piston rings, and valves need to be ground.

When the Tractor Runs Badly or Stops at Work

(1) Fuel tank may be empty, or fuel tap may be only partly opened. This keeps float chamber short of fuel when engine is working hard and causes irregular running.

(2) Tractor too hot, and oil fumes coming out of crankcase. This may be due to no water in radiator, not enough lubricating oil in sump, or a damaged bearing. If hot engine stops suddenly it may mean engine has seized. Sometimes a seized engine will free itself when it is cool and more oil has been added. Do not pour cold water into radiator if engine is very hot.

(3) Leak intake in manifold, due to loose nuts or broken gasket, can make mixture weak, and cause uneven firing.

(4) Black smoke in exhaust gases shows mixture is too rich. This causes irregular firing. Setting of adjustable jet may have slipped, due to vibration. Choke may be partly closed, or air cleaner may be dirty or overfilled with oil.

(5) Flooding of carburettor, causing rich mixture and irregular firing, may be due to float being punctured.

(6) Dirt or water in jet or fuel pipe will cause irregular running, and may cause the tractor to stop completely. First, clean out the fuel

pipes and sediment bulb, and then take out the carburettor jets and clean them by blowing through them. If the fuel in the tank contains any water the tank must be drained and washed out with a small quantity of fresh fuel. Dirt can also jam the needle valve of float chamber, and cause flooding and rich mixture, which brings misfiring.

(7) Turn starting handle to test compression. Lack of resistance in one cylinder may be due to a valve being stuck in open position. This will cause irregular firing. Free valve stem with paraffin to dissolve carboned oil. Another possible cause of open valve is broken valve spring.

(8) Plugs may be misfiring. Short out each plug in turn while engine is running. Use screwdriver with wooden handle, and connect blade first with metal body of engine, then with sparking plug top. If shorting a plug affects running, that plug is likely to be all right. If shorting has no effect, plug is defective. Take out defective plug, clean flakes of carbon from points, or, if points are too wide, bend outer electrode till gap is correct.

(9) Points of contact-breaker may be dirty, or out of adjustment. Take out breaker parts and clean points by rubbing with fine emery paper. Reassemble and adjust breaker until points open to a gap as wide as the thickness of a postcard. Broken carbon brush in magneto causes uneven firing, or complete stoppage. Carbon particles on inner surface of distributor cover cause uneven firing. Cracks in the inside of the cover can collect carbon particles which provide paths for the current. Temporary repair is to gouge out the cracks and drill small holes to obstruct path of current, but new cover is only permanent repair. Coil may be defective or may have broken-down condenser; magneto may be defective through broken-down condenser, or broken-down armature: expert attention required.

(10) Rubber coverings of sparking plug leads may be perished, and spark may be jumping across to metal of engine. Wrap rubber binding and adhesive tape round faulty places until leads can be replaced.

DIESEL TRACTORS

Some of the points in the sequences relating to vaporizing oil engines apply also to Diesel engines, but the following lists cover points peculiar to Diesel engines.

When the Tractor Fails to Start

(1) Decompressor pulled out.
(2) Air leaks in pipe line.
(3) Air bubbles in fuel system. Bleed the fuel lines.
(4) Injector needle stuck.
(5) Injector holes blocked.
(6) Broken injector spring. Fit replacement.

When the Tractor Runs Badly or Stops at Work

(1) Engine too hot. Injection timing may be incorrect or some of the injectors may be faulty.

(2) Engine knocks. Injection timing is incorrect, or there is air in the fuel system.

(3) Exhaust emits excessive black smoke. Maximum stop screw out of adjustment. Air cleaner is dirty, or delivery valves are worn.

(4) Engine misfires. Injector pipe is broken, delivery valve springs are broken, or pump plunger springs are broken. May be air in system.

(5) Engine starts and immediately stops. Governor idling setting is incorrect, or there is air in system.

(6) Engine does not give full power. Injection timing is incorrect, or there are broken delivery valve springs or the air cleaner is dirty.

(7) Engine idles unevenly. Injection timing incorrect. There is an air leak in governor diaphragm. The fuel pump plungers and barrels are worn. Manifold gasket leaking and needs to be replaced.

ACCESSORIES AND CONVERSIONS

THERE are many attachments and accessories which, though they are not essential to the operation of the tractor, increase its range of usefulness. Some of these improve the efficiency of the engine or the hitching of the implements or the grip of the tractor on the soil; some make the tractor more versatile; and others, by adding to the comfort of the driver, make longer hours of work possible.

An example of the aids to engine efficiency is the oil filter which can be fitted to some popular makes of tractor not already having a replaceable cartridge-type filter. One such attachment uses suction from the induction pipe to cause the oil to circulate from the sump, through the filter, and back to the sump. When the cartridge becomes choked it can in some models be cleaned with paraffin and petrol; in others it has to be replaced by a new one.

SPECIAL DRAWBARS

Some tractors have only a fixed drawbar, with no range of hitching height, and only a small range of lateral adjustment. This restriction can cause much extra work when various implements of different heights have to be used, and when it is desired to offset an implement sufficiently to be clear of a standing crop or to work near to a hedge. Several extension drawbars, or complete alternative drawbars, are available to increase the adjustability. Some are made by the tractor manufacturers and some are proprietary accessories, but many have been designed by users and made up on the farm.

Some of them allow as much as 36 in horizontal adjustment and 16 in vertical adjustment. Some of them incorporate also a swinging drawbar. Any attachment to fasten a drawbar to three-point linkage points needs to be considered very carefully before it is used, because it is essential that the drawbar load shall not be transmitted in such a way as to strain the linkage.

ROAD-BANDS

When a spade-lugged tractor has to travel on the highway it must, to fulfil the Road Traffic Regulations, be fitted with wheels presenting a smooth continuous surface to the road. This is reasonable, because the indentations made in the road by the lugs provide pockets in which water can collect, and, indeed, percolate through the fracture made in the top layer of the surface. If this water freezes it will burst open the

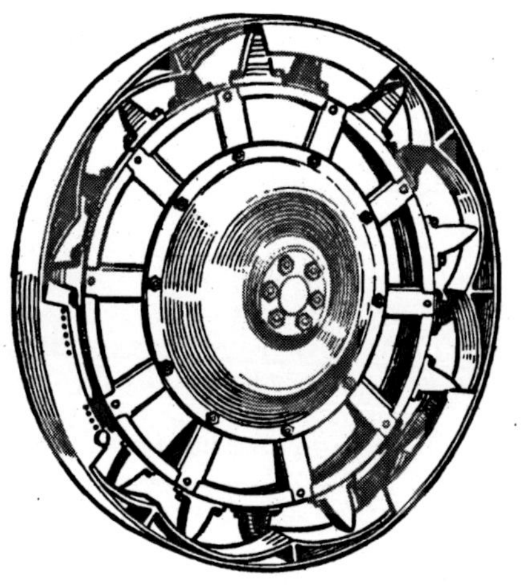

The Tamkin-Bird road band. To fit this band the tractor is run on to a block so that the inside lugs of the wheel are on the block. Then the road-band is pushed over the outside lugs and given a twist in the direction opposite to the forward drive of the wheel. The tractor is then driven off the block and driven forward. The wheel tightens itself within the road-band, and then a steel peg is pushed into one of the holes behind the lug to fix the tyre. Taking out the peg and driving the tractor backwards loosens the band.

road material, and the resulting damage will be far greater than the original mark made by the spade lug.

It would take too long to take all the spade lugs off the rear wheel before the tractor goes on to the road, and so detachable road-bands or iron over-tyres are made by the tractor makers and by other manufacturers. These bolt on to the rim of the wheel in such a way as to enclose the spade lugs. They are made in two or more sections so that they can be put on without the tractor being jacked up. One part is put on, and then the tractor is driven forward until that part of the road-band is in contact with the ground.

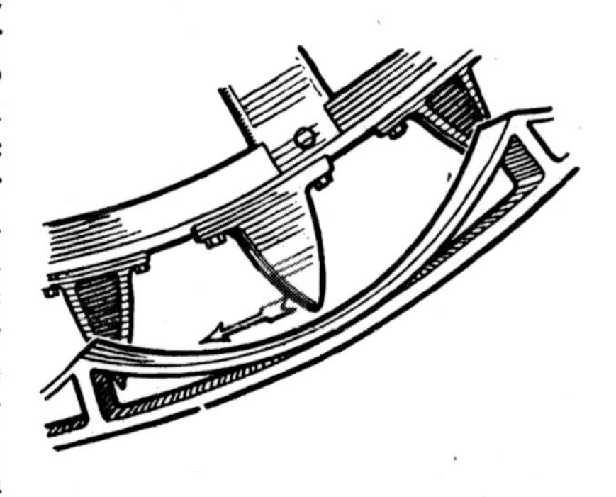

RETRACTABLE LUGS

Although it takes a very much shorter time to get a road-band on or off than it does to take off all the spade lugs, or replace them, valuable cultivation time is lost every time the tractor has to cross the road from one field to another; and many attempts have been made to design a wheel with retractable spade lugs or strakes.

The trouble in many of these devices has been that although they work well enough when they are new and clean, they become almost unworkable when their mechanism is rusty and caked with mud. In at least one make, however, this difficulty has been overcome, and the strakes on the wheel can, reliably, be adjusted easily from the working position, in which they project beyond the rim, to a position in which their edge is flush with the rim.

CONVERSION TO VAPORIZING OIL

In countries where petrol is cheap, it is an economical and convenient fuel because it does not dilute the lubricating oil harmfully, and no separate tank for a starting fuel is needed. Where there is great difference in price between petrol and vaporizing oil, vaporizing oil should be used in large engines employed on continuous steady duty, such as those in tractors. Most engines designed to run on petrol can be converted to run satisfactorily on vaporizing oil.

The compression ratio of most petrol tractor engines is 6·0 to 1, and therefore these engines need considerable modification if they are to run smoothly on vaporizing oil. It is, of course, quite possible to run a petrol engine on vaporizing oil with no structural modifications whatever: all that is necessary is to fill the petrol tank with vaporizing oil and to place enough petrol in the float chamber to start the engine. The engine is left running until it begins to falter, then the tap to the tank is turned on immediately. Alternatively, a small auxiliary tank for petrol for starting can be fitted. The combustion of the vaporizing oil in arrangements of this kind will be very incomplete and the engine may soon be damaged by detonation and uneven firing, and dilution of the lubricating oil.

An arrangement as crude as the above is not to be recommended, and the minimum reasonable alteration is that the carburettor shall have a cowling to conserve the heat in the induction pipe. Complete vaporizing of fuel is possible providing sufficient protection is given to the vaporizer against the cold winds and fan draught, and the cooling water is maintained at a temperature not lower than 195° F.

Conversion kits which are on sale are vaporizers which fit between the manifold and the present carburettor, providing a system of air-fuel mixture heating. To achieve smooth running when the conversion has been fitted and the engine is being run on vaporizing oil, it is usually necessary to retard the ignition. This reduces the thermal efficiency of the engine, but it has the subsidiary effect of keeping the engine warm. This latter point is important because when the engine has been converted to run on paraffin it must be kept warmer than is necessary when it runs on petrol; however well the fuel is carburetted, a high cylinder wall temperature is needed to keep the mixture as a vapour during the induction and compression strokes.

The other method of preventing detonation when the heavier fuel is being used is to lower the compression ratio of the engine. The simplest way to reduce compression ratio is to place an additional thickness of gasket between the cylinders and the cylinder head. An extra asbestos-copper gasket can be used for this, but these gaskets have only a thin copper rim at the edges through which heat can be transferred from the cylinder head to the cylinder block and be dissipated. When only one gasket is present, the heat probably passes from head to block sufficiently well, but a double thickness can cause serious insulation. To get over this difficulty a metal gasket can be purchased which gives quick heat transference. These gaskets are made up of thin laminations, and therefore can be adjusted to the thickness required.

The lowering of the compression ratio and the retarding of the ignition reduce not only the efficiency of the engine but its power output. One method of taking this factor into consideration when converting an engine is to put in a larger-bore sleeve and a larger piston. This restores to some extent the power of the engine. For many services, however, the loss of power has no noticeable ill effect, since most engines are built with a reserve of power above the demand usually expected on the tractor.

It is well to change the thermostat in the water cooling system, and replace it by one which will maintain a higher water temperature, since a vaporizing oil engine should have a higher working temperature than a petrol engine.

ELECTRIC LIGHTING

Most tractors are sold with basic electrical equipment included in the standard price. Electric lighting is necessary for night ploughing and is useful when the tractor outfit has to be driven home at the end of a winter day's work. When only lighting is required, a generator alone, with no storage batteries, can be used satisfactorily. Since the speed of the engine on a tractor is governed, the dynamo is driven at constant speed, and, even without a regulator, the light does not fluctuate seriously. With a magnetic regulator, which adjusts the field current according to the speed of the dynamo and the load upon it, the light remains fairly constant even when the speed of the engine

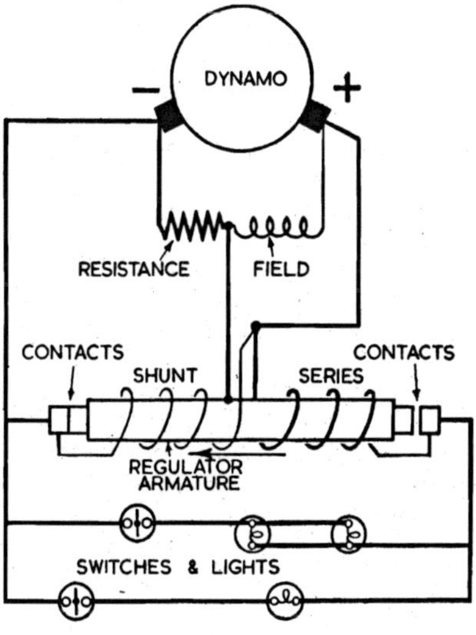

A wiring diagram of the C.A.V. dynamo lighting system using no battery. The contacts each end of the barrel-type regulator armature open or close according to speed of dynamo and load being taken from it, and they regulate the field current so as to maintain a light of constant intensity within quite a wide range of engine speeds.

varies right down to idling speed. Alternatively, the lighting system can employ batteries which are charged from some outside source.

If the lighting system is to be used both for night work in the fields and for road work either in carting or in driving the tractor to or from the fields, two different sets of lamps will be needed. The set for work in the fields will have powerful lamps to throw white light at front and rear, or one of the high-slung flood lamps which can be suspended from a standard erected on top of the tractor. The set for road work will consist of two white side lights, at the tractor's extreme edges, nearside and offside, to show its width to oncoming traffic, and a red rear light which should also have a white side window to illuminate the tractor's rear registration number plate. If a trailer is being towed, the rear light and number plate should be on the back of the trailer, and if the trailer is very much wider than the tractor it is well to have some additional side lamps fixed either on the trailer or on extension brackets on the tractor, to show the overall width.

The electric wire cable for the trailer light can be fitted with an adaptor to fit into the lamp socket of the tractor rear light. The bulb from the tractor rear light can be taken out and transferred to the trailer rear light.

Floodlights for Night Ploughing

Success in night ploughing depends mainly on careful adjustment of the lamps; detailed preparation for the job before night falls; and on keeping the driver warm.

One method of lighting the field for work at night is to have several paraffin pressure lamps, using a mantle, placed around the headland of the field. These pressure lamps give 5,000 candle power, and each one burns for forty hours on six pints of burning oil; it weighs 27 lb and in use it is placed on a tripod which allows it to be swivelled in the direction needed. Smaller oil-wick lamps can be slung on the tractor itself, and it has been found that a lamp fore and aft of the tractor and one shining along each headland make night ploughing fairly easy.

Another kind of illumination for work in the field at night is one of the electric floodlights that are carried high on a lamp standard built on to the tractor. This floodlight, operated from the tractor's dynamo and battery equipment, sends its rays downwards and gives a general light over a large area.

Lighting on the Tractor

Ordinary headlamps of the motor-car type, such as are fitted to many tractors, can make work reasonably easy if they are carefully adjusted.

If sufficient electrical power is available on the tractor, two headlamps should be fitted, one having its reflector adjusted to give a very wide beam, and the other a narrower, penetrating beam. If, however, the battery or dynamo is not large, it is better to have one really power-

An Allman speedometer mounted on the stub axle of a tractor. It is driven from the tyre and can be put in and out of action by a lever. (See Chapter 14.)

(Below) The gearbox, flexible drive and (inset) dial of a Smith Tractometer which shows engine speeds and forward travelling speeds, and registers the number of hours work the tractor has done. (See Chapter 14.)

A tractor hour recorder, developed by English Numbering Machine Ltd., which can be fitted to the generator body of a tractor and be driven by a gear wheel attached on the inner side of the generator belt pulley. (*See Chapter* 14.)

A cab on a Rotary Hoes Ltd. Platypus 30 *Bogmaster tractor.* (*See Chapter* 15.)

David Brown Cropmaster tractor fitted with detachable cab. (See Chapter 15.)

The Reekie tractor cab, fitted with unsplinterable glass windshields. The doors and screens can be removed in hot weather. (See Chapter 15.)

Fordson wheeled tractor fitted with a winch. (See Chapter 15.)

On the left is a Diesel delivery valve which has been in use with clean fuel. On the right is a valve that has to deal with fuel not adequately filtered. Abrasive matter in the fuel has caused vertical scores on the collar. (See Chapter 17.)

ful headlight than two poor ones. The best place to mount a single headlamp is on a bracket fixed near the bottom of the radiator, not more than 2 ft from the ground level, and the best place for the rear lamp is on a bracket behind the driver's seat. Experiment will show the angle at which the rear lamp lights up the plough to the best advantage.

But it will be found that this single lamp cannot possibly light up all the mouldboards and coulters well enough for plough adjustments to be made while the plough is working. The light illuminates the top of the plough very clearly, but the coulters and the bottoms are deep in the shadow thrown by the upper membranes of the frame of the plough. However, to have more than one lamp at the rear would be too complex, and so the operator must make up his mind that he will have to dismount from time to time and take an electric torch to the plough to make sure that all is going well.

The best way to find out exactly where to mount the lamps is by actual trial in darkness. Two lamps connected to the lighting set by flexible wires should be mounted loosely, one at the front of the tractor, one at the back, and both tractor and plough should be taken into the field and left in the furrow till darkness comes. In the dark it will soon be found that if the front lamp is held too high, it has to be pointed steeply downwards if it is to light up the ground immediately in front. At this angle it will throw no light into the distance; all that will be seen is a brightly lit-up patch just in front of the tractor, and this white patch makes the distance all the blacker. What is more, the steep angle of the rays will illuminate the ground too evenly and give a flat lighting.

On the other hand, if the lamp is too near the ground, much of the light will shine upwards and so be wasted. Only trial will determine the best height and place for the lamp, so that its light throws into relief the outline of the furrow wall immediately ahead of the tractor, and at the same time sends a soft glow into the distance, enabling the driver to see when he is coming towards the end of the bout.

The same method of trial must be used to find where to fix the lamp at the back of the tractor so as to light up the plough. As soon as the best positions have been found, marks should be made on the tractor so that in the morning iron brackets may be made for carrying the lamps. The bolt holes in the brackets should be made much bigger than the bolts in order to provide some latitude for movement, so that the final adjustment can be made again in the dark, before the nuts are tightened on the bolts of the brackets.

Some guiding system must be thought out to prevent the driver getting lost on the headland when he tries to get back to the furrow. An arrangement of hurricane lamps or white pegs or sticks with white paper attached to them or reflecting discs provides the best guide. It is true that when these devices are used the driver has to leave the tractor at each turn, and put the lamps or pegs in a new position, and this decreases his output; but experience has shown that he welcomes these breaks, especially on cold nights, and they give him an opportunity to look at the plough.

Many tractors, although they are not at present fitted with lighting outfits, can have standard dynamos, batteries and lamps fitted by the local agents. For some makes of tractor, however, this will not be possible, and the farmer will have to improvise.

Headlamps can usually be found at motor scrap dumps, and a very simple bent iron bracket will suffice for mounting these lamps on to the tractor. The same dump may yield a dynamo. On some tractors the dynamo could be fixed up to be driven as a three-point drive from the fan belt. If the fitting of a dynamo proves to be impossible, the lamps will have to be powered only by a battery. A motor car lighting and starting battery will keep these two lamps, with bulbs of 36 watts, alight for about twenty hours. If it can be managed, two batteries ought to be at hand so that one can be on charge while the other is being used on the tractor, or it may be possible for the farmer to arrange with his local garage for the hire of charged batteries. The battery should be mounted in a wooden box, which need only be roughly made, and this box ought to be mounted on pieces of rubber or cloth, so that the battery will be insulated from some of the vibration. It is better to have a separate switch for each of the two lamps, so that, during most of the time, the rear lamp may be switched off to save the battery.

HOUR RECORDER

Another accessory for a tractor is an hour recorder, to help in deciding when maintenance and overhaul jobs are to be done. Some of these meters have a clockwork mechanism which is set in motion each time the tractor is started and then stops when the tractor stops, but a type which is really an engine revolution counter is quite accurate enough. The engine is running at somewhere near its rated speed for most of the time it is in operation, and therefore the number of revolutions of the engine can be related directly to the number of hours of operation. In one accessory kit pack, the meter is attached to a strap which can be put round the generator body. The meter has a driving gear which meshes with another gear which comes in the pack and has to be fastened behind the driving belt pulley of the generator. The figures which appear on the dial represent hours of work.

When the engine is running faster than average the counter will register more hours than the engine has actually worked, but that will be desirable, because the wear and need for maintenance will be correspondingly greater. Similarly, when the engine is running slowly an hour's work will be recorded as less than an hour on the counter.

An engine revolution speed indicator, fitted with suitable dials, can show the forward speed of the tractor in the various gears and this is particularly useful when spraying is being carried out. It can be used also to indicate the correct engine speed for belt work and power take-off work. If a revolution counter is incorporated, the meter can be used also as an engine servicing guide.

There are further types of hour recorder which are true time clocks. These are usually electric clocks which are driven from the battery and are set in motion either when the ignition key is switched on or whenever oil pressure in the lubricating system is developed.

Methods of tabulating the readings of hour meters have been discussed in the chapter on maintenance.

COMFORT AND SAFETY

THE opportunities which tractors provide for getting work accomplished quickly are taken full advantage of on most farms, but the opportunities they give for the work to be done in comfort are not fully exploited. It is not always realized that in farm work, as in factory work, comfort of the operator can improve the output of work. Tiredness in field work is often brought about by wind and weather and by noise and vibration.

COMFORT

The Cab

One model of an American wheeled tractor has a completely enclosed coachbuilt cab with a padded seat. The cab is lined with sound-absorbing material and is air-conditioned, with a hot water radiator for the winter and an air circulation device for the summer. The floor of the cab has two trapdoors through which the drawbar hitch can be adjusted.

The driver can be totally enclosed, except for the gap for the implement trip cord. Safety-glass windows, near the operator's feet, like the windows in aircraft, allow a view on to the ground immediately ahead of the tractor. A radio receiver and a cigarette lighter and ash tray are fixed in the cab.

This saloon-tractor is designed to pull trailed implements, and its defect is that access has to be provided to the drawbar hitch and the trip mechanism of the implement. Hydraulic control of the implement, either through lift arms which can be outside the cab, or by remote control through oil lines, could do away with this need for access. With hydraulically operated implements straightforward ploughing and cultivating can be done with very few stops for attention to the implement. The few adjustments that do call for a stop are made much more quickly and easily by a man who has stepped from a warm cab than by one with cold fingers or wearing gloves.

Some row-crop operations are difficult to perform when the driver is enclosed in a cab, and it is well for the cab on a general-purpose tractor to be removable, but to make any form of heating effective the cab must have a floor, and the joints must be reasonably airtight. Considerable protection is, however, afforded by the various types of canopies sold as accessories for fitting to ordinary tractors. Some of them are made of plywood, some of metal, and some of canvas on an iron framework.

In some tractor cabs celluloid side screens can be detached to let in more air in summer, and in others the canvas sides can be rolled up. Opening the sides in this way also relieves the amplification of sound caused by the closed box. In a tractor in good mechanical condition and having an efficient silencer this is not serious, but the amplification can make a noisy tractor very unpleasant. Even so, the amplified noise may be worth putting up with on a cold wet day. On a warm, dry day the driver will be glad to roll up the sides.

Exhaust Pipe

Most tractors are fitted with good silencers, or special silencers can be bought as optional equipment. Silencers must, of course, reduce slightly the power given by the engine, but in practice it is difficult to detect any loss in pull or increase in fuel consumption.

A tractor with an open exhaust system can be converted quite easily by fixing to it a silencer, either bought new specially for the job, or a scrap one from some large old car. It must be a large silencer; a small one will not let the gases get away quickly enough. Some welding will probably be necessary to make gastight joints.

It is best for the outlet from the exhaust pipe to be at the rear of the tractor and to be as near the ground as is practicable. The fumes from a high vertical exhaust pipe in front of the driver can be injurious to health. The harmful carbon monoxide is heavier than air and sinks downwards through the atmosphere and is blown back by the movement of the tractor. Many tractor operators have headaches caused by exhaust fumes. Vertical exhaust pipes are less liable to damage than pipes taken right to the back of the tractor, but carefully controlled medical tests have shown that a high exhaust outlet in front of the tractor driver is extremely harmful. Moreover, exhaust noise is much more noticeable when the outlet is just in front of the operator's face.

Vibration

Vibration is bad for the nerves, and it is difficult to do anything to make an out-of-balance tractor smooth, beyond taking the play out of any slack bearings that may have been adding to the vibration. Tractors are rigid and unyielding, and have no spgrins to damp vibration.

The next best thing to eliminating vibration is to insulate it from the driver. Most tractors are provided with curved metal seats of the same shape as has been employed on horse-drawn implements since the last century. These metal seats are more comfortable than they look but it is sometimes better to replace them by a motor cycle saddle or by one of the special spring seats sold for agricultural machines.

SAFETY

Fatigue in the driver is one of the causes of accidents, and for this reason alone attention to comfort is worth while. There are, however, many other avoidable causes of injuries to operators, and there are more accidents than there need be. When bulls and some other animals are to be looked after there is always a certain amount of risk because their actions are unpredictable, but a tractor can be thoroughly understood and if it is handled with skill it will play no tricks.

Knowledge and thought can forewarn against any danger likely to arise. For example, if the driver is wearing rubber boots and there is mud on the platform and control pedals of the tractor, he can tell himself that the coefficient of friction between rubber and wet soil is so small that he must step very carefully when he climbs on to the tractor, and that when he works the clutch or brake his foot must be squarely on the pedal so that it does not slip off sideways.

In the same way, his knowledge of the science of mechanics will tell him that a wheeled tractor has, necessarily, a high centre of gravity, otherwise it would be no use for row-crop work nor for deep ploughing. He will know therefore that it is dangerous to attempt extreme sideling work, and that there is a danger of the tractor toppling sideways when it is driven too near the crumbling side of a ditch.

Another cause of overturning is the use of one back wheel brake instead of both. The purpose of having independent operation of the brakes, one on each rear wheel, is to help the tractor to turn when it is running slowly on a narrow headland, but to check the tractor when it is running fast on the roadway or from field to field, the brakes must be applied on both rear wheels simultaneously. Some tractors have a catch for interlocking the two brake pedals. If only one rear brake is applied it may swing the tractor so sharply to one side that it will overturn.

Starting and Hitching

Thought will tell the driver that a tractor, with its large engine and low gear ratio, moves almost irresistibly. Therefore it should be allowed to start moving only when the operator is in a position to assume complete control over it. Before any attempt is made to start the engine of a tractor either by hand crank or electric starter, the operator must satisfy himself that the gear lever is in its neutral position, lest the tractor should move off when the engine starts.

Most cases of the driver being run down by the tractor occur, however, not when the engine is being started but when implements are being connected to the tractor. The driver must remain on the seat of the tractor while he is backing the tractor towards the implement. He should never try to work the clutch while is he off the tractor. If a second man is available he can hold the hitch of a trailed implement ready for the clevis to engage the tractor drawbar, or can help to engage the arms of a unit attachment implement, but this second man must not kneel or crouch at the job, or he may himself get run over. He must be on his feet, ready to move away quickly if the tractor comes back too far.

A help to the safe and easy hitching of implements by a driver who has no helper to hold the implement clevis while he backs the tractor is an iron hook on a long handle, to be used rather like the hooks used by shunters on the railway. This hook can be carried on the tractor. The driver backs the tractor until he is near the implement, then, still seated on the tractor, he leans back and, with the hook, lifts up the implement drawbar and holds it so that the clevis will engage with the tractor drawplate as he eases back the tractor the last few inches.

Broken wrists when starting tractors can be prevented if the driver will be content only to pull the starting handle up and not to try to swing it. He can keep his thumb lying alongside the fingers with the handle lying in the cup of the palm. Then, if the engine should backfire, its rotation will throw his hand away and his grip of the handle will be lost; whereas, if the thumb is twined round the handle it is almost impossible to let go.

When the tractor engine is running in the shed, the doors of the shed should be open to allow the poisonous exhaust fumes to escape.

The tractor should not be refuelled when the engine is running, nor indeed when the exhaust pipe is extremely hot, even when the engine has been stopped.

When the tractor is on the highway extra care must be taken, because it is then not only the driver's life that is being risked. When a pneumatic-tyred tractor is running down a hill on the road, it should not be allowed to coast. One or other of the gears should be engaged so that the engine turns and provides an even braking effect. If the tractor coasts and attains too high a speed, and then the brakes are applied, the wheels may lock, and slide, and the tractor get out of control. On the highway the drawbar pins holding a trailer to a tractor should have retaining cotters in place. A tight-fitting pin, without cotter, is particularly liable to work its way out, because it is too close a fit to be able to drop back each time the pull is relieved, and so it steadily climbs out.

Power Shafts

Power take-off drives should generally be stopped before the driver dismounts from the tractor, and at any rate no adjustments to implements should be attempted until it is quite certain that the power drive is disengaged.

Guards and shields on power take-off shafts should always be firmly in position while the outfit is at work.

Power take-off shafts should be run as straight as possible. Run like this they are more efficient as a drive, and they are safer because they are less likely to break. If they do break, the driven broken end may swing wide and injure the operator.

Tractors drawing trailed implements are more prone to rear when they are working uphill than when they are on the level. But even on the level it is possible for a hidden tree root, or other serious obstruction in the ground, to lock the implement so that the engine, unable to drive the tractor along, will start to wind the tractor up, pivoting on the back axle. Usually the tractor will come to rest when the drawbar hits the ground; but it is possible for a tractor in these circumstances to rear badly enough for the operator to be injured. Therefore, if the driver suspects hidden roots, he should sit on the tractor in such a position that he can drop his foot, or hand, on the clutch lever at the very first sign of a halt in forward movement. Another precaution that can be taken is to use a wooden drawbar pin which a sudden check to the implement will break, or to use a spring-loaded drawbar hitch.

When the filler cap is being taken off a hot radiator, for water to be added, the operator should stand well back so that his face is not over the opening. Steam or a spray of boiling water may gush out of the radiator.

Precautions against fire must be taken. Canister fire extinguishers are very efficient, but it must be remembered that the various types are not universally effective for all kinds of fire. Some are designed to deal with petroleum fires, and others to deal with fires of carbonaceous material such as straw. Some send out a spray of foam and some send out a heavy gas which dilutes the air surrounding the fire and smothers it. Two types should be kept in readiness, and it is worth while to keep two, clearly marked, clipped to the tractor, on the assumption that when a fire occurs in threshing or other work in the field or yard a tractor is likely to be somewhere near and can act as a fire engine.

Another possible danger is that of the driver's clothing getting caught in spade lugs or in the sprockets of tracks. Flapping overcoats ought not to be worn on the tractor.

It is unwise to give a passenger a lift on a tractor. Even on a level highway it is not easy for a second person to find a safe place to sit or stand on the tractor. On a bumpy field, the risk of the passenger falling off is very real indeed, and if a trailer or implement is being towed the person who falls off may be killed.

If a tractor is to work regularly on hillsides it is worth while to set the wheels at a greater track width than normal, for all operations that allow the wide setting. This helps the stability of the outfit. Dual rear wheels which increase the width of the tractor will have the same effect.

Implements

Implements, as well as the tractor itself, present hazards to the operator. The risk here occurs generally when adjustments are being made or blockages are being cleared. The golden rule is to disconnect all drives to the working parts before attention is given to them. For instance, no attempt should be made to clear grass from the cutter bar of a mower until all possibility of the knife being put into motion by movement of tractor or mower has been removed. Power shafts should be disconnected and ground wheel drive should be put out of gear.

THE TRACTOR SHED

A REALLY dry and warm tractor garage will save a lot of time and energy in starting the tractor in the morning, and the same conditions, coupled with adequate lighting, will encourage the spending of more time on greasing and general maintenance.

It may not be possible, or even desirable, to have a garage just for the tractor itself, for there are several of the more modern complicated machines and implements that need just the same sort of storage and attention as the tractor. This is particularly true of the self-propelled machines and of engine-driven implements. On the other hand, there are implements such as cultivators and harrows for which warm, dry storage is not nearly as important; they can quite well go in an open-sided implement shed, and be brought into the workshop-garage only when they need repair.

Building and Fittings

What is needed is a well-built, well-lighted building large enough to take the tractor, the combine harvester or any other machine of that sort, and perhaps the motor car or lorry, and to leave plenty of room for a bench, an anvil and a forge. There must be room, too, to work at the bench and other equipment even when all the machines are in the building, for it is on wet days, when none of the implements can be out on the land at their jobs, that most of the work in the building will be done. It is worth fitting a good slow-combustion stove. This will make work more comfortable, and also will do away with the need for letting the water out of radiators on winter nights against freezing.

For motor car work a pit is a great help, but for tractor work a hoist is much more useful. The tractor engine and gear casings are much higher off the ground than are motor car parts, and a pit is rarely needed. If a little more room is wanted under the tractor it is much better to hoist the tractor up. Moreover, the hoist is extremely useful for supporting one half of the tractor when the tractor has been 'split' to get at the gearbox.

The most convenient hoist is a pulley block running on an overhead railway, but one of sufficiently high capacity, say 2 tons, needs a heavy and expensive construction for the gantry. A cheaper substitute is a set of shear legs, a large tripod to straddle the work to be lifted; at the apex of the tripod is a pulley block. The pulley block ought to be of the Weston differential type, which, by mechanical advantage, allows heavy weights to be lifted easily. Also the action of the pulley is irreversible, so that the weight remains firmly held at the point to which it has been lifted.

Another alternative is a good high-lift hydraulic jack, which can lift a tractor sufficiently for many underneath jobs to be done.

A smooth concrete floor, or a closely boarded floor with no gaps in it, saves time when nuts and bolts and small parts are dropped on to the floor and have to be found, but a spade-lugged wheeled tractor with no road-bands on it, or a track-layer with no pads on the tracks, will smash a concrete floor or tear up a board floor. Therefore a lane of dirt floor must be left alongside the concrete or boards of the main part of the building if such a tractor has to be accommodated.

A concrete floor should slope down to a drainway, so that waste water will not make the building damp. If the building has a wooden floor, a good big sink, with waste pipe, should be fitted.

If possible water should be laid on. If not, a large rain-water butt should be fixed outside for the roof drainings. Indeed, it is a good thing to have soft rain water at hand for radiators even if there is also a piped supply of water inside.

Where electricity is available, several three-pin points should be fitted round the walls in addition to the main lighting system. These will be useful for such tools as portable electric drills and emery wheels and a tyre pump. For the tyre pump a universal pneumatic unit can be used, which will do also such jobs as pressure greasing and spray painting. The compressed air is useful also for cleaning radiator grills. The three-pin points will be useful for connecting a trickle battery-charger, and they are the best kind of junction for a hand lamp, which, for safety, should always have its bulb encased in a wire cage well earthed through three-core wire.

NECESSARY TOOLS

The tools that will be needed depend so much on how far the farmer is going to do his own overhauls that it really is better to start with only a few hand tools and to buy the more complicated and special items only when the need arises. For example, if the farmer restricts his tractor overhauls to decarbonization and gets the more serious repair jobs, like gear renewal, done by the implement agent, he will not need the special wheel pullers and many other aids almost essential for major overhauls.

The equipment ought to include a strong vice for the bench, a good set of open-ended and socket spanners, a hack saw, tin snips, pliers with a wire-cutting edge, some good hammers, a grindstone, several cold chisels, punches and screwdrivers, some files, and a soldering outfit and blow-lamp.

A hand electric drill is a very useful tool indeed; and if a good big heavy-duty model is bought, able to take drills up to ½ in, it can, when more convenient, be temporarily fitted into a bench stand, with a feed arm, and this will make it into a pillar press drill.

Measuring Instruments

Some simple measuring instruments ought to be purchased, even though many jobs can be managed perfectly well without any measurements being made. A few measurements usually save time rather than waste it, because cutting and drilling by guesswork so often have to be corrected later. For metal work, a simple steel rule marked in inches, or inches and centimetres, is sufficient. Folding rules which shut down to 6 in or less can be bought, but it will be found that a single scale is easier to read accurately on a flat surface, because its whole length is in contact with the surface. Even with a thin, single-section steel rule, it is possible to make an error in measurement unless the eye is exactly over the division line which is being read. The thicker the rule the more serious the error can become. Flexible steel tape measures which roll up into a pocket case are convenient for many purposes, but they cannot be used for marking out. The engineer's steel single-section rule is rigid enough to be used as a straight edge for scribing lines on metal surfaces which are to be worked upon, and to provide a bearing surface for a knife when cutting soft material such as gaskets.

For marking out metal, a round scriber is normally used. This is a steel rod knurled to provide a good grip for the hand and tapered to a point at one end so that it will scratch metal. If the scriber is thin and its point is a very gradual taper, it is possible to draw a line fairly near to the guiding edge of the rule, provided the scriber is held so that it tends to undercut the rule. If the scriber is held upright, there will be a space between the rule and the point of the scriber.

A steel try-square should be part of the equipment of the tractor shop. It consists of a thin rectangular blade fixed exactly at right-angles into a rectangular length of thicker steel. The edges of each arm are parallel, and so the try-square can be used for the inside or outside of angles equally well. Centre punches, such as are used for starting a drill position, will also prove useful to mark the boundaries of scribed patterns. For example, a piece of metal may be centre-punched at the point of intersection of two straight lines. For this work, both centre punches and scribers have to be kept sharp.

If much accurate work has to be done, it will be found that a level surface plate is needed as the datum for making measurements and marking out work to be cut or drilled. Another use for the surface plate is for finding the high spots on metal which is being filed or otherwise made flat. To do this, a thin film of red lead and oil is spread on the surface plate and the object being examined rubbed in contact with it. The dabs of colour transferred to the object will show where the high spots are. A surface plate is made of cast iron, and is planed and carefully scraped until it is dead flat. The fine surfaee must be preserved, and therefore the plate must never be used roughly or hammered upon. For some purposes, a piece of plate glass will serve in place of a surface plate.

Callipers and cylinder-pressure gauges may be needed if overhauls are undertaken.

FUEL INSTALLATIONS

It is well for the fuel storage installation to be near the tractor shed, and the lubricating oils can best be stored actually inside the shed. Storage should be adequate. In a busy week one tractor can use eighty gallons of fuel, and in most parts of the world it is well to have at least three weeks' supply at a time. The quantity to be kept in store depends also on price differentials offered by the suppliers for various amounts at each delivery. Petrol must be stored underground or in special buildings. Diesel oil and vaporizing oil can be kept in tanks mounted above ground. The tanks can be placed on brick piles, but there should be a layer of wood or bituminous material between the top of the bricks and the bottom of the tank. If this is not done the part of the tank in contact with the bricks will rust. The easiest way to feed the fuel from the tank into a tractor is by gravity, so that no pump will be needed, and therefore the bottom of the tank should be near the level of the filler cap of the tractor, but some few inches lower, so that although the greater part of the fuel can be drawn off through a flexible hose pipe into the tractor tank the last few inches in the storage tank can be drawn off only into a container placed at a level lower than the bottom of the tank. This gives a reserve and a reminder that more fuel must be ordered.

Outdoor tanks should be protected from rain that would collect around the filler cap and also perhaps get into the spout of the hose pipe. It is well to protect the tank as much as possible from the direct rays of the sun. A thatched roof can be very a satisfactory arrangement. If the tank is alongside the tractor shed it is convenient to let the protecting roof for the tank be a lean-to. The fuel supplier's tanker lorry will pump the fuel through a hose-pipe into the filler cap of the storage tank and room must be left between the lower edge of the lean-to roof and the top of the tank for the hose from the wagon to be manipulated.

With vaporizing oil or petrol tanks can be of galvanized mild steel, but a tank which is to contain Diesel oil should not be treated inside in any way. Diesel fuel reacts with galvanizing material and some other coatings and forms undesirable compounds in the fuel. Diesel oil tanks should be tilted towards the rear and there should be a drain plug at the lowest point for sludge to be drawn off, once a week. This requirement makes it usual for Diesel tanks to be sited above ground rather than underground, because it is easier to arrange access to the drain point.

Tanks for underground siting need an opening about 4 ft 6 in wide, 6 ft long, 5 ft 6 in deep, for the 300-gallon size, but Diesel tanks must have space for the drain cock to be used to draw off a few pints to remove sediment. Two concrete supports are put at the bottom of the opening, and the tank is lowered into position. The foot valve of the pump and the suction pipe are screwed on, jointing compound being used for external threads. Great care has to be used in making the joints between the suction pipe and the base of the pump, and the foot valve and the pipes must be kept free from dirt. If the valve has dirt in it, the pump will fail to hold up the fuel when it has been out of use for some days. The space round the tank is filled with sand, and a layer of concrete about 3 in thick is run over the top to form a cover and to encase the flange. A 300-gallon tank installation needs a cart-load of sand and 3 cwt of cement.

Most underground storage tanks have a dip-stick. Some have also a recording device for indicating how many pumpfuls have been taken out; on the usual size of piston type pump one complete stroke delivers one gallon.

Since Diesel fuel is heavier and more viscous than vaporizing oil, it will hold dirt in suspension longer, and therefore more care must be taken in storing and handling Diesel fuel than vaporizing oil and petrol. A storage and settling tank, with either gravity or pump supply to the engine tank, eliminates unnecessary agitation of the fuel during tractor filling, and after the storage tank has been filled continuous settling is obtained so that fuel accidentally contaminated during delivery is clarified as the dirt settles out in the tank.

Diesel tanks should have a slope of not less than ½ in per foot of length, and with the outlet 3¼ in above the bottom. This arrangement provides that clean fuel will be drawn from this level, while the water and dirt settle on the bottom.

The gravity outlet or pump must have an oil resistant flexible hose. If a straight nozzle or pipe is used instead of a valve at the end of the hose, then some arrangement must be provided so that the outside of the cotton-covered hose can never come in contact with the fuel. A long nipple may be used, or a bar may be welded across the outside of the pipe several inches from the end so that it can be inserted into the filler opening for only a short distance. A large filler opening should be fitted and it should have a fine mesh strainer. For outdoor installations the filler opening itself may be adequate as a vent, but for indoor installations a separate vent leading to the outside must be used.

When the fuel supply nears the bottom of the tank the rate of withdrawal should be reduced to keep down the agitation of the sediment.

When the oil company's delivery man arrives with a supply of Diesel fuel in the tanker, he connects the tanker discharge hose to the filler opening on the owner's storage tank. The connection at the end of the truck discharge hose must be clean before the fuel is discharged, and the screen in the filler opening on the owner's storage tank should also be clean. After the supply has been delivered, the cap should be placed back on the filler opening at once. The fuel should be allowed to settle for 24 hours before any of it is drawn off.

If there is no Diesel storage tank, drums will have to be used, and it is well to mount them on their sides on a rack with the back end lower than the front. If a tap is fitted to a drum almost all the fuel can be drawn off without disturbing the sediment which has collected at the low end at the back of the drum. This means that a tap must be provided for each drum so that it will not be necessary to disturb the fuel after it has once settled. Another method is to stand the drums on end and insert a pump with an outlet reaching to within a few inches of the bottom of the drum, but with this arrangement care is needed when the pump is changed from an empty drum to a full one to see that no dirt from the top of the drum falls inside while the pump is being inserted.

With either method of dealing with drums it is best to allow a quantity of fuel to be left in each drum. After the fuel has been used from several drums, the small amount remaining in each one can all be poured into a single drum and be allowed to settle. Only a very small quantity of fuel need be discarded finally as dirty and unsuitable for use.

Diesel fuel should always be withdrawn from drums by a pump or through a tap. Drums should never be tilted, either on end or on their sides so that the fuel can be poured over the edges. Dirt will be washed into the fuel and any sediment will be mixed up again with the fuel from which it had settled.

STORING LUBRICATING OIL

Lubricating oils stored on the farm in bulk are usually retained on the premises for a long time before they are used up, and there are plenty of opportunities for contamination since quite small quantities are removed each time a barrel or drum is tapped.

Barrels exposed to the elements and left in an upright position are likely to draw in moisture from rain, dew, or melting snow which collects on the barrel head. Even when the bungs are drawn fairly tight, the suction created by the cooling and contracting of the barrel contents after sunshine is sufficient to do this.

Inside storage is best whenever possible, if only to keep the oil warm so that it will flow easily into containers when wanted. If inside storage is not feasible, the first essential is to make sure that the bungs are hammered tight. Laying the barrels on their sides is much the most satisfactory method for avoiding moisture contamination when they must be stored outside or under a cover that only partially protects them. When it is unavoidably necessary to store barrels outside in a nearly upright position, they should be tipped slightly to one side, in such a way that the bungs in the barrel head are on the high side of the tilt in the region where no water could collect.

Lubricating oils must be transferred to the tractor in clean containers. Being more viscous and less volatile than petrol or paraffin or Diesel oil, lubricating oil hangs around the insides of measures and attracts dirt. A measuring can which has been used for a thick lubricating oil has quite a thick coating of oil left on its inner surface. If a can in this condition is left exposed to dust-laden air it will become quite dangerous for use for further oil until it has been very carefully cleaned and dried.

Therefore, it is better to have the oil store inside the tractor shed or to arrange for the drums to be covered by a cabinet which will hold the delivery pump and any cans which have to be used. This cabinet must be provided with a dustproof door.

EFFICIENT WORKING

TRACTOR fuel efficiency is a measure of the amount of useful work done on the farm for every gallon of tractor fuel used, but fuel is not the greatest expense in connection with a tractor, and fuel efficiency is not the only factor that matters. There are, in fact, many other points – like some of those in cattle judging – that are difficult to express in figures, though they are very real to the operator. Examples are smoothness of running, ease of starting, and general reliability. A new tractor has a high score on all these points, and if it is well handled from the very first day of its life, the score will keep high for a long time.

BREAKING IN

Breaking in a new tractor is usually complete after 25 hours of work. During these first working hours the tractor should pull only light loads and the governor control should be set at no more than three-quarters open. In vaporizing oil engines, ¼ pint of engine lubricating oil should be added to each gallon of petrol put into the starting tank of the tractor. This oil provides some lubrication for the valves and pistons immediately the engine starts, which is valuable because in a cold new engine with tight bearings the proper circulation of oil from the sump may not begin for some seconds after the engine has started.

There is no need to add oil to the paraffin fuel, because the paraffin will not be used until the engine is thoroughly warm and the sump oil is circulating normally. Alternatively, an upper cylinder lubricant can be used in all the fuel used, and also a suitable additive in the lubricating oil in the sump.

In petrol engines and Diesel engines too much oil must not be added, because the one fuel is in use all the time and considerable carbon deposit may be formed. A ¼ pint of lubricating oil in four gallons of petrol or Diesel oil is a suitable mixture.

When the running-in period is complete the oil should be drained out of the engine sump while it is still hot. When the oil has ceased to drip the drain plug should be replaced and ½ gallon of thin lubricating oil poured into the engine. Then the engine should be turned over a few times by hand and the drain plug removed again. This thin oil will flush out any particles of metal worn off the bearing surfaces during their bedding in.

At this time, too, the cylinder head bolts should be tightened, for the cylinder head gasket, like the bearings, beds down during the running-in period.

Nuts on the land wheel retaining studs, and on spade lug bolts, should be tried with a spanner and tightened if loose. All other visible nuts and screws should be tested and tightened, for sometimes in new machinery a nut is held from metal-to-metal contact by a flake of paint or oxide. When this flake has shaken loose, or when some uneven metal surface has flattened with use, the nut develops a tendency to lose its firm grip.

Breaking-in completed, the tractor can now be given its full load.

When the tractor is run in, it is well to find whether any adjustments can be made to reduce its consumption of fuel. In general this attention will consist in a spark ignition engine of adjusting the carburettor and advancing the ignition, but in a Diesel engine reducing the use of the excess fuel device for starting is as much as ought to be attempted. The timing of the injection certainly should be left alone.

In a spark ignition engine, weakening the mixture of fuel and air may save fuel. The maker's setting is of necessity a compromise and is arranged to allow for unskilled operating of the tractor. Therefore an operator who is prepared to take extra care in the driving can sometimes hit on a more economical setting.

If the temperature of the engine cooling water is maintained at only slightly under boiling point the engine will run on a weaker mixture than the usual setting provides at that temperature.

THE CARBURETTOR

Where the carburettor has an adjustable jet the resetting should be made when the engine is hot and the tractor is pulling its usual load. In a paraffin-burning tractor the adjusting should be done when the engine is running on paraffin, not petrol. The governor control lever should be set fully open.

The jet should first be opened wide. This will probably cause the exhaust gases to look black, and to smell of partly burnt fuel. Then the jet adjusting needle should be turned down slowly until the engine either slows down or begins to misfire. When this point has been reached the jet should be screwed back just a little, and the carburettor will then be at its most efficient. It will be noticed that the exhaust gases are now invisible and nearly odourless.

Too small a jet opening, and therefore too weak a mixture, is not economical of fuel, and it causes the cylinder valves to become burnt.

If the carburettor is not of the adjustable jet type, the main jet should be taken out and a note made of the figures that will be found marked upon it. These figures should be quoted to the manufacturer, and a jet one size smaller ordered.

Some drivers use a variable jet adjustment to provide a richer mixture to help keep the engine from stalling while it is warming up. This is not a good practice. Such frequent use wears the screw, and there is always the risk that the driver will forget to re-set the jet when the engine has got warm.

The need for opening the jet during the warm-up period can be avoided by closing the radiator shutter and spending more time in warming up the engine before the tractor is put to work. A tractor engine should not be put under load until the intake manifold is hot, so that it can run on its normal mixture. Indeed a good test for carburettor setting is to start the tractor and, after the engine has run for about half a minute, see if it can be driven, even unloaded, without the engine stalling. If it can, the carburettor mixture is too rich.

If the manifold has a heat regulator, it must be adjusted to the fuel being used. It should be set in the hot position when using vaporizing oil and in the cold position for petrol.

Some tractor carburettors have three adjustments: idling speed adjustment, idling speed mixture adjustment, and the main variable jet for mixture adjustment at working speeds. None of these adjustments should be altered until the engine has been brought up to proper operating temperature.

The idling speed adjustment is a stop screw which regulates how far the throttle will close when the governor control lever is set in the idling position. This is set properly at the factory, but after a tractor has been in service for a few years wear occurs, permitting the throttle to close so far that the tractor stops when idling. Such trouble can be corrected by screwing in the idling speed stop screw about a turn or more. For a petrol engine this should be enough to make the engine idle, at a speed of from 350 to 450 r.p.m.; for a vaporizing oil engine it is well to adjust the engine to idle at a somewhat higher speed so that the engine temperatures will not drop too low.

To adjust the idling mixture valve, set the throttle in the idling position and turn the needle valve screw in until the engine begins to run unevenly, then back it off until satisfactory idling is obtained. On some carburettors the idling mixture adjustment screw controls the air, and when turned 'in' the mixture becomes richer. If the running of the engine is not affected when this idling needle valve screw is opened a turn or two, it indicates that the carburettor float level is too high for best operation or that the float valve is leaking or that the area around the throttle valve is restricted by intake manifold deposits.

On some makes of tractor, the work jet adjustment needle is too awkwardly placed for adjustment to be made when the engine is on field work. If so, the engine can be pulling a belt load or we can adjust it running at full speed with no load. The spark should be retarded, if possible. When the adjustment has been made without load, it is often necessary to open the needle valve a little, not over one-eighth turn per trial, if the tractor tends to stall when the load is applied. It is best to adjust for no load and then keep opening the carburettor slightly until the engine pulls the load.

Some carburettors have no accelerating pumps and the engine cannot be expected to have great accelerating capacity. However, when it is pulling a power take-off machine, such as a combine or pick-up baler, more responsive pick-up for suddenly encountered loads can be given by the engine if a slightly richer mixture adjustment is used.

Surveys show that most tractors are operated only about 30 per cent of the time at full load, and it is well to bear this in mind when carburettor adjustments are made. When any tractor is to be operated at light load for a considerable time, it may be advisable to set the load mixture somewhat weaker than normal. Some carburettors automatically give weaker mixtures at part loads although most do not.

Whether a new setting has been made or whether the maker's setting has been accepted, care should be taken that the proportion of fuel and air it provides is not upset by other factors. For instance, a blocked-up air-cleaner will cause an increased suction of fuel through the jet. Also, damage to the float chamber mechanism can make the mixture become too rich, by allowing the fuel to stand too high in the jet, so that it is drawn too easily out of it. It can also cause the carburettor to be flooding continuously with consequent loss of fuel. The damage may be merely a sticking float needle, or it may be that the needle seating is worn out of true, or it may be that the float itself is punctured and heavy with fuel that has entered it through the puncture.

The needle seating can be ground in gently with fine emery. The float, made of thin brass, can be mended by careful soldering after the fuel has been boiled out from inside it. The boiling must not be done by flame, or an explosion may result. Immersion in hot, nearly boiling water will vaporize the fuel and eject it. The immersion in hot water will also show where the puncture is, since bubbles will be seen coming from the hole. An alternative method of ejecting the fuel is to make a second small hole in the float and blow out the fuel. Then seal both holes with a touch of solder.

SAVING FUEL

Fuel sediment bulbs and fuel strainers should be cleaned before too much dirt accumulates and restricts the flow of fuel. Erratic running can waste time and fuel.

Decarbonization and the grinding-in of valves, and the fitting of new valve springs, will help an engine to give out more power for the fuel it consumes. The magneto and sparking plugs and high-tension leads also must be kept in good order so that every charge of fuel shall be ignited and give power.

Much fuel can be lost through leakage, splashing and spilling. To avoid leakage, the fuel pipe unions on the tractor should be kept tight. Any cans or drums that have become battered until the fuel can seep through their rims should be discarded or repaired.

If bulk fuel storage is not available, or if tractors have to be refuelled in the field, a sound, good-sized funnel should be used for filling the tractor tanks. If ordinary 2-gallon petrol cans are used, it is important that the can, when it is being emptied, should be held with its opening uppermost. Then the air can flow into the can steadily, and the fuel it replaces will flow out of the can just as steadily. If the can is held with the opening at the bottom, the fuel will come out with sudden ebbing and flowing, and will often splash outside the funnel. If, however, the funnel is large enough the can, or even a 5-gallon drum, can be inverted and can rest inside the funnel so that the operator has only to steady the drum and does not have to take its weight. To hold a 5-gallon drum at arm's length is tiring; with the tiredness comes inability to hold the drum steady.

The funnel must have a filter to keep back dirt and water. This is most important where the drum is inverted and is inside the funnel, because however carefully the can has been wiped before the pouring started it is nearly sure to have upon it some particles of soil.

Some air vent must be provided in the funnel so that air can escape smoothly from the tank as the fuel goes in. On some funnels this is achieved by making the stem fluted instead of exactly circular so that it can never be a close fit against the bung-hole of the tank; on another it is arranged by allowing the funnel to rest not against the edges of the bung-hole but on the body of the fuel tank.

Bulk Fuel Storage

A lot of handling of the fuel, with consequent possible loss through spilling, can be done away with by using an underground bulk storage tank with a hose delivery pipe. The need for clean storage of Diesel fuel brought a quick changeover to bulk storage on farms in Britain, and the amount of fuel handled in cans is now very small. Details of bulk storage installations are given in the chapter on tractor sheds.

It is best for the tank and pump to be placed so that the tractor can be brought to the pump without being bumped along a yard or metalled road, yet with the opening to the storage tank against a roadway hard enough to bear the tank lorry which brings the bulk supply.

Whatever way of filling the tractor tank is used there is a difficulty in that the operator while he is pouring in the liquid cannot see to what level the fuel in the tank has reached, and he may go on pouring in fuel to the funnel when the tank has become quite full. This leads to much waste; therefore, along with the use of a hose or funnel should go the use of a dip-stick.

Graduated Dip-stick

This dip-stick can be made either of wood or metal. It can be graduated by starting with an empty tank and filling up by stages from a 2-gallon petrol tin used as a measure. After each 2 gallons has been put in, a mark must be made at the point reached by the liquid on the dip-stick. When the filling is completed these marks can be made permanent and the relative figures can be set against them. It is well to have the figures running from top to bottom of the tank rather than from bottom to top, so that a direct reading will give the number of gallons which still can be added.

It is, of course, necessary to graduate the measuring stick experimentally all the way along the range because the intervals along the scale are not equal since the tank is not rectangular.

FULL LOADING

Light running of the tractor with the implement not in effective work must be kept as low as possible. In ploughing, careful planning of the layout of the bouts and the order in which they are to be done can often save a lot of headland running. The turns should be made as short as possible with a non-stop sweep.

When the tractor is in work the work should be as hard as it possibly can do. Much of the fuel used in a tractor engine is used only to move the tractor itself over the ground, and the tractor has to be propelled whether the work it is doing is heavy or light. If the outfit were running round the field with the plough lifted out of work, the tractor would still be consuming more than half as much fuel per hour as it would if the plough were turning furrows to the full depth of the tractor's power. Therefore the load being drawn by the tractor ought to be made as great as it can pull without the engine labouring below its governed speed, or the driving wheels slipping on the soil.

Ploughing is the heaviest continuous job carried out by tractor, and so it is particularly important that it should be done efficiently. If the plough load is not great enough it must be increased. The first way that comes to mind for increasing the plough load is to set the plough deeper. But this is not always a useful way of employing the power. The deeper ploughing may not give a bigger crop nor make the land any cleaner than would shallow ploughing. Another way of increasing the quantity of work done is to increase the forward speed of the tractor by changing into a higher gear. This works all right in such operations as harrowing and rolling, but with our present designs of ploughs high-speed work does not produce well-turned furrow slices. The other way to give the tractor a full load is to widen the strip of work the tractor-drawn implement is doing. If the tractor is pulling a plough that cuts two 9 in furrows, and is pulling it too easily, it is likely that it would pull a three-furrow 9 in plough set only a very little shallower than the two-furrow one. Changing from a two-furrow plough to a three-furrow means that half as much again work is done in the same time, with the same expenditure of labour, with very little extra consumption of fuel, and with probably less wear of the tractor engine.

Experiments to investigate the magnitude of loss of efficiency due to incorrect loading have been made by the National Institute of Agricultural Engineering and reported in an N.I.A.E. paper 'Influence of Engine Loading on Tractor Field Fuel Consumption'. The results indicate that the careful selection of the correct gear ratio for a particular job can greatly influence the consumption of fuel, at any rate as far as the spark ignition petrol engine used in the test is concerned.

When the load on a spark ignition engine is reduced by the normal process of regulating the throttle, the ignition timing being fixed, the charge becomes progressively diluted by residual exhaust gases; this reduces the speed of burning and the heat losses through the cylinder walls are therefore increased. This effect can be remedied by increasing the ignition advance as the load is reduced; the indicated efficiency then stays constant over a wide load variation. With many tractors there is no manual control of the ignition timing, and even when such a control is provided the possible variation of ignition advance is small. The indicated efficiency therefore normally falls with a reduction of load. Although a reduction of bearing pressures takes place with a reduction of load, the consequent decrease in frictional forces is small enough to be negligible. The frictional loss therefore becomes a rapidly increasing proportion as the load is reduced and the brake thermal efficiency must always suffer through this fall in mechanical efficiency.

Decrease in engine speed is generally advantageous. Friction and pumping losses increase slightly more than linearly with speed and so an increase in mechanical efficiency is gained at slower speeds. Although the hot gases are in contact with the cylinder walls for a longer time at a slow speed, this loss is counteracted by the increase in heat losses due to increased turbulence occurring at high speeds and so over the normal speed range the heat losses from the cylinders increase only slightly as the speed is reduced. The volumetric efficiency also benefits from a speed reduction as the velocities of the gases in the passages are lowered.

Therefore, for any given power requirement from an engine with no direct control over ignition timing or mixture strength it is advantageous from the point of view of fuel consumption to obtain that power with as high a torque load as possible and a correspondingly low engine speed. This is achieved in a normal tractor by a combined use of a stepped gearbox and a hand control of the governor setting, and a good tractor driver always selects the highest gear possible and then reduces the governed speed to give the required forward speed. The best speed is dictated by the operation being performed. This entails a greater amount of work for the driver in altering the gear or governor setting more often. Moreover, a feeling of security is given by the larger reserve of drawbar pull available in lower gears. It should be noted that a reduction in the engine speed does not reduce the maximum drawbar pull of the tractor, unless the speed is governed below the speed at which the maximum crankshaft torque is developed, and this is improbable in normal work.

Substantial savings are possible only with a tractor having a fairly wide range of gears, and from the point of view of fuel economy it certainly is well to have such a selection of gears. The other necessary characteristic is that the tractor engine should be flexible and develop its maximum torque at a low speed so that the engine speed can be appreciably reduced without affecting the drawbar pull of the machine.

Given these conditions, fuel can be saved on a light load by shifting to a higher gear and slowing the speed of the engine down to give the desired travel speed, but since some engines when operated below normal speed and heavily loaded do not receive sufficient lubrication for safe protection, it is advisable not to run the engine under load at extremely low speeds.

It is instructive to make a field test to determine how much fuel can be saved by selecting as high a gear as is reasonable. To make this test, fill the fuel tank and operate the tractor at full throttle and in the lower gear for a given time, then measure the amount of fuel used. Repeat the test in the higher gear but with a slower engine speed so as to give the same rate of travel. Accurate results can be obtained if each test is run for 4 or 5 hours.

Use of a Trailer

The efficiency of the tractor itself is not the only thing to be considered in discussing the economy of tractor operations. The power must be used to good purpose, and no tractor time must be wasted. When the weather is favourable for work on cultivation or harvest, the tractor must be in the fields on the job for as long as possible each day. A service trailer helps towards this aim, and with ingenuity the same trailer can be made to serve several purposes. For instance, a root trailer, with its side-boards detached and raves or ladders fixed at its ends, can be used for haymaking and harvest, and the same trailer can be converted for servicing a tractor.

Two or three 40-gallon drums can be placed on the trailer platform horizontally. So that they shall not roll, they can rest on two wooden brackets and can be secured by iron bands passing over the drums and bolted to the platform of the trailer. A convenient way to fill the tractor tank from the fuel drums is to use a 1 in semi-rotary pump and a short length of petroleum-resisting flexible hose. The pump can be mounted on to an iron suction pipe long enough to reach nearly to the bottom of each drum.

A locker also should be mounted on to the trailer platform, large enough to hold plough shares and spares, tins of petrol and lubricating oil and water, and a chain. There will still be room to carry a bicycle or motor bicycle so that the tractor driver can get back from the field to his meals. It must be pointed out, however, that attempts to save tractor wear can be taken too far. Calculations show that reckoning ordinary rates of wage per hour and ordinary costs of tractor operation per hour it pays better for a man to ride to and from the field for his meals than it does to walk, because the journey will be so much quicker.

A drawbar should be made on the back of the trailer so that a trailed plough or other implement can be hitched on and towed along the road. Pick-up implements for direct attachment must be carried on the trailer.

With this outfit a tractor can tow to the field all the equipment it will need for a week's work. The frame for the drums, and the locker, can be made easily detachable so that the trailer can be converted quickly back to a flat platform type.

TRACTOR IN ROUGH USE

Although it is well to be as careful as possible with a tractor, and not submit it to any avoidable strain, there are occasions when it does pay to use the tractor for rough work which otherwise would need much hand labour. Such jobs as these cannot be reckoned by ordinary standards of tractor efficiency. An example of this kind of task is the removal of trees from old orchards and coppices. The tractor can be used either by direct traction or through a winch. In direct traction snatch must be avoided as far as possible and yet best use must be made of the available pull, so the hitch must be high on the tree and low on the tractor.

Tracklayers are best for direct traction work, but if a wheeled tractor must be used it is well to try to contrive a front hitch for the haulage cable. Since the cable must make a steep angle from the tractor hitch to the point of attachment high up the tree, the pull will tend to lift the tractor off the ground. If the rear end of a wheeled tractor is lifted it will lose its driving wheel grip, but if the front is lifted, no driving grip will be lost. It is to be noticed that this tendency for the front wheels to lift is not as dangerous as is front wheel lift when an ordinary load is being towed by too high a hitch at the rear drawbar.

The best angle for the cable or chain is about 50° to ground level. A wheeled tractor with an attached winch and a sturdy anchor sprag can uproot quite large trees, and can take up so much of the stump and

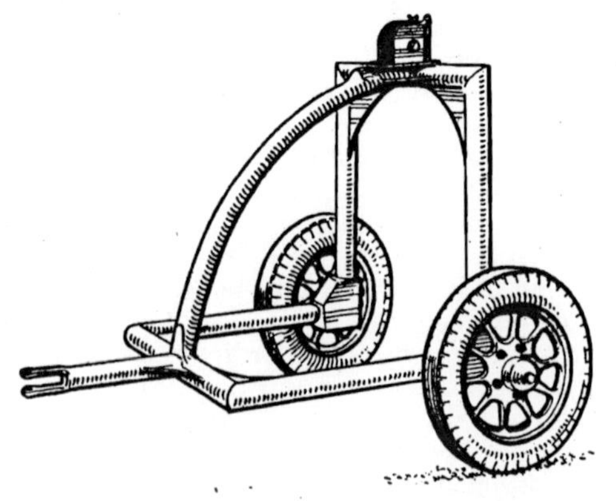

Fairholme log-hauling attachment for a tractor. The cable from a tractor winch runs on the pulley in the bracket on the upper cross-piece.

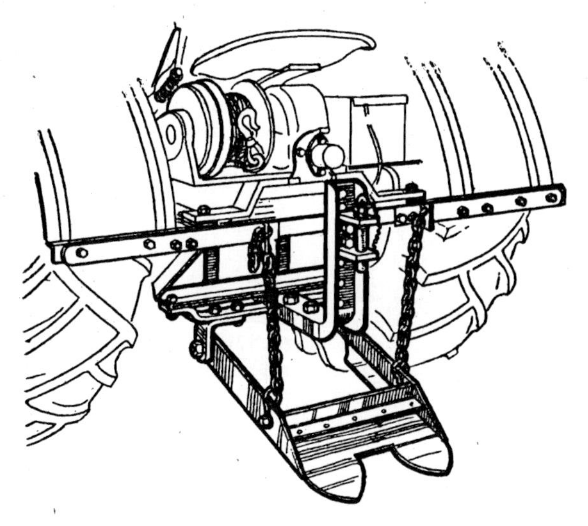

A sprag or anchor to resist the pull of a mounted winch.

roots that little further work is needed to prepare the land for ploughing and cultivation.

The truth is that if a farmer is prepared for a little quicker depreciation in order to increase the range of work that the tractor can do, the high tractive effort available can be put to all sorts of uses, without the need for any very expensive accessories. In the tree pulling mentioned above, if only small tree roots are to be dealt with the only accessory needed is a stout chain, but taking up the load with a jerk ought to be avoided as much as possible. It is true that a running pull at a slack chain will wrench out roots that have resisted steady pulling — and the transmission gears of a tractor seem to stand nearly anything. Nevertheless, one savage tug at a tree probably deteriorates a tractor as much as many hours of heavy ploughing.

A set of pulley blocks and steel cable can be used to help the tractor in pulling small trees. If the cable is run round the pulleys in such a way as to allow the tractor to move forward, say, four times as far as the end of the cable attached to the tree moves, the pull exerted on the tree will be approximately four times as great as the normal pull of the tractor. This is better than using the momentum of the tractor to provide the extra pull by snatch.

If much heavy tree pulling is to be done, however, it is well worth while to get a winch, either portable or attached directly to the tractor. The directly-attached winch can be driven through a belt from the tractor pulley, or it can be driven from the power take-off.

A winch and a good sprag-anchor increase the range of work of a tractor enormously, for they make it cover all the uses of the old steam traction engine. A tractor and winch can be used for loading felled timber on to wagons, for making mole drains at a far greater depth than they could be made by direct hauling, and for such handy little jobs as hoisting heavy loads into lofts through pulley tackle. A sprag or anchor let down from the back of the tractor enables greater pulls to be sustained when the tractor is used with a winch for such jobs as timber hauling and mole draining.

WHEEL SLIP

Whatever the gear or speed, we must avoid giving the tractor a heavier load than it can comfortably manage. Overloading uses the engine at too near its maximum output and will cause it to wear out quickly. It is likely that overloading will bring wheel slip, with consequent waste of fuel. Sometimes the slip is bad enough to bring the tractor to a standstill, but often a pneumatic-tyred driving wheel can be making many more revolutions than it need, without the slip being apparent to the driver. Wheel slip is a direct loss of power and therefore a direct loss of fuel.

It is worth while making a rough estimation occasionally to find how much the driving wheels are slipping when the tractor is dealing with a given load. First paint a mark on the wall of one of the tyres, or on the metal rim near the tyre; this will make it easy to count the number of revolutions of the wheel. With the implement in work the distance the tractor travels while the driving wheel makes ten revolutions is marked on the ground by pegs, and then measured. Next, the implement is unhitched, or is lifted out of work if it is a unit-principle outfit, and the tractor is run free in the same gear as before, and the distance of travel for ten revolutions of the driving wheel is measured again. The difference between these two distances, divided by the distance travelled without load, and multiplied by 100, will give roughly the percentage of slip. If the percentage is more than 18, then something ought to be done to reduce it, for fuel is certainly being wasted.

It is worth mentioning here that the speed of a tractor can be estimated roughly by walking alongside the tractor and counting the number of steps, reckoned at 35·2 in each, taken in 20 seconds. Then mark off one decimal place. For example, 32 steps in 20 seconds would be 3·2 miles an hour.

Drawbar pull can be estimated only by using some kind of dynamometer. The simplest kind is merely a spring balance. The more accurate sort are hydraulic; the pull works through a linkage which causes oil to be compressed in a cylinder. The pressure of the oil can be measured and is proportional to the pull. It is impossible to estimate drawbar pull without instruments.

One way of reducing slip is, of course, to reduce the load of work; but it often happens that wheel slip occurs long before the work load is great enough to put a full demand on the engine. In such cases, if the grip of the driving wheels can be improved, the tractor will be able to deal with a heavier load more economically. Indeed, improving the grip of the wheels can sometimes actually relieve the engine by reducing the loss of power due to the rolling resistance of the wheel passing over the ground. When tyres are slipping so much that they are on the point of digging in, their rolling resistance often becomes so high that the engine is stalled. The grip of the driving wheels can be improved by adding ballast or by fitting strakes.

Spade-lugged wheels rarely slip seriously without digging themselves in, and calculation is rarely necessary. Moreover, an inspection of the impressions left in the soil by the lugs of the wheels when the tractor is pulling its load will give sufficient guide to whether any excessive slip is taking place. If the impressions retain about the same shape as the lugs which have formed them, then the wheel is gripping satisfactorily; but if the imprint is much enlarged there is too much slip. The grip of steel wheels can sometimes be improved by ballast to cause deeper penetration of the lugs into the soil. A frequent cause of slip is, however, worn spade lugs; as well as losing their total length for penetration, the worn lugs become rounded instead of wedge-shaped.

When the driving wheels are gripping well, the driver must avoid taking undue advantage of their grip by increasing the load to such an extent that the engine is overworked. With a little practice and thought, he will be able to judge this by listening to the rhythm of the engine; when the speed of the engine seems to have been brought down and the exhaust note indicates that the engine is labouring, the driver should reduce the load or change to a lower gear.

PLANNING FOR EFFICIENCY

This chapter on efficient working must not be closed without a reference to the matter of general efficiency in tractor operations. We have seen how to make the tractor yield the greatest possible amount of work for the smallest possible quantity of fuel, but we must not forget that fuel can be saved also by saving unnecessary tractor hours, and that a saving of this kind can also bring lower depreciation and longer life to the tractor itself. Efficient tractor management can save fuel, time, money and spare parts, and it can bring more and better work for every hour the tractor runs. It is, in fact, a part of farm management. Good farming consists in getting the best crops for a given expenditure of land, labour, fertilizers and tractor fuel and oil.

On the field it is well to spend a little time working out a scheme of campaign. For example, in ploughing, time is well spent in deciding carefully how the lands are to be laid out and how the rigs are to be opened, and, in irregular fields, deciding which of the corners to veer out in short runs. Then when work begins it should be done carefully without haste, so that the lands are marked out accurately, so that the rigs will be parallel. If they are not parallel, the finishes also will be out of parallel and to complete the ploughing unnecessary short ends will have to be veered out, using the plough at only part of its full working width, and making many empty runs to get back into position.

In drilling seed or spreading fertilizer it may pay to calculate from rate of sowing and length of field where to set bags of seed or manure along the headland at places where the hopper of the drill will become empty; but here is a case where a zeal for planning must be tempered by caution. On a damp day, bags of artificial fertilizer left scattered along the headland may come to much more harm than they would if they were left all in one place on a cart or trailer covered by a tarpaulin sheet.

The great thing is to picture every operation as a whole, not as separate jobs. Looking at things in this way it is much easier at, say, harvest time, to plan so that just at the time when a trailer has been filled, an empty one returns from the rickyard or grain store.

To get full value from a tractor it must be made to work usefully for as many hours in the year as possible. The trouble is that the number of days in the year when useful tractor work can be accomplished is not 365. Taking Sundays into account and the days on which bad weather makes work impossible, or undesirable from the point of view of the well-being of the soil, it is unlikely that there are more than 200 days on which anything like 8 hours of useful work can be obtained from the tractor.

One way to keep the number of hours per year as near as possible to the theoretical maximum is to employ a second driver to keep the tractor going while the regular driver has his meals and during overtime hours which are beyond the fatigue limit of the regular driver. A disadvantage of this is that it breaks the very desirable principle of keeping one tractor to one man. The relief drivers will not understand the tractor as thoroughly as the regular drivers do, and it is much more likely to go wrong during these relief times than it is during the ordinary work; and breakdowns are a high price to pay for the relief working.

Some mounted implements make the use of a service trailer rather difficult; indeed it is often worth while to use a trailer large enough for the implement to be deposited upon it for carriage to the field so that the trailer can be attached to the tractor without the mounted implement being in the way.

It is important not to lose time through faulty operation of an implement or through inefficient planning of the work which may bring too great a proportion of idle running time. For example, serious loss of time can occur through ploughing being set out in lands that are too wide. If, for example, a field 100 yards long is set out in lands 50 yards wide, the tractor will spend one-third of its time running along the headlands and only two-thirds ploughing when nearing completion of the land. It is important in ploughing a field to begin with the triangular pieces or pykes, then plough the straightforward lands, before ploughing a complete headland right round all sides of the field. Ploughing the pykes after ploughing the straightforward lands leaves no room for turning on unploughed land when the pykes come to be ploughed.

A plough on which the polished wearing parts have been allowed to rust will tend to choke. These parts should be greased whenever the plough is not in use. Even when all these precautions are taken, choking is sometimes unavoidable. In these cases it is well to stop in time, that is, before the choking material becomes tightly packed. If this is done, the plough can be quickly cleared with a paddle made of a $\frac{3}{4}$ in steel bar with a 4 in × 4 in spade at one end, and an open pointed hook about 5 in across at the other. The rubbish can be hooked out of the plough, and the spade end of the paddle can then be used to scrape the choked parts clean.

Rust will also cause unnecessary friction, and this brings an unproductive load. A more frequent cause of unproductive load in any implement is incorrect adjustment of the various components. Implement adjustment and care is all part of tractor economy.

Keeping the tractor engine in good order is a sound step towards making the best use of tractor fuel. Leaking valves or gummed piston rings can cause the engine to consume much more fuel than is warranted by the output of power, and faults in the ignition system can make the engine difficult to start. Difficult starting, which in the case of a vaporizing oil engine often involves emptying and refilling the carburettor chamber several times, wastes fuel directly, and it can waste it indirectly by causing the driver to keep the engine running while he is loading a trailer or adjusting his implements.

Altogether it can be truly said that good tractor maintenance and careful management to avoid idle running and inefficient loading can make a worthwhile contribution towards a high standard of farming by ensuring full value from every gallon of fuel and every hour of working time.

Lightning Source UK Ltd.
Milton Keynes UK
UKOW05f1652030913

216452UK00001B/91/P